The CISO 3.0

This isn't just a book. It is a roadmap for the next generation of cybersecurity leadership. In an era where cyber threats are more sophisticated and the stakes are higher than ever, Chief Information Security Officers (CISOs) can no longer rely solely on technical expertise. They must evolve into strategic business leaders who can seamlessly integrate cybersecurity into the fabric of their organizations.

This book challenges the traditional perception of CISOs as technical leaders, advocating for a strategic shift toward business alignment, quantitative risk management, and the embrace of emerging technologies like artificial intelligence (AI) and machine learning. It empowers CISOs to transcend their technical expertise and evolve into business-savvy leaders who are fully equipped to meet the rising expectations from boards, executives, and regulators. This book directly addresses the increasing demands from boards and regulators in the wake of recent high-profile cyber events, providing CISOs with the necessary skills and knowledge to navigate this new landscape.

This book isn't just about theory but also action. It delves into the practicalities of business-aligned cybersecurity through real-life stories and illustrative examples that showcase the triumphs and tribulations of CISOs in the field. This book offers unparalleled insights gleaned from the author's extensive experience in advising hundreds of successful programs, including in-depth discussions on risk quantification, cyber insurance strategies, and defining materiality for risks and incidents. This book fills the gap left by other resources, providing clear guidance on translating business alignment concepts into practice.

If you're a cybersecurity professional aspiring to a CISO role or an existing CISO seeking to enhance your strategic leadership skills and business acumen, this book is your roadmap. It is designed to bridge the gap between the technical and business worlds and empower you to become a strategic leader who drives value and protects your organization's most critical assets.

Security, Audit and Leadership Series

Series Editor: Dan Swanson, Dan Swanson and Associates, Ltd., Winnipeg, Manitoba, Canada

The scope and mandate for internal audit continue to evolve each year, as do the complexity of the business environment and the speed of the changing risk landscape in which it must operate. The fundamental goal of this exciting series is to produce leading-edge books on critical subjects facing security and audit executives and practitioners. Key topics addressed include leadership, cybersecurity, security leadership, privacy, strategic risk management, auditing IT, audit management and leadership, and operational auditing.

Agile Enterprise Risk Management: Risk-Based Thinking, Multi-Disciplinary Management and Digital Transformation
Howard M. Wiener

University Auditing in the Digital Era: Challenges and Lessons for Higher Education Professionals and CAEs
Sezer Bozkus Kahyaoglu and Erman Coskun

Agile Audit Transformation and Beyond
Toby DeRoche

Information System Audit: How to Control the Digital Disruption
Philippe Peret

Next-Generation Enterprise Security and Governance
Edited by Mohiuddin Ahmed, Nour Moustafa, Abu Barkat, and Paul Haskell-Dowland

Auditing Information and Cyber Security Governance: A Controls-Based Approach
Robert E. Davis

Evidence-Based Cybersecurity: Foundations, Research, and Practice
Pierre-Luc Pomerleau and David Maimon

Operational Auditing: Principles and Techniques for a Changing World
Hernan Murdock

The Security Leader's Communication Playbook: Bridging the Gap between Security and the Business
Jeffrey W. Brown

The CISO 3.0: A Guide to Next-Generation Cybersecurity Leadership
Walt Powell

For more information about this series, please visit: https://www.routledge.com/Security-Audit-and-Leadership-Series/book-series/CRCINTAUDITA

The CISO 3.0

A Guide to Next-Generation Cybersecurity Leadership

Walt Powell

CRC Press
Taylor & Francis Group
Boca Raton London New York

CRC Press is an imprint of the
Taylor & Francis Group, an **informa** business

Designed cover image: Shutterstock Image ID 2490883849

First edition published 2026
by CRC Press
2385 NW Executive Center Drive, Suite 320, Boca Raton FL 33431

and by CRC Press
4 Park Square, Milton Park, Abingdon, Oxon, OX14 4RN

CRC Press is an imprint of Taylor & Francis Group, LLC

ISBN: 978-1-032-82351-5 (hbk)
ISBN: 978-1-032-84007-9 (pbk)
ISBN: 978-1-003-51078-9 (ebk)

DOI: 10.1201/9781003510789

Typeset in Times
by Apex CoVantage, LLC

Dedicated to my family – Lindsay, Rivers, and Axel

Thanks for all the love and support

Contents

Foreword

In recent years, the role of the chief information security officer (CISO) has undergone a transformation unlike anything we have seen before. What was once primarily a technical position focused on safeguarding systems and data has evolved into a strategic role that is central to business operations, risk management, and executive leadership. This shift has not been gradual; it has been driven by sweeping changes in regulatory, technological, and threat landscapes. As CISOs, we are no longer just stewards of cybersecurity – we are business leaders, fiduciaries, and critical advisors to the boardroom.

Walt Powell's *The CISO 3.0* arrives at a pivotal moment. With the Securities and Exchange Commission (SEC) Cybersecurity Rule redefining material disclosures and courts recognizing the fiduciary obligations of directors and officers regarding cybersecurity oversight, the stakes for CISOs have never been higher. The modern CISO must bridge the gap between security and business strategy, articulating risk in financial terms that resonate with executives and shareholders. All this has occurred against the backdrop of the rapid adoption and evolution of artificial intelligence following the release of ChatGPT.

What sets the CISO 3.0 apart is its clear, actionable roadmap for embracing this evolution. It challenges readers to step into a forward-thinking role, where cybersecurity is viewed not as a cost center but as a business enabler, risk is quantified and treated like any other board-level priority, and capabilities, not just tools, define program success. Walt draws upon decades of experience to offer deeply pragmatic insights, empowering current and aspiring CISOs to transform their programs and careers.

Reflecting on my journey, the fundamentals of business alignment and leadership remain unchanged. However, the world has demanded that we continuously elevate our abilities. Walt brilliantly captures this next step: Encouraging CISOs to embrace their evolving roles as architects of resilience, innovation, and value within their organizations.

If you are a security leader ready to face tomorrow's challenges – regulations, AI-driven threats, and cyber resilience – this book is your guide. Walt Powell has charted the path to becoming a CISO 3.0, a leader capable of navigating today's complexities while preparing for what lies ahead.

Welcome to the next generation of cybersecurity leadership.

Matthew K. Sharp
Author of *The CISO Evolution: Business Knowledge for Cybersecurity Executives*

Acknowledgments

Writing this book has been a collaborative journey, and I am deeply grateful to the incredible individuals who have contributed their guidance, insights, and support along the way.

First and foremost, my heartfelt thanks to *Matt Sharp* for his invaluable help and guidance throughout this process. Your wisdom and encouragement have been instrumental in shaping this book.

I am immensely grateful to *Dan Swanson*, whose guidance and generosity in sharing resources have been a constant source of inspiration. Your wealth of knowledge has been a cornerstone for much of the content in this book.

Special thanks to *Peter Gregory* for his insightful review of the proposal and for the advice he shared, both directly and through his excellent book, *The Art of Writing Technical Books: The Tools, Techniques, and Lifestyle of a Published Author.* Your expertise has been an anchor in navigating the writing process.

To my amazing Field CISO team – *John Candillo, Aaron McCray, Justin MacDonald, Steve Allison, Tom Sinnott*, and our director *Buck Bell* – thank you for your insights, for the peer reviews, and for being such a phenomenal group of collaborators. Your collective experience has enriched this book beyond measure.

A heartfelt thank you to *Dr. Larry Burke* for the partnership, comprehensive book review, and the depth of insights you provided. Your assistance has been invaluable in refining this work.

To *Douglas W. Hubbard* and *Richard Seiersen*, authors of *How to Measure Anything in Cybersecurity Risk*, thank you for transforming the way I think about cyber risk. Your groundbreaking work has directly influenced many of the ideas presented here.

I also want to express my appreciation to *Kevin Richards* and the team at *X-Analytics® Secure Systems Innovation Corporation* for sharing their innovative approaches to risk quantification. Your work continues to inspire.

A special acknowledgment to *Gary McIntyre*, a great Canadian and friend, whose thought leadership and guidance around cyber resilience were invaluable. Your expertise and perspective have had a meaningful impact on this book, and I am deeply grateful for your contributions.

Lastly, a sincere thanks to my family – *Lindsay, Rivers, and Axel* – for your unwavering love and support. This book is dedicated to you, and I couldn't have done it without your encouragement.

To all those mentioned in this book and everyone who has played a part in this journey, thank you. Your contributions have made this work possible and meaningful.

Author

Walt Powell is an experienced executive coach and CISO advisor who has extensive experience working with countless CISOs and developing cybersecurity programs. Walt helped pioneer the role of Field CISO and is a founding member of the Global Security Strategy Office at CDW. Walt now leads a team of Field CISOs, composed entirely of former executives, who bring a wealth of experience and knowledge to their clients, underpinned by unique insights gained from contributing to and learning from the strategies of hundreds of chief information security officers (CISOs) and chief information officers (CIOs) across every size of organization and vertical. Walt and his team leverage this wealth of knowledge and experience to provide executive coaching, support, and mentorship to elevate other CISOs, their programs, and organizations, sharing lessons and providing strategic guidance that would typically take several careers to acquire.

Prior to his role at CDW, Walt was the owner and vCISO at Left Brain Security, which is now Left Brain Security Media. He has served as an award-winning cybersecurity leader, advisor, architect, and pre-sales engineer and has also served as a professor of networking and security at Wright College. Walt firmly believes in the importance of giving back to the industry, which is why he taught in CISSP and CISM boot camps and contributes as a certification exam development committee member for numerous organizations. He holds an impressive array of professional certifications, including CISSP, CISM, CCISO, Carnegie Mellon CISO, and the Stanford Advanced Cybersecurity Certificate, and numerous technical and sales certifications from leading cybersecurity firms. Walt also leads a cybersecurity book club, which is being launched as a podcast.

A proud Mensa member and futurist, Walt is deeply invested in exploring the implications of emerging technologies on cybersecurity. He actively contributes to the cybersecurity community by writing and speaking at industry conferences such as BSides, CypherCon, and CrowdStrike Fal.Con; sharing white papers; and authoring articles on critical security topics. Beyond his professional life, Walt is a former professional musician and multi-instrumentalist who cherishes spending quality time with his children, traveling, and learning new languages.

1 Introduction

1.1 PURPOSE

CISOs are at a crossroads. The old ways of leading security teams and managing risk are fading fast. Today, you need more than technical skills: you need to be a business strategist, a risk guru, and a trusted advisor. You need to speak the language of the boardroom, not just the server room.

The CISO 3.0: A Guide to Next-Generation Cybersecurity Leadership is your guide to navigating this new world. It is a roadmap for current and aspiring CISOs who want to step up their game and become the leaders their organizations need. We'll challenge the old ways of doing things, pushing you to embrace a risk-based approach that aligns with business goals and delivers real value.

This isn't just another checklist of security controls or a dry recitation of compliance requirements. It's about empowering you to go beyond your technical comfort zone and become a true business leader.

Pause for a moment and reflect: How aligned are your current cybersecurity strategies with your organization's broader business objectives?

1.1.1 Target Audience

This book is your roadmap if you're a cybersecurity professional aspiring to a CISO role or an existing CISO seeking to enhance your strategic leadership skills and business acumen. We assume you have a foundational understanding of cybersecurity principles. Still, our primary focus is on developing the broader business and leadership capabilities required for success in the growing CISO role.

Consider this: Do you feel confident in your ability to communicate cybersecurity risks and strategies in a way that resonates with your organization's executive leadership and board of directors?

If you've ever felt like your technical expertise isn't translating into the boardroom influence you desire, or if you're struggling to secure the budget and buy-in needed for your security initiatives, this book is for you. It's designed to bridge the gap between the technical and business worlds, empowering you to become a strategic leader who drives value and protects your organization's most critical assets.

1.1.2 Scope of the Book

This book isn't about theory: it's about action. We explore the practical side of business-aligned cybersecurity through real-life stories and examples. We'll address the challenges CISOs commonly face, such as securing budgets, articulating the value

DOI: 10.1201/9781003510789-1

of security programs, prioritizing initiatives, managing risk acceptance, and determining if risk is well-managed. Beyond that, we'll guide you through developing an executive presence, securing necessary resources, and understanding strategic decision-making and risk management.

The book's scope encompasses a wide array of topics vital for the modern CISO:

- Strategic leadership
- Business alignment
- Risk management and governance
- Meaningful metrics development
- Effective board-level communication
- Emerging technologies and their impact on cybersecurity
- Incident response and crisis management
- The future of cybersecurity leadership

The book's structure allows you the flexibility to either progress sequentially through its chapters or reference specific sections as needed to address particular challenges you encounter along the way.

As you explore these topics, consider how you currently approach each area and identify opportunities for growth and development in your own CISO journey.

The book distinguishes itself through several special features that enhance its value for readers. It emphasizes real-world application, incorporating numerous real-life stories and examples that illustrate the effective application of business acumen in cybersecurity. The book also offers unparalleled insights gleaned from the author's extensive experience advising hundreds of successful programs, including in-depth discussions on risk quantification, cyber insurance strategies, and defining materiality for risks and incidents. It provides practical guidance, equipping CISOs with actionable advice, tools, and strategies that can be implemented immediately to enhance their leadership skills and improve their security programs. As a time-saving resource, the book condenses lessons and strategies learned from countless programs into a wealth of knowledge that would otherwise take years to acquire. It also serves as a cost-effective alternative to formal education, delivering targeted and practical knowledge directly applicable to the CISO role at a fraction of the cost of traditional programs. Finally, to further support your learning and implementation of these skills, the book offers companion resources and templates available for download at www.CISO30.com, providing additional tools and guidance to apply the concepts and strategies discussed in the book to your own organization. In short, we hope that *The CISO 3.0* is an invaluable resource for current and aspiring CISOs seeking to elevate their leadership skills, drive business value, and navigate the complexities of modern cybersecurity.

1.2 WHY WRITE THIS BOOK NOW?

The cybersecurity industry is in a state of flux, shaped by changes in regulations, advancements in technology, and an increasingly complex threat environment. With recent regulatory actions, such as the Securities and Exchange Commission's (SEC's) cybersecurity disclosure rule, and high-profile breaches, such as SolarWinds and Uber, CISOs are facing unprecedented scrutiny. This moment calls for a new kind of leader – one who can blend cybersecurity expertise with broader business strategies. The pressure is mounting, and the need for a practical approach to navigating these challenges has never been greater.

That's where *The CISO 3.0* steps in. This book isn't just a response to the current challenges: it's a forward-thinking guide designed to help CISOs become the strategic leaders their organizations require. It provides the tools and insights needed to thrive in today's cybersecurity landscape, helping you lead with confidence and clarity.

The timing couldn't be more appropriate. As the CISO role evolves, the resources available to support this transition must keep pace. *The CISO 3.0* addresses this gap, offering a hands-on, actionable roadmap to help CISOs translate business alignment concepts into real, measurable outcomes. It's not only about understanding the "why" of aligning cybersecurity with business objectives but also about conquering the "how".

This book is your opportunity to stay ahead of the curve, enhancing your skills and perspective in line with the evolving demands of cybersecurity leadership. It's designed to empower you as a strategic partner, enabling you to drive business value, manage risks, and lead your organization into a secure and successful future.

1.3 MOTIVATION AND RELEVANCE

I was inspired to write this book as a way to give back to the security community, which has shaped so much of my career. Over the years, I've seen firsthand how the role of the CISO has evolved, and I felt a strong responsibility to share what I've learned to help others navigate the growing demands of this position. Working closely with CISOs and their teams, I've noticed that expertise often remains isolated or closely guarded, which is why I wanted to break down those barriers and make valuable knowledge more accessible. As William Gibson once said, "The future is already here, it's just not very evenly distributed". This book is my effort to level the playing field by providing the tools and insights that can help CISOs become more effective leaders.

The need for *The CISO 3.0* is driven by the real-world impact of cyber incidents. With the cost of data breaches soaring into the millions and cyberattacks ranking as a top global risk, the role of the CISO has never been more important. My goal is to equip CISOs with the practical skills and knowledge they need to navigate these challenges. This includes understanding finance, economics, and risk, as well as tackling complex topics like cyber insurance. I also focus on helping CISOs develop modern strategies for resilience so they can ensure business continuity in the face of ever-evolving threats.

One of my core motivations is to educate and empower. After years of working with organizations that face similar security challenges, I wanted to create a resource to address these issues directly. Writing this book allows me to reach a wider audience, including emerging leaders in the field. It's also my way of documenting the evolution of cybersecurity, highlighting the changes we've seen, and offering insights into where we're headed. I explore leadership strategies tailored specifically for cybersecurity teams, providing valuable takeaways that can help readers excel in the industry.

As someone who enjoys storytelling, I saw writing this book as an opportunity to share my experiences in a structured and meaningful way. I aim to take complex technical concepts and turn them into an easy-to-follow guide for both current and aspiring CISOs. I also recognized gaps in the existing literature, particularly around cybersecurity leadership, and wanted to fill those gaps with resources that address specific challenges the industry faces today.

In short, my motivation for writing this book comes from a genuine desire to educate, contribute to the field, share my experiences, and fill crucial knowledge gaps. I believe these are all essential steps toward advancing cybersecurity leadership.

1.4 BOOK STRUCTURE AND ORGANIZATION

The CISO 3.0 is structured as a comprehensive guide for CISOs on their transformative journey. The book is organized into four sections, each building upon the previous ones to provide a complete overview of the CISO 3.0 role:

- **Part 1: The Evolving Role of the Security Leader.** This section traces the historical evolution of the CISO role, from its technical origins to its current strategic leadership position. It highlights the need for a new breed of security leader, the CISO 3.0, who can seamlessly integrate cybersecurity with business strategy.
- **Part 2: Business and Risk Alignment.** This part delves into the practicalities of aligning cybersecurity with business objectives. It explores the "languages of business", including accounting, finance, economics, and risk, providing CISOs with the tools to communicate effectively with executives and the board.
- **Part 3: Risk Treatment.** This section is divided into two subsections. The first, "Transfer, Avoid, and Accept Risk", explores cyber liability insurance and risk financing, offering insights into strategic risk transfer and acceptance. The second, "Risk Mitigation", guides readers through creating a modern, risk-quantified security strategy roadmap, focusing on proactive risk mitigation.
- **Part 4: Bringing It All Together.** The final section focuses on integrating the concepts and strategies from the previous parts to build meaningful metrics and communicate effectively with executives and the board. It explains the value of data storytelling and visualization, empowering CISOs to create compelling narratives that resonate with their audience.

Each chapter is packed with real-world examples, actionable advice, and unique insights. It provides a roadmap for CISOs to evolve into strategic business leaders who can effectively protect their organizations and contribute to their long-term success.

1.5 AUTHOR'S BACKGROUND AND PERSPECTIVE

I've spent the last two decades working in the cybersecurity field, and this time has provided me with the opportunity to develop and lead numerous security programs. I've had the privilege of serving as the inaugural Field CISO at CDW, and before that, I was the owner and vCISO at Left Brain Security. Throughout my career, I've worn many hats, including cybersecurity leader, advisor, architect, and pre-sales engineer. I'm also passionate about education and have served as a professor of networking and security at Wright College. At this point, I've held nearly every type of role on both sides of the negotiating table.

Currently, I lead a team of Field CISOs at CDW, all of whom are former executives. Together, we consult with CISOs and CIOs, helping them navigate the challenges of security leadership. Last year alone, our team had the opportunity to engage with over 500 organizations, gaining valuable insights into best practices across diverse security programs, industries, and company sizes. It's these insights I hope to bring to you.

I believe in continuous learning and hold a variety of professional certifications, including CISSP, CISM, CCISO, Carnegie Mellon CISO, and the Stanford Advanced Cybersecurity Certificate. I'm also an active member of the cybersecurity community, contributing through writing, speaking at industry conferences, and facilitating discussions in a cybersecurity book club-turned-podcast.

As a Mensa member and futurist, I'm fascinated by the impact of emerging technologies on cybersecurity. When I'm not immersed in the world of cybersecurity, I enjoy spending time with my children, exploring new places, and learning languages.

This book is a culmination of the lessons and strategic guidance I've gleaned throughout my career. Through my observations, I've identified a maturity curve among CISOs and outlined activities that can foster growth along that path. My hope is that *The CISO 3.0* will serve as a roadmap, empowering CISOs to strengthen their skills, navigate the complexities of modern cybersecurity leadership, and ultimately contribute to the advancement of the profession.

We hope that *The CISO 3.0* becomes an indispensable resource for modern CISOs navigating the complexities of cybersecurity in a digitally transformed world, equipping them with the knowledge and skills to align cybersecurity strategies with business objectives, communicate effectively with the board, and drive innovation while managing risks effectively.

Whether you are just beginning to work toward a CISO role or are an experienced security leader looking to enhance your strategic impact, this book is designed to provide you with the insights, tools, and strategies you need to succeed. My intent is not only to educate but also to inspire, helping you become the kind of leader who can effectively bridge the gap between technical expertise and business strategy, driving both security and organizational success. By sharing the knowledge I've gained over decades in the field, I aim to equip you with the wisdom and foresight to navigate the future of cybersecurity leadership with confidence and competence.

1.6 SELF-ASSESSMENT: ARE YOU READY FOR CISO 3.0?

Reflect on your current role as a CISO or aspiring CISO. Are you ready to take on the challenges and opportunities of the CISO 3.0 era?

Consider the following questions to gauge your readiness for the CISO 3.0 transformation:

- Do you possess a strong understanding of your organization's business objectives, financial metrics, and risk appetite?
- Can you effectively communicate cybersecurity risks and strategies in a way that resonates with the board and executive leadership?
- Are you comfortable leveraging data and analytics to make data-driven decisions about security investments and risk mitigation?
- Can you translate technical security concepts into business terms that resonate with nontechnical stakeholders?
- Can you read and interpret cyber insurance policies?
- Do you have a strong understanding of risk management principles?
- Can you quantify and prioritize risks effectively?
- Are you open to new technologies and approaches to security, such as AI and Zero Trust?
- Can you prioritize your security roadmap strategy?
- Are you able to justify your program, budget, and investment requests to leadership?
- Do you understand what constitutes a material risk, and if so, can you determine which of your organization's risks are material?

If you answered "yes" to most of these questions, you're well on your way to becoming a CISO 3.0. If not, don't worry, this book is designed to help you develop the skills and mindset needed to thrive in this challenging yet rewarding role.

The journey to becoming a CISO 3.0 starts now. Commit to embracing the challenges and opportunities that lie ahead. Let this book be your guide as you navigate the complexities of modern cybersecurity and evolve into the strategic leader your organization needs. As you embark on this adventure, I encourage you to actively engage with the material, take notes, and reflect on how you can apply the concepts and strategies discussed to your own organization. Don't hesitate to reach out to me directly to collaborate, ask questions, or seek advice.

You can reach me at:

CISO30.com
Linkedin.com/in/WaltPowell
TheCISO30@gmail.com
Welcome to *The CISO 3.0* – your guide to becoming the next generation of cybersecurity leadership.

Part 1

The Changing Role of the Security Leader

2 What Is a CISO 3.0?

2.1 INTRODUCTION

In this chapter, we will take a trip through time, tracing the evolution of the chief information security officer (CISO) from its beginnings in the early 2000s to the present day. We'll witness how the role has transformed from a primarily technical expert, troubleshooting firewalls and antivirus software, into a strategic business leader, orchestrating the complex symphony of cybersecurity and organizational goals.

We'll look into the distinct phases of this evolution, starting with the technically driven CISO 1.0, who focused on implementing security controls and safeguarding the IT infrastructure. Then, we'll explore the rise of the CISO 2.0, a more strategic figure who sought to align security practices with regulatory guidance, standards, and frameworks. Finally, we'll arrive at the CISO 3.0, the next-generation leader who seamlessly integrates cybersecurity into the very fabric of the business strategy, leveraging advanced technologies like AI to manage risks and drive business value proactively.

This chapter will examine the dynamic nature of the CISO role and its importance to organizations of all types and sizes. It's a story of adaptation, innovation, and strategic vision, where the CISO emerges not only as a protector but also as a catalyst for growth and resilience in the digital age.

2.2 BRIEF HISTORY OF CISOS

In the early days of cybersecurity, it wasn't uncommon for security to be handled by a team of one. Sometimes that person was someone with a real passion for security, but just as often, the "security person" was simply an IT professional who had been roped into helping with a project or incident, and over time, they became the de facto security expert. I remember meeting many of these early pioneers, people who had no formal security background but found themselves on the front lines, figuring things out as they went along. These individuals had to take on responsibilities that stretched far beyond their original IT roles. They were often the ones responsible for both setting up servers and securing them, juggling infrastructure management with fending off cyber threats. These makeshift security roles laid the groundwork for what would eventually become the formalized CISO position.

The story of the CISO is one of remarkable evolution, mirroring the rapid transformation of the digital landscape itself. In the nascent days of the Internet, when Steve Katz assumed the very first title of CISO at Citigroup in 1995, the cybersecurity world was a vastly different terrain. There were no stateful firewalls or web gateways, and even antivirus was novel and rare. Those possessing even a rudimentary understanding of security tools were considered experts. If you knew anything at all about cybersecurity, including that Internet connectivity was a risk, then you were generally considered a thought leader who was way ahead of the curve. The turn of

DOI: 10.1201/9781003510789-3

the millennium marked a turning point. The Y2K scare and the subsequent bursting of the dot-com bubble exposed the vulnerabilities inherent in IT systems and the interconnectedness of global supply chains. Organizations, suddenly awakened to the looming cyber threats, sought out IT professionals who could navigate this new and dangerous situation. The CISO, once a rare and specialized role, was becoming an indispensable figure in organizational structures.

Back then, cybersecurity wasn't a boardroom priority. It was tucked away in the IT department and viewed as a technical issue, not a strategic one. Early IT programs frequently suffered from poor user experience and were commonly viewed through the lens of the socially awkward "IT Guy" stereotype portrayed on shows like *Saturday Night Live* and the *IT Crowd*. That very perception helped drive what became the first iteration of the CISO role.

The CISO 1.0 era was characterized by a focus on technical proficiency and compliance. The primary responsibility was to implement and maintain security controls, acting as a digital gatekeeper against the rising tide of cyberattacks. As the digital world expanded and evolved, the CISO's role followed suit, transitioning into the CISO 2.0 phase. This era saw a greater emphasis on business alignment, framework adherence, and qualitative risk assessment. The CISO 2.0 was no longer just a tech expert but a strategic partner, working alongside executive teams to integrate security into the broader business strategy.

I remember when I first started in this field, we were practically making it up as we went along. There were no fancy next-gen tools, just a bunch of us trying to figure out how to keep the bad guys out. It's amazing to see how far we've come. Today, we stand at the cusp of the CISO 3.0 era, a time of strategic leadership and business acumen. The CISO 3.0 is not merely reactive but proactive, aligning closely with financial goals and business risks. They leverage cutting-edge technologies like AI, ML, and decision science to anticipate and mitigate threats, transforming cybersecurity from a cost center into a value driver. The CISO 3.0 is a skilled leader, bridging the gap between technology and business, and their role is more critical than ever in today's complex and interconnected world.

2.3 CISO 1.0

The role of the CISO has undergone a sizable transformation over the years. In the early 2000s, CISO 1.0 was predominantly technically focused, their primary mission being to reduce disruptions and safeguard the new IT infrastructure. CISO 1.0s were responsible for ensuring the proper installation, configuration, and maintenance of security technologies. CISOs in the 1.0 era were often viewed primarily as technical experts rather than strategic partners to the executive team and were usually perceived as a back-office function. Consequently, they were frequently excluded from high-level discussions and decisions.

2.3.1 Role Tasks of the CISO 1.0

The CISO 1.0 were often chief architects who focused primarily on deploying and managing security infrastructure such as firewalls, intrusion detection systems, antivirus solutions, and maintaining the technical bulwarks of organizational security. The CISO 1.0 provides high-level project and program management, as well as ticket and case management.

2.3.2 Challenges and Limitations of CISO 1.0

CISO 1.0 faced a great many challenges. Security was often seen as part of IT, leading to conflicting priorities between CISOs and CIOs. The former needed to reduce risk, while the latter focused on service uptime. CISO 1.0s often had no formal budgets for implementing and maintaining security, which meant funds were not predetermined and subject to change. This financial unpredictability, coupled with a lack of stakeholder buy-in, meant that even though CISOs were constantly working on projects, those projects struggled to stay on time as they weren't viewed as a business priority.

Security in this era was often reactive rather than proactive. CISO 1.0s found themselves responding to tickets and requests, overwhelmed by project work, and generally too understaffed to focus on strategy.

The need to drive security projects without first aligning strategically to business goals is what leads many organizations to refer to cybersecurity as the "department of NO". The idea that security is a blocker that prevents progress and innovation is a common challenge for CISO 1.0s and is just another reason CISOs struggle to earn a seat at the table for broader business discussions.

2.3.3 The Legacy of CISO 1.0

CISO 1.0 represents a critical phase in the evolution of the CISO role, marking the beginning of the CISO's involvement in organizational security. It laid the technical foundation upon which future strategic expansions would be built, ultimately shaping the CISO role into what it is today. As you can see, the limitations and challenges of CISO 1.0 highlight the need for a more strategic approach to security, setting the stage for the emergence of CISO 2.0 and beyond.

Many of you may read the description of the CISO 1.0 and see yourselves in those tasks and challenges, or you might know a CISO who fits that mold. While nearly every CISO in the early 2000s was a 1.0, it's not uncommon to still encounter CISO 1.0s today. In fact, I still come across plenty of organizations with one-person security teams, where security is embedded within IT, and the IT lead has to wear both hats. This is especially true for organizations just beginning to build out their security capabilities.

What's important to understand is that these CISO versions aren't simply historical markers, they also represent levels of maturity for both the CISO and the overall security program. Think of it like software versions: Some companies are running the latest and greatest, while many others are stuck on legacy versions that are no longer supported. The same holds true for CISOs. Plenty are still operating as CISO 1.0s or 2.0s, struggling to keep up with modern challenges because they haven't yet started down the path to version 3.0 of the role.

That's the whole purpose of this book: To help CISOs upgrade to the newest version of their role. The version model is a guide to understanding where you are now and where you can go next.

2.4 CISO 2.0

Most CISOs today are a CISO 2.0, which is compliance-driven with a focus on assessment and framework alignment. CISO 2.0s evolved to become more strategic by creating artifacts like program roadmaps and risk registers to help communicate departmental goals. CISO 2.0s also began to work more closely with other departments, such as risk management, compliance, and legal, to ensure that the organization's security strategies aligned with these departments' objectives. This collaboration helped CISOs gain a better understanding of the organization's overall risk profile and ensured that security risks were appropriately managed across the organization.

2.4.1 Role Tasks for CISO 2.0

When the CISO 2.0 originally emerged, it felt like a shift toward true leadership, transcending beyond mere technical expertise to embody a more people-centric managerial role. These professionals not only oversaw the security team and worked to establish a security program that was aligned to some kind of best practices. The CISO 2.0 was primarily a shift toward compliance alignment, where CISOs focus on ensuring adherence to relevant regulations and industry standards such as Health Insurance Portability and Accountability Act (HIPAA), Sarbanes-Oxley (SOX), and Gramm–Leach–Bliley Act (GLBA). The rise of compliance and regulations was a game-changer for CISOs. Suddenly, we had a playbook, a set of rules to follow. But it also meant a whole new set of challenges. Navigating the maze of compliance requirements while still keeping the lights on was no easy feat. I've seen many CISOs struggle to find that balance, but the ones who succeeded emerged as true leaders, not just techies.

CISO 2.0s also introduced risk assessment, moving toward a qualitative analysis that evaluates risks based on controls assessment frameworks. Those risk-based assessments created an information security baseline that CISOs could use to steer the organization toward security best practices and limit policy deviations.

CISO 2.0s also proactively shaped the security of their organizations by prioritizing the prevention of attacks over reactive measures. This strategic approach was complemented by their responsibility to adopt modern cybersecurity technologies, ensuring the organization stays ahead of emerging threats. They also took charge of incident response in an effort to minimize damage from any breaches.

CISOs in this phase are generally working to drive better security awareness within the organization. Through the promotion of best practices and awareness of potential threats, they challenge the employees to be a part of the organization's security posture. In essence, CISO 2.0 encapsulates a broader, more compliance-focused approach to cybersecurity leadership, which is a major step forward from the purely tactical approach of the CISO 1.0.

2.4.2 Challenges and Limitations of CISO 2.0

CISO 2.0s also face a number of major challenges. While creating strategic roadmaps is a big step forward from 1.0 to 2.0, determining what should be part of those strategies is often a big challenge. Generally, the drivers are purely external, gaps in regulatory compliance, standard frameworks, insurance underwriting guidance, or contractual obligations are the most common inputs. While many people will refer to these 2.0 strategies as "business alignment", there is usually very little introspection or coordination with business goals and priorities. 2.0 is more like compliance or framework alignment. We'll discuss what real business alignment looks like and what the differences are when we discuss what it is to be a CISO 3.0.

Even after a strategy is drafted, most security programs are seen as a cost center rather than a value driver and fail to earn budget and buy-in. Most CISO 2.0s struggle to articulate the return on investment (ROI) of their programs and find it difficult to prioritize which strategic initiatives and risks to tackle first. Many find it even harder to justify those choices and their budget requests to boards and executive leaders. Rightly so, most 2.0s have piles of controls they can't rationalize that are duplicative, underutilized, misconfigured, or understaffed, or are just shelfware.

The core problem for many CISOs centers around a persistent lack of command, which undermines their effectiveness and places them in a position of accountability without authority. Despite the title of "chief", many CISOs lack control over critical areas including budget allocation, policy enforcement, and resource deployment. This absence of authority forces CISOs into a continual cycle of seeking approval and "selling" the importance of cybersecurity investments to other executives, which not only delays responses to potential threats but also limits strategic alignment. The situation is exacerbated by the difficulty of translating abstract cyber risks into relatable terms for non-security-minded executives, often leaving CISOs in a position where their efforts to safeguard the organization are not fully understood or supported. Teams look to leaders who can make decisive calls, yet the CISOs' lack of authority reduces their credibility, leading to weakened confidence internally and externally.

I often hear from CISOs and have seen articles detailing how CISOs lack authority and would be able to better defend their organizations if they could act more unilaterally. The thought is that CEOs and chief financial officers (CFOs) typically control their domains, but the CISO remains responsible for security outcomes without the necessary authority, making it challenging to foster trust with the board, peers, and their own team. But it's a fallacy to assume that other C-levels enjoy full autonomy or operate without similar accountability challenges. In reality, all executives, including the COO, CFO, and even CEOs, must advocate for their initiatives and align them with broader business goals, often justifying their decisions to the board, stakeholders, or even to each other. I find for most CISO 2.0s it's not a lack of authority that holds them back but rather a failure to communicate cybersecurity in a way that aligns with business priorities. Many CISO 2.0s find themselves having to repeatedly "sell" security because they fail to speak the language of the business and end up selling security concepts instead of business initiatives, and sell individual security projects instead of selling their entire program, strategy, and approach. By emphasizing security jargon and individual projects over the overarching value of a robust cybersecurity strategy, they miss the chance to frame their goals in terms that resonate with executives focused on ROI, risk management, and strategic alignment.

Another interesting anomaly for 2.0 security leaders is that they often get the responsibility and accountability of the CISO role without the actual C-suite title. It is very common for me to meet the top security leader at even Fortune 500 organizations whose title is VP, director, or even manager. Beyond that, when security leaders are given the "C" title, it is commonly a second-tier "C" that reports to another C-suite executive like the CIO and has their communications filtered and proxied to the CEO, CFO, and board of directors. For years, I've seen CISOs striving for full C-suite status and for the right to present to the board. Still, even when security leaders finally earn that seat, it's been my experience that most are not prepared and don't possess the necessary skills to take on the fiduciary duties and legal responsibilities associated with the true C-suite.

For example, CISOs have gotten away with board presentations for years, which would get other C-level executives fired. Imagine if a Chief Financial Officer (CFO) walked into the board meeting and said, "I can't tell you in dollars and cents what our last quarter's earnings were, but I can tell you it's a 'green up arrow'", or if the CFO said, "we can't tell you exactly where all our assets are, but we do have a project underway that will hopefully allow us to, track most of them within the next three years". The CFO would be run out of the building, but that is *EXACTLY* how CISOs present their programs, with qualitative red, yellow, and green charts that don't communicate the impact to the business. That is why CISO 2.0s tend to be second-class executives and have their messaging gatekept by other executives.

This disconnect from the board and C-suite leads to one of the biggest challenges for CISO 2.0s, which is risk ownership. Who owns the risk, and who should own the risk? Well, that depends not only on who you ask but also on how you define risk. The problem starts when you explore risk mitigation from a RACI point of view. A RACI chart defines who is Responsible, Accountable, Consulted, and Informed about a task, and CISO 2.0s commonly find that they are accountable but not responsible for risk.

Here's an all-too-common true story. A CISO 2.0 discovers, as part of their risk assessment, a set of legacy operating systems (OS) running on production equipment. The OS is out of support and cannot be patched. This is a major vulnerability because there are known exploited vulnerabilities for that unpatched version of the OS. The CISO would like to have the system operations team patch these devices, but that would require a restart, which would momentarily interrupt production. The department head says, "They cannot tolerate a reboot, and he will accept the risk". That scenario makes the department head the risk owner, who would be responsible but not accountable for the risk. Consequently, if an incident and breach were to occur, the CISO would be on the hook for the impact even though the department head "accepted the risk". What the department head actually accepted was a vulnerability, and we'll discuss this common misconception and how to avoid this pitfall in Chapter 6.

So, if the stakeholders don't own the risk, who should? Many believe it's the CISO, and often, when I talk to board directors, they will say things like, "We hired a CISO so that they can own the risk". However, if we look at the RACI chart again, the CISO should be responsible, while the entire business is accountable. Meaning the board of directors and the executive management owe a fiduciary duty. We'll discuss these duties at length in Chapter 5, but for now, just take note that the board of directors should provide risk oversight.

Let's take a look at another very common real story. Imagine a CISO who created a strategy and then worked that plan, implementing and maintaining security controls for 2 years. They have been reporting progress to the board, including their compliance with both the payment card industry Payment Card Industry Data Security Standard (PCI-DSS) and the NIST cybersecurity framework (CSF) controls framework. One day, they are hit by a ransomware attack, and a segment of their systems that runs their loyalty card program was encrypted and ultimately taken offline. Subsequently, 250,000 miscellaneous records were stolen, and the loyalty card program was down for 16 days. Ultimately, a $2.5 million ransom was paid, and another $2 million+ was spent on investigation, remediation, and recovery costs. Following the incident, the CISO was let go because the incident "was their fault" since they owned the risk.

This event didn't have to cost the CISO their job, and there are a number of things they could have done in advance to avoid that fate. We'll explore those concepts in depth throughout the rest of the book. However, for now, we'll say that the CISO should have been regularly reporting the company's current risk exposure, determining if that exposure was aligned with the company's appetite, and sharing the plan to buy down the risk instead of reporting PCI-DSS and NIST CSF compliance. The CISO should have been earning buy-in that the outstanding risk was acceptable and then set expectations for the probability of material impacts. Then, when that event happened and incurred $4.5 million+ in material impact, the CISO could determine that $2 million of that was offset by cyber insurance, and another $1.4 million was transferred to the commercial property policy, leaving $600K in losses, which was within expectations. In that scenario, the board "owns the risk", and the CISO keeps their job.

2.5 ROAD TO THE CISO ROLE

The path to becoming a CISO is rarely a straight line. Some start in the trenches, honing their technical skills, while others come from the business side, bringing a strategic mindset. I've seen both paths lead to success, but the modern CISO needs to be a hybrid, comfortable in both worlds. It's like being a bilingual diplomat, able to speak the language of both tech and business (Table 2.1).

The ideal CISO possesses a combination of technical and business skills. Modern organizations require leaders who can bridge the gap between these two worlds. These hybrid CISOs not only understand the technical intricacies of cybersecurity but also possess the skill to translate these complexities into business terms.

The modern CISO is not solely a technical expert or a business strategist but a blend of both. They can "walk in both worlds", understanding cybersecurity's technical challenges while also speaking the language of business leaders. The modern CISO can translate complex technical concepts into actionable insights for executives and, conversely, convey business priorities to the cybersecurity team.

As we conclude this section on CISO 2.0 after dissecting the responsibilities, challenges, and transformations of CISOs through various stages, it's evident that the role has grown beyond its initial scope, which presents new challenges for would-be security leaders. If you find yourself or your program resonating with any of the

TABLE 2.1
Technical versus Business Centric CISOs

The Technical CISO: The Smartest Head Architect

The technical CISO typically starts as a brilliant cybersecurity expert, often with an extensive background in areas like network security, encryption, and threat detection. They are the individuals who excel in understanding the intricate details of cybersecurity technologies, vulnerabilities, and exploits. Their ascent to the CISO role is often based on their technical prowess and ability to architect robust security solutions.

Challenges of the Technical CISO:

Limited Business Understanding: One of the primary challenges faced by technical CISOs is their limited understanding of business operations, strategy, and management. They may struggle to connect cybersecurity initiatives with the broader goals and objectives of the organization.

Communication Barriers: Technical CISOs may find it challenging to convey complex cybersecurity concepts to nontechnical stakeholders. Effective communication with executives, board members, and employees becomes crucial, but it can be a significant hurdle.

Risk Management: Technical CISOs may prioritize technical solutions without considering the broader risk landscape. This can lead to a disconnect between cybersecurity investments and the organization's overall risk profile.

The Business-Centric CISO: The MBA in the Cybersecurity World

On the other side of the spectrum, the business-centric CISO often starts with a strong business background, frequently holding an MBA. They understand organizational strategy, finance, and management principles, and their path to the CISO role is driven by their ability to align security with business objectives.

Challenges of the Business-Centric CISO:

Limited Technical Proficiency: Business-centric CISOs may lack the technical expertise needed to understand and address the intricacies of cybersecurity threats and solutions. This can lead to critical blind spots in security strategies.

Technical Communication: Communicating effectively with the technical cybersecurity team can be a challenge for business-centric CISOs. They may struggle to grasp the technical nuances and may not be able to evaluate the effectiveness of security measures.

Risk Oversimplification: Business-centric CISOs may oversimplify cybersecurity risks, potentially underestimating the complexity and severity of emerging threats.

descriptions or challenges discussed, you're not alone. These are prevalent issues in many organizations, reflecting the common hurdles that numerous CISOs encounter. This widespread nature of the challenges is precisely what inspired the creation of this book. If you identify with these problems, this book is tailored for you. The ensuing chapters will serve as an illustrative guide, offering insights and strategies to navigate these complexities. We aim to empower you to overcome these challenges and aid in your transformation into a next-generation CISO 3.0. This process won't just elevate your role: it will also elevate the security posture and strategic capability of your entire organization. As you turn each page, you'll gain the tools and knowledge not only to adapt but also to thrive and lead in the new world of cybersecurity leadership.

2.6 CISO 3.0

2.6.1 Why CISO 3.0?

We are entering a new era, one that demands a different kind of security leader. CISO 3.0 marks a significant shift in the role of the chief information security officer, moving beyond a purely technical focus to become a fully integrated partner for business strategy, risk management, and forward-thinking technological innovation. This transformation is essential to meet the increasingly complex challenges of today's threats and to rise to the growing demands of regulators and the board. To thrive in this environment, the CISO 3.0 must meet and overcome the limitations and challenges that current CISO 2.0s face. For most CISO 2.0s, cybersecurity is embedded in IT and seen as a back-office function; the move to 3.0 includes a shift toward security being a front-office function that is a competitive differentiator that supports business growth.

I introduce the concept of CISO 3.0 as a paradigm shift in the role of the CISO. One meant to elevate the role to a truly strategic, business-aligned position where CISOs become full-fledged members of the C-suite, facing the board and actively contributing to creating and communicating business value. The CISO 3.0 will be instrumental in enabling innovation and growth by aligning to business goals.

Now is the time because, like it or not, the role of a CISO is already changing. Maxim Kovalsky, managing director, Cybersecurity and Privacy Services at Grant Thornton Advisors LLC said in the National Association of Corporate Directors (NACD) October 2023 newsletter, "One important immediate consideration for boards as a result of the new SEC regulations is a re-examination of the role of the CISO. The people in these positions have traditionally been hired for their technical expertise, and they often don't report to the board or the CEO. The regulations put the CISO in a more prominent reporting role that requires the ability to engage with boards on a regular basis . . . The question for boards to ask is, 'How do you make the role of the CISO a true executive?'" Maxim added, "Elevating the role of the CISO should be on any board's shortlist in preparation for compliance with the SEC disclosure rules". It doesn't get much clearer than that! Furthermore, that guidance is being given directly to board members via their primary communication channels. In the next chapter, we will explore the regulations and some of the enforcement actions that emphasize the need for next-generation CISOs to understand and accept their fiduciary duties.

This CISO transformation is timely not only because of regulatory pressure but also due to the growing consumer demand for security. We're at a moment where consumers are beginning to care about security and privacy in the same way they now care about safety features in their cars. Think back to the 1970s, before Ralph Nader's campaign to make seat belts mandatory. At that time, car buyers didn't give much thought to seat belts, but today, most people wouldn't even consider buying a car without advanced safety features. We're entering a similar era for cybersecurity.

Regulators like the Securities and Exchange Commission (SEC) and Federal Trade Commission (FTC) are shining a spotlight on security and privacy in digital products, and this is driving consumer expectations. People are starting to demand more transparency and accountability in the way their data is handled, and companies that fail to prioritize this will fall behind. The rapid rise of generative AI services has only accelerated this trend. As AI-powered products become more common, consumers are becoming increasingly aware of the potential risks and are demanding stronger protections.

The role of the CISO is also broadening. Many CISOs I talk to are being tasked with additional leadership responsibilities, such as taking on the roles of data privacy officer (DPO) or chief risk officer (CRO) and overseeing product security, fraud, or even physical security. The consolidation of responsibilities is likely to continue, and the increased expectations from executive leadership are driving this transformation even further. As the scope of the CISO's responsibilities expands, so does the need for next-generation CISOs who can handle these complex roles. So, it stands to reason that CISOs don't have long before this next-gen evolution becomes a table-stakes expectation.

2.6.2 So, What Is a CISO 3.0?

This next evolution in cybersecurity leadership demands a blend of technical acumen, business savviness, and visionary leadership. The CISO 3.0 must be fluent in the language of business, accounting, finance, economics, and risk management, allowing them to communicate effectively with executives and the board. Their role goes far beyond protecting systems; they align security initiatives with the organization's broader goals, demonstrating how cybersecurity can drive business value. CISO 3.0 represents a forward-thinking approach where most strategic drivers are internal, based on what effectively reduces risk for the organization. External pressures are meticulously weighed against the internal cost–benefit analysis. This means that pressures like framework compliance and cyber insurance requests take a backseat to the understanding of what buys down the most risk for the organization. CISOs in this phase should be leveraging third parties to help provide assurance and validation of their program and findings to executives and the board.

A deep understanding of risk management is central to the CISO 3.0 role. This new generation of CISOs recognizes the nuances of risk, from technical vulnerabilities to broader business risks, and can effectively quantify and prioritize these risks to guide decision-making. Quantitative risk management becomes a cornerstone, allowing CISOs to assign monetary values to threats and focus resources on the most critical areas. They lead by shaping a risk strategy that informs both the program and tactical measures, ensuring efforts are always directed toward reducing the greatest risks to the business.

Strategic leadership is another key trait of the CISO 3.0. They focus on buying capabilities instead of tools, understanding how to shop for, build, and maintain capabilities that are business-level choices and, whenever possible, are tool-agnostic or outsourced. This is a departure from many CISO 2.0s who may have piles of redundant controls and shelfware they are unable to rationalize.

In terms of technology, CISO 3.0 isn't just keeping up; they are driving the adoption of innovations like AI, machine learning, and automation. They understand how these technologies can enhance security operations, allowing for proactive threat identification and real-time incident response. Equally important, they are aware of the risks these technologies introduce and work to safeguard the organization against AI-driven threats. This ability to balance cutting-edge technology with risk awareness sets them apart as leaders.

Modern security approaches like identity-centric security, zero trust, and continuous monitoring are also central to the CISO 3.0's toolkit. They understand that by focusing on user identity, real-time data analysis, and adaptive security measures, they can ensure that only the right people have access to sensitive information and that risks are addressed before they escalate. The CISO 3.0 also leads efforts to unify security strategies, bringing together risk management elements into a cohesive program that provides better visibility and quicker responses to threats.

One of the key traits of the CISO 3.0 is their adaptability. Continuous monitoring, real-time insights, and a proactive security approach allow them to stay ahead of evolving threats. They understand that vulnerabilities must be addressed as soon as they are identified, and incidents should be mitigated in real time. Automation and AI are leveraged to streamline processes, enabling the security team to focus on strategic initiatives while maintaining a vigilant defense.

The CISO 3.0 is also a master of decision science. They use data to drive informed decision-making about risk management, translating complex technical risks into business impact and effectively communicating these insights to stakeholders.

Above all, the CISO 3.0 is a skilled communicator. They have the ability to translate complex technical risks into clear, actionable insights that resonate with both technical teams and nontechnical stakeholders. Whether presenting to the board or collaborating with business units, they foster a culture where cybersecurity is seen as a shared responsibility across the organization. This leadership helps build trust and ensures that security is woven into the fabric of the company's strategic goals.

2.6.3 How to Become a CISO 3.0

The journey from CISO 1.0 to CISO 3.0 is a transformation that requires more than just mastering new skills: you also have to change the way you view cybersecurity's role within the organization. CISO 1.0 focuses on protecting systems, but as businesses evolve, security must become integral to operations. Moving to CISO 2.0 involves aligning security with compliance and risk management frameworks, like NIST or ISO, but it remains reactive, driven by external pressures. The true transformation to CISO 3.0 happens when cybersecurity is fully integrated with business strategy, driving outcomes like growth and innovation. A CISO 3.0 communicates in business terms, turning technical risks into discussions about financial impact and ROI, making security a key enabler of business success.

Undertaking this journey requires several key transformations. First, it's essential to expand your knowledge of the business. You need to understand how the organization generates revenue, manages risk, and delivers value to customers. Engage with leaders in finance, operations, and marketing to learn how cybersecurity can protect not only digital assets but also the overall business model. Develop a deeper understanding of risk, not only in terms of vulnerabilities but also in how different threats can disrupt business operations. As a CISO 3.0, you must be able to quantify risk in a way that resonates with executives, using data to demonstrate the business implications of security decisions.

Leadership also plays a significant role in this transformation. As you transition, you will need to focus on building cross-functional relationships and influencing decisions at the executive level. You'll need to position yourself as a trusted advisor to the board and C-suite, contributing to strategic discussions rather than just providing technical updates. As a leader, you'll also need to foster a culture of security across the organization, ensuring that everyone understands their role in protecting the business.

The journey from CISO 1.0 to CISO 3.0 isn't always linear, and it doesn't have to follow a strict progression. While it might seem like you have to move gradually from CISO 1.0 to 2.0 and then to 3.0, the reality is, you can leap straight to 3.0. I frequently hear people suggest the "Crawl, Walk, Run" approach, and while that may make sense for certain projects with very long time horizons, it's always better to skip straight to running if you can. Think about it this way, many of the skills you learn while crawling don't necessarily translate to running. You end up spending time mastering things that don't get you closer to your real end goal. Meanwhile, your competition might be sprinting ahead while you're still figuring out how to crawl. This is one of those moments where you can jump straight to running as a CISO 3.0, regardless of where you're starting from.

Ultimately, this is about transforming yourself and your program into a true strategic asset for your organization. Whether you're currently a CISO 1.0 or 2.0, the leap to CISO 3.0 is not just possible; it's necessary to stay ahead in today's fast-paced, risk-driven world. In this new phase, CISOs are not just adapting to change, they are driving it. They help enable innovation and business growth while safeguarding the organization's most valuable assets. The CISO 3.0 is the future of cybersecurity leadership, prepared to meet the challenges of tomorrow and turn them into opportunities for strategic advancement.

2.7 CONCLUSION

As we wrap up this chapter, it's clear that CISO 3.0 is not just a title but a reimagining of the entire position, aligning it with strategic business objectives, sophisticated risk management, and advanced technological capabilities. This progression is a response to the increased pressure of regulators and consumers, as well as the increasing expectations for CISOs to be business-savvy leaders. The role will now encompass a deep understanding of business drivers, an adeptness in decision science, and a grasp of cutting-edge technologies like AI.

If you see your own experiences or your program reflected in the discussions of CISO 1.0s and 2.0s, know that you're in good company. These are common challenges and considerations many in the field are navigating. The subsequent chapters of this book will examine each of the topics outlined here in greater depth. From the nuanced strategies of quantitative risk management to fiduciary duties and board-level communications, you'll find detailed discussions and guidance aimed at empowering you to navigate these complexities. This book is designed to be your companion as you work toward modern cybersecurity leadership, transforming you into a next-generation CISO 3.0 ready to lead your organization into a secure and strategic future.

2.7.1 Key Takeaways

1. **The Evolution of the CISO Role**
   - The role of the CISO has evolved significantly over time, marked by distinct phases:
     - **CISO 1.0:** Primarily focused on technical tasks such as firewall configurations and antivirus solutions. These professionals operated as technical experts without direct strategic involvement.
     - **CISO 2.0:** Shifted toward compliance, emphasizing framework alignment and collaboration with other departments. This phase introduced strategic roadmaps but often lacked deep integration with business goals.
     - **CISO 3.0:** Represents the current and future iteration of the CISO as a strategic leader, integrating cybersecurity into core business strategies, enabling innovation, and fostering resilience.
2. **Core Competencies of a CISO 3.0**
   - **Strategic Business Alignment:** The CISO 3.0 collaborates across departments, aligning cybersecurity initiatives with organizational objectives and financial goals.
   - **Quantitative Risk Management:** Emphasizes the use of data and analytics to quantify risks, enabling informed decisions about prioritization and resource allocation.
   - **Technological Leadership:** Adopts advanced tools such as AI and automation, balancing their benefits with associated risks.
   - **Advanced Communication:** Bridges the gap between technical teams and business leaders, presenting cybersecurity concerns in relatable and actionable terms.
   - **Decision Science Expertise:** Applies decision-making frameworks to balance competing priorities and optimize risk-reduction strategies.
3. **The Challenges and Opportunities of Transitioning to CISO 3.0**
   - **Challenges:**
     - Limited authority in some organizations, requiring CISOs to advocate and communicate effectively.
     - Overcoming siloed thinking to integrate cybersecurity into broader business processes.
     - Avoiding inefficiencies caused by redundant or misaligned security tools.
   - **Opportunities:**
     - Positioning cybersecurity as a business enabler rather than a cost center.
     - Proactively driving innovation while ensuring organizational resilience.
     - Building stronger stakeholder trust through transparent and impactful communications.

4. **Practical Steps to Transition**
   - **Learn the Language of Business:** Develop fluency in finance, operations, and strategic planning to align cybersecurity initiatives with business priorities.
   - **Focus on Risk Quantification:** Shift from qualitative assessments to measurable impacts, enabling precise and defensible resource allocation.
   - **Enhance Cross-Functional Collaboration:** Build relationships with departments such as finance, operations, and legal to ensure comprehensive risk management.
   - **Lead Cultural Change:** Foster a culture where cybersecurity is everyone's responsibility, integrating it into daily operations and decision-making processes.
   - **Invest in Continuous Learning:** Stay updated on emerging technologies, regulatory changes, and best practices to maintain relevance and leadership.

2.7.2 Reflection and Exploration Questions

Understanding the Concepts

1. How do the tenants of the CISO 3.0 redefine the traditional role of a CISO?
2. How does the CISO 3.0's approach to cybersecurity differ from earlier iterations?

Real-World Application

3. What real-world examples can you think of that demonstrate the qualities of a CISO 3.0?
4. Have you encountered situations where a lack of strategic alignment hindered cybersecurity efforts? How could a CISO 3.0 have addressed those challenges?

Personal and Professional Growth

5. How might you apply the main takeaways from this chapter to your own career or organizational role?
6. What steps can you take to strengthen your ability to communicate cybersecurity strategies to nontechnical stakeholders?

Further Exploration

7. What steps can you take to begin your transition to the CISO 3.0?
8. How might emerging trends, such as AI and regulatory shifts, shape the role of the CISO in the future?
9. What additional skills or resources do you think are critical for becoming a successful CISO 3.0?

3 The Evolving Regulatory Landscape

3.1 INTRODUCTION

The regulatory environment surrounding cybersecurity is advancing at a rapid pace and, with it, the responsibilities of today's CISOs. The role of the CISO has transformed into a position that intersects business leadership with regulatory compliance. As a result, understanding the legal landscape has never been more critical for security leaders.

In this chapter, we'll look into the modern regulatory framework, exploring key laws and regulations that are reshaping corporate governance and cybersecurity accountability. These include landmark legal decisions like Caremark, emerging SEC regulations, and rules from the Federal Trade Commission (FTC) and New York Department of Financial Services (NYDFS), all of which set the stage for the future of cybersecurity leadership. We will examine some of the major regulatory shifts, focusing on how CISOs can navigate these changes, anticipate risks, and stay ahead of potential legal pitfalls.

At the core of these changes is an increasing emphasis on transparency, accountability, and the alignment of cybersecurity efforts with broader business strategies. Regulators are no longer just focused on technical failures; they're scrutinizing how well companies manage their risks, communicate with stakeholders, and adhere to their fiduciary duties. From the expansion of Caremark principles into cybersecurity to the implications of the SEC's cybersecurity disclosure rules, it's clear that we need a modern, business-aligned approach to cybersecurity governance. Beyond that, as the legal landscape shifts, we must move to meet it because the CISO is now expected to play a part in managing these obligations at the highest levels of corporate leadership. The stakes are high, but so are the opportunities for CISOs to solidify their role as indispensable leaders in this new era of corporate accountability.

3.2 CAREMARK LAWSUITS

Caremark lawsuits, originating from the landmark *In re Caremark International Inc. Derivative Litigation* case in 1996, represent a specific kind of shareholder derivative action. These suits aim to hold corporate directors accountable for failing to oversee the company's most critical risks. In many ways, Caremark cases form the backbone of modern corporate governance when it comes to the fiduciary duties of directors. The original Caremark decision established that directors could face personal liability if they neglected to provide "reasonable" oversight of company operations, especially concerning risks that are essential to the organization's survival and success.

Since then, the concept of Caremark liability has evolved to mirror shifting views on risk within corporate governance. Over the years, this has expanded to cover areas like cybersecurity. This progression in legal thinking underscores the growing

DOI: 10.1201/9781003510789-4

demand for rigorous board-level vigilance over "mission critical" areas that, if mishandled, could endanger the entire organization.

Today, understanding Caremark lawsuits is fundamental for corporate governance, as these cases outline directors' duties to actively protect the company's vital interests. The principles of duty and oversight introduced by Caremark have become core to board responsibilities, guiding directors in managing complex, evolving risks.

We'll dive deeper into Caremark cases and the fiduciary duties they establish in Chapter 5, following our discussion of the board of directors' role.

For now, just know that directors owe a duty of loyalty, which includes the key concepts of "reasonable" oversight over "mission critical" risks.

3.2.1 Defining the Reasonable Standard and Mission Critical Risks

While it is possible for shareholders to sue board directors, Caremark established a high bar for director liability. A successful lawsuit requires a plaintiff to demonstrate that the board acted with *knowing* bad faith, meaning they *utterly failed* to implement a system for monitoring critical risks or completely disregarded glaring red flags. "Reasonable" oversight, which was laid out in the *Marchand (Blue Bell Ice Cream) vs Barnhill* case, doesn't equate to guaranteeing perfect execution or oversight. It signifies a good faith effort to establish a system for identifying and mitigating the biggest risks.

Mission critical risks are those that could pose an existential threat to a company's financial health or core operations. The question of whether cybersecurity is one of those mission critical activities has been the focus of several recent legal battles.

***Marchand v. Barnhill* (2019) (Bluebell):** This case, although not directly related to cybersecurity, highlighted the potential application of Caremark to this area. The court faulted the board for lacking a systematic approach to food safety oversight, a concept that could be extended to cybersecurity preparedness.

In the *Boeing Case,* Shareholder lawsuits following a series of high-profile 737 Max crashes alleged the board failed to oversee safety protocols adequately. While the case ultimately settled, it expanded mission criticality and highlighted the potential vulnerability of directors who neglect critical aspects of a company's operation, which could include cybersecurity.

McDonald's Corporation Shareholder Derivative Litigation (2020) represents another meaningful development. Here, the court allowed a Caremark claim to proceed against certain officers, not just directors, suggesting an expansion of fiduciary duties beyond the boardroom in cases of critical risk. This decision underscores the potential for broader accountability within a company's leadership structure.

Several recent cases illustrate the growing application of Caremark principles to cybersecurity more directly. In *Firemen's Retirement System of St. Louis on behalf of Marriot International, Inc. v. Sorenson* (2022), often referred to as the *Marriot* case, the court dismissed a Caremark claim against directors for a data breach. Nevertheless, the court acknowledged the increasing importance of cybersecurity and the potential for future claims to succeed, particularly if the board exhibits a blatant disregard for data security.

Only a month after the Marriot case, shareholders brought a Caremark claim against T-Mobile for a data breach in *T-Mobile US, Inc. Shareholder Derivative Litigation* (filed November 2021). This lawsuit alleged breaches of fiduciary duty by the T-Mobile board in connection with cybersecurity failures. The case centered around two key data security incidents: A data breach impacting 54 million customers in August 2020 and a Federal Communications Commission (FCC) fine levied in February 2021 for data security weaknesses. The lawsuit specifically focused on whether the T-Mobile board exhibited a "knowing" failure to oversee cybersecurity risks, as defined in Caremark. The T-Mobile shareholders aimed to differentiate their claim from the Marriott claim, which was dismissed because the plaintiffs couldn't demonstrate that the board ignored known legal violations ("red flags"). The argument hinged on the fact that no established legal standards for data security existed at the time. The T-Mobile case argued that the FCC investigation and resulting fine established a "red flag" that the board "failed to heed", potentially demonstrating a breach of fiduciary duty. The suit was ultimately dismissed because the plaintiffs couldn't establish standing. Yet, it still highlights the possibility for cyber breaches and related fines to constitute a breach of fiduciary duty. Courts are increasingly scrutinizing how boards address data security risks, and the presence of FCC fines or other legal violations could be used to support claims of director negligence in future cases.

The most high-profile cases in recent history involved a cyberattack on SolarWinds, a software company. The court, while ultimately dismissing the Caremark claim, recognized the potential for future success in similar cases if directors fail to implement a *reasonable* system for overseeing cybersecurity risks. This decision further emphasizes the developing definition of "mission critical" in the modern era, potentially encompassing robust cybersecurity measures.

The Caremark doctrine has become a powerful tool for holding corporate leadership accountable for cybersecurity preparedness. The courts are increasingly recognizing cybersecurity as a "mission critical" risk, and the standard for "reasonable" oversight is likely to adapt as cyber threats continue to evolve. The extension of fiduciary duties to officers beyond the boardroom demonstrates a broadening scope of potential liability for C-level officers including the CISO. Companies that fail to prioritize cybersecurity and establish appropriate oversight mechanisms may find themselves facing not only financial losses from breaches but also legal challenges from shareholders and, potentially, employees.

3.3 NAVIGATING THE SEC'S CYBERSECURITY RULE

The Securities and Exchange Commission (SEC) took a major step toward increased transparency by enacting its final rule on cybersecurity risk management, strategy, governance, and incident disclosure on *July 26, 2023*. This landmark rule, often referred to simply as the SEC's cyber rule, fundamentally alters cybersecurity disclosures for public companies. The rule's core provisions take effect in two parts:

Incident Disclosure on Form 8-K: Public companies must begin filing disclosures about material cybersecurity incidents on Form 8-K within *4 business days* of determining the incident's materiality. This requirement became effective in mid-December 2023.

Annual Disclosures on Form 10-K: Companies must include a full overview of their cybersecurity risk management strategy, governance, and incident history in their annual filings on Form 10-K. This requirement also took effect in mid-December 2023.

The SEC's rule creates a new level of accountability for public companies and their cybersecurity posture. The SEC has a well-established track record of addressing cybersecurity through enforcement actions, even before the new cyber rule was enacted. These actions underscored the growing importance of cybersecurity as a critical risk for public companies. Notably, the SEC brought an enforcement action against SolarWinds and its CISO, Tim Brown, for disclosure failures related to the Sunburst cyberattack. It's important to clarify that these charges stemmed from *unrelated disclosure regulations* in place at the time, as the SEC's cyber rule wasn't yet in effect during the SolarWinds incident. This enforcement action serves as a potent illustration of the SEC's commitment to holding companies and their leaders accountable for cybersecurity transparency.

The final rule mandates two primary disclosure avenues for public companies:

Form 8-K: Companies must now report any material cybersecurity incidents within 4 business days of determining their materiality. This report, filed on Form 8-K, should detail the nature and scope of the incident, as well as its potential financial impact.

Form 10-K: Annual filings on Form 10-K now require a more thorough overview of a company's cybersecurity risk management strategy, governance, and incident history. This includes details about the company's cybersecurity policies, procedures, and risk assessment processes.

The SEC's rule necessitates closer collaboration between CISOs and legal teams. Traditionally, CISOs have sometimes been excluded from the development of filing language. This practice, however, needs to be reevaluated. CISOs are central to identifying and assessing the materiality of cyber incidents. They will need to gather and analyze data to determine the nature and scope of the incident, as well as its potential financial impact. This information will be vital for legal teams to accurately draft the disclosures for both Form 8-K and Form 10-K filings.

The SEC's lawsuit against SolarWinds CISO Tim Brown underscores the heightened accountability placed on CISOs under the new rule. The SEC alleged that Brown, along with SolarWinds itself, failed to disclose the severity of the Sunburst cyberattack adequately. This case is a reminder that CISOs can be held personally liable for the accuracy and completeness of risk disclosures.

3.3.1 Materiality: The Heart of the Matter

The concept of *materiality* lies at the core of the SEC's rule, but the SEC doesn't provide a definitive definition of materiality within the cyber rule itself. The responsibility falls on organizations and their officers to determine what level of risk and events constitute a material cybersecurity incident. This determination requires careful consideration of both quantitative and qualitative factors.

The SEC's cyber disclosure rule represents a major change in how companies should approach cybersecurity, and it marks a new era of accountability for public companies and their officers. The SEC's rule necessitates a shift in how companies approach cybersecurity governance. CISOs will need to adapt by fostering close collaboration with legal teams. This collaboration will be important for ensuring accurate and timely disclosures of both cyber incidents on Form 8-K and ongoing cybersecurity risks in Form 10-K filings.

Now more than ever, businesses are under pressure to prioritize cybersecurity, committing to strong risk management programs that go beyond the basics. Effective detection, prevention, and response capabilities are necessary not only to reduce cyber risks but also now to ensure that disclosures are both timely and accurate.

This rule also takes investor protection to a new level. With more transparency around a company's cybersecurity posture and history of incidents, investors can make better-informed decisions about where to place their money. This added clarity empowers investors, allowing them to assess risks and opportunities with greater insight. At the same time, this will bring new legal challenges. Companies that fall short of compliance could face lawsuits from investors and scrutiny from regulators. This makes it critical for businesses to fully understand the rule's requirements.

Navigating the complexities of cybersecurity governance in the face of the SEC's rule requires a multifaceted approach. Understanding the nuances of materiality, fostering collaboration between CISOs and legal teams, and prioritizing robust cybersecurity programs will all be important for companies seeking to prepare for this new level of accountability.

3.4 NAVIGATING THE FTC SAFEGUARDS RULE

The Federal Trade Commission (FTC) has a longstanding commitment to protecting consumer data, often taking enforcement actions against companies that fail to meet adequate data security standards. In 2021, the FTC advanced its efforts with the finalization of its Safeguards Rule, officially titled "Standards for Safeguarding Customer Information". This rule sets mandatory cybersecurity standards for certain financial institutions under the FTC's oversight.

The FTC Safeguards Rule aligns cybersecurity initiatives closely with consumer protection, drawing some notable parallels to the SEC's cyber disclosure rule. Here's a look at how these rules compare:

Focus: The SEC's rule is rooted in transparency, requiring companies to disclose cybersecurity risks and incidents openly. The FTC Safeguards Rule, by contrast, focuses on the proactive implementation of a comprehensive data security program aimed specifically at protecting customer information.

Covered Entities: The SEC's rule applies strictly to public companies registered with the SEC, while the FTC Safeguards Rule casts a wider net. It includes financial institutions like lenders, mortgage brokers, and payday lenders, as well as nonbank entities that handle customer financial information, such as investment advisors and tax preparers. Data brokers, which gather and sell consumer financial data, are also covered under the FTC's jurisdiction.

By comparing these rules, we see two complementary approaches to data security: The SEC's focus on transparency and risk awareness, and the FTC's emphasis on concrete security measures to protect consumer information. Together, these standards reflect an evolving regulatory landscape that prioritizes both corporate accountability and consumer privacy.

3.4.1 Requirements of the FTC Safeguards Rule

The FTC Safeguards Rule outlines a set of core requirements that covered institutions must implement to safeguard customer information. These requirements focus on establishing a robust information security program that addresses the following key areas:

Designation of a Safeguards Program: The rule mandates the appointment of a qualified individual to oversee the information security program and ensure its ongoing effectiveness. Basically, this means organizations need a CISO or at least someone who functions as a CISO.

Risk Assessments: Organizations are required to assess potential vulnerabilities and threats to customer data regularly. These assessments should take into account the sensitivity of the information, possible attack methods, and the potential consequences of a data breach.

Data Security Controls: The FTC Safeguards Rule doesn't prescribe specific security controls. However, it emphasizes the need for a layered approach that includes access controls, data encryption, data disposal procedures, and ongoing monitoring activities.

Incident Detection and Response: Organizations must have a documented incident response plan to effectively identify, contain, and recover from data security incidents. This plan should include procedures for notifying law enforcement and affected consumers, as well as a process for investigating and remediating the root cause of the incident.

Employee Training and Awareness: The rule highlights the importance of employee training in raising awareness of cybersecurity threats and best practices for data security.

While the FTC Safeguards Rule doesn't mandate specific technologies or solutions, it does set the expectation that organizations implement a "reasonable program" to protect customer data. By examining past FTC enforcement actions and leveraging industry best practices, organizations can build a robust data security program. This proactive approach helps companies not only meet compliance standards but also mitigate the risk of breaches and potential penalties.

While the SEC and FTC rules focus on different areas of cybersecurity, they work together to encourage stronger cybersecurity practices for public companies and financial institutions. Understanding and aligning with both sets of requirements will be crucial for organizations moving forward.

Even if you aren't a covered entity under the FTC Safeguards Rule, you can still benefit from reviewing consent orders that resulted from FTC enforcement actions. These orders reveal what courts typically regard as "reasonable" and can serve as a guide for creating a defensible set of controls. To help with this, I've developed a framework called the Secure Care Standard, which crosswalks controls recommended by recent consent orders with those in the FTC Safeguards Rule. You'll find an example of this in Figure 3.1, and the full version is available for download at CISO30.com. The Secure Care Framework enables you to pinpoint controls that appear in multiple regulatory or enforcement sources, allowing for a quick assessment of your current control environment and identification of potential deficiencies.

SecureCare Standard

Framework for reasonableness, defensibility and minimum standard of care

16 CFR Part 314 | 17 CFR Parts 229 | 17 CFR Parts 232

Category	Sub Category	Security Control Activity Category	Required Control	Requirements	Description	Chegg	Drizly	CAFEPRESS	FTC Safeguard Rule	FSEC Cyber Rule	SEC Broker Dealer Cyber Rule
				CI = covered information	Plain English	1/25/2023	1/9/2023	4/23/2022	6/9/2023	Est 10/2023	Est 10/2024
Governance											
Governance	Governance	BOD Oversight	Board of Directors Oversight	Designate a senior member of your personnel responsible for direction and oversight of the Qualified Individual	Designate oversight for CISO	V C	IV C	II C		Proposed Item 106(c)(1) of Regulation S-K	
Governance	Governance	Policy & Reporting	Policy & Reporting	No more misrepresentation to consumers in published Security Policy	Stop misrepresenting policies over information collection and usage or security measure	NAP	I	I			
Governance	Governance	Business Continuity / Disaster Recovery	BC/DP Plan	Proposed Item 106(b) of Regulation S-K would Amend Form 10-K to **require disclosure, as applicable, of whether: The registrant has business continuity, contingency, and recovery plans in the event of a**	Firms should have BC/DR plan, policies, and controls, and test at least annually						Proposal section B.1.e
Governanace	Cybersecurity Risks and Incidents Disclosure	Disclosure & Reporting	Incident & Risk Disclosure & Reporting		Report Cyber incidents and material risks on 8-k, 10-k, SOR, etc.					Proposed Item 106(d) of Regulation S-K	See paragraph (c)(1) of proposed Rule 10 See paragraph (c)(2) of proposed Rule 10 See paragraph (b)(1)(vi)(A)(4) of proposed Rule 10.
Governance	Data Governance	Data Lifecycle	Data Retention Policy	Delete or destroy all Covered Information that is not being used or retained in connection with providing products or services to Corporate Respondent's customers, and provide a written statement to	Mandated Deletion and Data Minimization	NAP	II		16 CFR 314.4(c)(6)(i)		
Governance	Data Governance	Data collection, storage and retention	Data Warehousing	Policies and procedures to minimize data collection, storage, and retention, including data deletion or retention policies and procedures;	Policies and procedures to minimize data collection, storage and retention	V E.4		II.2, II.E.3			
Governance	M&A	Due Diligence	M&A Due Diligence	Assess, prior to the acquisition of any entity that maintains, processes, or transmits Covered Information ("Acquired Entity"), the effectiveness of that entity's safeguards to	Perform pre-aquistion due diligence over cyber security controls	V G					
Security											
Security	Governance	Comprehensive Information Security Program	Comprehensive Information Security Program	establish and implement, and thereafter maintain, a comprehensive information security program ("Information Security Program") that protects the security, availability, confidentiality, and integrity of Covered Information under Respondent's control. To satisfy this requirement, Respondent must, at a minimum:	Establish, Implement and Maintain, a comprehensive Information Security Program to protect security, availability, confidentiality and integrity of Covered Information	V	IV	II (includes privacy and excludes availability)	17 CFR 314.4(c) 17 CFR 314.4(g)	17 CFR Parts 229	17 CFR Parts 232
Security	Governance	CISO	CISO	Designate a qualified employee to coordinate and be responsible for the CISP	Designate a CISO	V C	IV C	II C	16 CFR 314.4(a)	Proposed Item 106(d) of Regulation S-K	
Security	Governance	Annual BoD Reporting	Annual BoD Reporting	Provide the written program and any evaluations thereof or material updates thereto to	Provide BOD updates at least annually	V B	IV B	II B	16 CFR 314.4(i)		
				establish, maintain, and enforce written policies and procedures that are reasonably designed to address the Covered Entity's cybersecurity risks In particular, the policies and procedures would need to address:	Firms should have some type of risk assessment program, aligned to a well-known framework like NIST or ISO, and it should be updated at least annually and include risks and responses including controls validation activities. Controls should be identified and implemented to prevent unauthorized access (e.g., 2-factor identification)						

FIGURE 3.1 Secure Care Standard.

3.5 NAVIGATING THE NYDFS CYBERSECURITY RULE

The NYDFS has established itself as a leader with its *NYDFS Cybersecurity Requirement.* This rule, enacted in 2017 and amended in 2023, sets a stringent standard for cybersecurity practices within the financial services industry operating in New York state.

Unlike the SEC's focus on disclosure or the FTC's emphasis on prevention, the NYDFS cybersecurity rule takes a more detailed, prescriptive approach. It mandates specific actions and controls while still requiring a risk-based strategy. Additionally, the rule applies to a wider range of financial institutions in New York. The NYDFS cybersecurity rule shares some common ground with the SEC's cyber disclosure rule and the FTC Safeguards Rule. Here's a breakdown of the NYDFS rule and how it compares to the others:

Focus: The NYDFS cybersecurity rule prioritizes the implementation of a *robust cybersecurity program* to protect consumer data and financial systems from cyber threats. It emphasizes a risk-based approach tailored to the specific risks faced by each covered entity.

Covered Entities: The NYDFS rule applies to a broad range of organizations operating in the financial services sector within New York, including

- banks
- insurance companies
- money transmitters
- investment advisors
- virtual currency companies.

This broader scope encompasses many organizations that wouldn't be covered under the SEC's rule (public companies) or the FTC Safeguards Rule (certain nonbank institutions).

3.5.1 Requirements of the NYDFS Cybersecurity Rule

The **NYDFS cybersecurity rule** outlines specific, mandatory requirements for covered entities, which are more detailed and prescriptive compared to the FTC Safeguards Rule. Here are the key aspects:

Comprehensive Cybersecurity Program: Organizations must create and maintain a written cybersecurity program that matches the level of risk they face. This program should cover areas like risk assessments, access controls, incident response, and data security monitoring.

Periodic Risk Assessments: Regular risk assessments are required to identify and prioritize vulnerabilities. These assessments should take into account evolving threats, the sensitivity of data, and the potential impact of a cyberattack.

Data Security Controls: The rule mandates several data security measures, including data encryption (both at rest and in transit), access controls, and multifactor authentication.

Incident Response Plan: Organizations must have a documented plan for responding to cybersecurity incidents. This plan should outline steps for identifying, containing, and recovering from incidents, as well as notifying regulators and affected consumers.

Third-Party Service Provider Management: The rule also extends cybersecurity accountability to third-party service providers. Covered entities are required to conduct due diligence to ensure these providers have adequate cybersecurity controls in place.

Independent Penetration Testing: Regular, independent penetration testing is required to uncover vulnerabilities in an organization's security posture before attackers can exploit them.

The NYDFS rule sets a high standard for cybersecurity in the financial services sector, requiring a robust cybersecurity program, frequent risk assessments, and strong data security controls, all aimed at protecting consumer data and financial systems from cyber threats.

For financial organizations, understanding the NYDFS rule alongside the SEC's disclosure requirements and the FTC Safeguards Rule offers a comprehensive view of the regulatory landscape. However, even if your organization isn't a financial institution or isn't operating within New York state, there are still valuable lessons to be learned from the controls mandated by NYDFS. These controls represent best practices that essentially form the baseline, or "table stakes", for a reasonable and defensible security posture.

For organizations not directly covered by NYDFS regulations, paying attention to these requirements can provide an additional guidepost on the path to reasonable and defensible cybersecurity practices. NYDFS regulations cover foundational areas like access controls, multifactor authentication, and incident response planning, which are universally applicable for mitigating cyber risk. By aligning with these standards, organizations can reduce their vulnerability to cyber threats while demonstrating proactive risk management, which is especially valuable in the event of a regulatory inquiry or litigation.

For non-covered entities, understanding these regulations shouldn't be about compliance but instead about establishing a security program that is both reasonable and defensible. Adhering to these best practices helps organizations frame their approach as one grounded in recognized standards. It's important to remember, though, that alignment with regulations like NYDFS is not a substitute for a thorough framework-based assessment using National Institute of Standards and Technology (NIST), Center for Internet Security (CIS), or other industry standards. These broader frameworks provide the structural foundation, while regulatory guidelines and consent orders offer direction for prioritizing critical gaps. Using regulations, enforcement actions, and consent orders as benchmarks can help guide and prioritize efforts, focusing resources where they're most impactful.

3.6 ENFORCEMENT ACTIONS AND LESSONS LEARNED

Cybersecurity enforcement is constantly developing, with regulators like the SEC, FTC, and NYDFS actively pursuing actions against organizations that fail to protect sensitive data adequately.

Here's a closer look at some recent enforcement actions and the valuable lessons they offer:

3.6.1 SEC Enforcement Actions

SolarWinds Supply Chain Attack (2023): The SEC brought a high-profile enforcement action against SolarWinds and its CISO for failing to disclose material information about the Sunburst cyberattack that compromised their Orion software platform. This case highlights the SEC's growing focus on *disclosure failures* related to cybersecurity incidents, even before the enactment of its cyber disclosure rule. (Note: The SEC's cyber disclosure rule came into effect later in 2023).

R.R. Donnelley & Sons Company (RRD) Cybersecurity Incident (2021): The SEC fined RRD over $2.1 million for disclosure and internal control failures related to cybersecurity incidents and alerts in late 2021. The SEC's order highlighted RRD's insufficient controls for escalating cybersecurity incidents to management and protecting company assets from cyberattacks. This case reinforces the need for effective internal controls and timely reporting of cybersecurity incidents to management responsible for disclosure decisions.

Lessons Learned: The SolarWinds case was a wake-up call for the entire industry. It showed us that even the most sophisticated organizations can be vulnerable and that CISOs can be held personally accountable for their actions. It's a sobering reminder that we need to be constantly vigilant and proactive in our risk management efforts.

We also learned that organizations must prioritize transparent and timely disclosure of cybersecurity incidents that could have a material impact on their business. Developing clear internal communication channels and protocols for incident reporting is fundamental. Cybersecurity measures should be integrated with financial reporting processes to mitigate the risk of fraud enabled by cyberattacks. The RRD case further underscores the necessity of robust internal controls and procedures for escalating and addressing cybersecurity incidents.

3.6.2 FTC Enforcement Actions

Global Tel Link Corp. Data Breach (2023): The FTC reached a settlement with Global Tel Link Corp. for inadequately securing customer data in the cloud, leading to a public exposure of personal information for over 650,000 individuals. This case showcases the FTC's focus on data security practices and the importance of implementing appropriate cloud security measures.

Drizly Data Security Failures (2021): The FTC imposed a $1.2 million penalty on Drizly, an alcohol delivery service, for failing to implement reasonable security measures to protect customer data following a data breach. This case underscores the FTC's broad reach in regulating data security practices, extending beyond traditional financial institutions.

Lessons Learned: Organizations must conduct thorough *security assessments* of cloud environments and implement robust *access controls* to safeguard sensitive data. Regular employee training on data protection best practices is also essential. The FTC's enforcement actions demonstrate that all organizations handling consumer data are subject to scrutiny.

3.6.3 NYDFS Enforcement Actions

EyeMed Vision Care Data Breach (2022): The NYDFS imposed a $4.5 million fine on EyeMed Vision Care for a data breach that exposed the personal health information of hundreds of thousands of customers. This case underscores the NYDFS's strict enforcement of its cybersecurity rule and its focus on *protecting consumer data* within the financial services sector.

Virtual Currency Entity Enforcement Actions (Ongoing): The NYDFS has been actively pursuing enforcement actions against virtual currency entities for violations of its Cybersecurity Rule and Anti-Money Laundering (AML) requirements. These actions highlight the NYDFS's focus on regulating the burgeoning cryptocurrency industry and ensuring robust cybersecurity practices within this sector.

Lessons Learned: Organizations subject to the NYDFS cybersecurity rule, or those operating in the financial sector more broadly, should meticulously follow the rule's requirements and invest in *robust cybersecurity programs.* Virtual and cryptocurrency businesses should be aware of the regulatory environment and ensure compliance with both cybersecurity and AML regulations.

Even for organizations that don't fall under the direct purview of the SEC, FTC, or NYDFS regulations, the principles and requirements outlined within these rules are increasingly becoming the de facto standard for cybersecurity best practices. Adhering to the guidance provided in these regulations demonstrates a commitment to a strong cybersecurity posture and can be a valuable asset in the event of a cyber incident or legal action.

3.6.4 Risk Quantification and Materiality: Cornerstones of Reasonableness

Establishing a "reasonable standard of care" in cybersecurity hinges on a data-driven approach that incorporates risk quantification and predetermining materiality thresholds. Quantifying potential losses associated with various cyber threats assists with the prioritization of cybersecurity investments and resource allocation. Similarly, *predetermining what constitutes a material cybersecurity risk* streamlines the incident response process and ensures timely and appropriate disclosures in the event of an incident. These proactive measures not only strengthen an organization's overall cybersecurity posture but also serve as powerful evidence of a well-reasoned and responsible approach to data security. We will discuss how to quantify risk and predetermine materiality later in the book.

3.7 IMPLICATIONS FOR CISOS: INCREASED ACCOUNTABILITY

The SEC has recently taken enforcement actions against several companies and individuals for cybersecurity disclosure failures. These actions highlight the increasing importance of disclosure for publicly traded companies and the importance of cybersecurity risk management for all organizations. The SEC's charges against SolarWinds' CISO and the criminal charges against Uber's CISOs demonstrate the potential for personal liability for CISOs in the event of security failures.

These cases make it clear that modern CISOs must be well-versed in legal and regulatory requirements, actively participate in the disclosure processes, and ensure the accuracy of information shared with stakeholders. The days of the CISO being blissfully unaware of disclosures and financial reporting are over. We're now in the spotlight, expected to be strategic leaders who can navigate the complex legal and regulatory landscape. It's a daunting responsibility, but it's also an opportunity to elevate our role and make a real impact on the business.

3.7.1 The Fallout of SolarWinds and Uber

The SolarWinds and Uber incidents, along with the enforcement actions taken against their CISOs, mark a major shift in the accountability landscape for cybersecurity leaders. The SEC's charges against the SolarWinds CISO for alleged misrepresentations and omissions in disclosures underscore the need for accurate and transparent reporting of cybersecurity risks and incidents.

While the SolarWinds case didn't directly involve fiduciary duty, it serves as a cautionary tale for boards. Despite the massive impact of the breach, the directors weren't held liable, mainly due to the specifics of the case and the complexities involved in attributing liability in such a widespread attack. Additionally, the court dismissed claims regarding the failure to disclose the breach's financial impact, as this happened before the SEC Cyber Rule mandated such disclosures. However, a fraud charge remains, tied to statements made on the SolarWinds website. This suggests that CISOs and their companies could still face fraud charges for lack of transparency in disclosing risks and incidents. Ultimately, the SolarWinds case highlights the steep financial and reputational costs tied to cybersecurity failures and emphasizes the need for rigorous oversight and risk management, even without explicit regulatory mandates.

In the Uber case, the CISO faced criminal charges for his handling of a data breach, emphasizing the potential for personal liability for CISOs in the event of security failures. These incidents go to show that CISOs must be well-versed in legal and regulatory requirements, actively participate in disclosure processes, and ensure the accuracy of information shared with stakeholders. CISOs are now expected to be strategic leaders who can navigate the complex intersection of cybersecurity, law, and business.

The SEC's recent enforcement actions also stress the need for solid internal controls and sub-certifications to guarantee the accuracy of disclosures. Companies are now expected to establish rigorous processes to identify and evaluate cybersecurity risks, communicate these risks to senior management and the board, and disclose material information to investors accurately and on time. This new landscape places a premium on transparency, proactive communication, and a comprehensive approach to cybersecurity risk management.

The fallout from these incidents has been far-reaching, impacting not only the individuals involved but also the broader cybersecurity community. The SolarWinds incident exposed the vulnerabilities of the software supply chain, prompting increased scrutiny of third-party vendors and their security practices. It also led to a heightened awareness of the potential for sophisticated, state-sponsored cyberattacks.

The Uber case brought to the forefront the ethical considerations surrounding data breaches and the importance of responsible disclosure. It also raised questions about the role of bug bounty programs and the potential for legal repercussions for security researchers. These incidents have spurred organizations to reassess their cybersecurity strategies, strengthen their incident response plans, and invest in more robust security measures. They have also prompted CISOs to seek legal counsel and insurance coverage to protect themselves from potential personal liability.

The SolarWinds and Uber incidents serve as cautionary tales for CISOs, highlighting the importance of proactive risk management, transparent communication, and ethical decision-making in the face of cybersecurity challenges. The lessons learned from these cases will shape the future of cybersecurity leadership and the role of CISOs for years to come.

The SEC's actions have far-reaching implications for CISOs and boards of directors. CISOs must now take a more proactive role in risk management and disclosure, while boards must ensure that they have the necessary expertise and oversight to manage cybersecurity risks effectively. The SEC's message is clear: With great power comes great responsibility, and CISOs and boards must rise to the challenge of protecting investors and the integrity of the markets.

3.8 CONCLUSION

To sum it up, the world of cybersecurity governance is changing fast. We've seen how major incidents like SolarWinds and Uber, along with the SEC cracking down, have put CISOs and boards in the hot seat. It's not enough anymore for CISOs to just be tech whizzes: they need to be fluent in the legal and regulatory side of things, too. The days of focusing purely on the technical bits are long gone. Now, CISOs have to be actively involved in disclosing information, making sure it's both accurate and transparent for everyone involved.

And the impact of these incidents? It's like a ripple effect, reaching far beyond the companies that were directly hit. The whole cybersecurity world is feeling the heat, with a spotlight on software supply chains, the looming threat of state-sponsored attacks, and the ethical handling of data breaches.

Organizations can't just sit back and wait. They need to take a comprehensive approach to cybersecurity governance. That means strong internal controls, putting numbers on cyber risks, setting clear thresholds for what's considered a major issue, and creating a culture where transparency and accountability are the norm. It's not just about avoiding legal trouble and financial hits: it's about showing a real commitment to data security. And let's face it, that's what everyone expects these days.

The SolarWinds and Uber debacles are lessons we can't ignore. They're shaping the future of cybersecurity leadership. CISOs and boards have a big responsibility to protect investors and keep the markets running smoothly. In this constantly changing field, having great power means shouldering great responsibility.

3.8.1 Key Takeaways

1. **The Changing Regulatory Landscape**
   - **Evolving Legal Environment:** The cybersecurity regulatory environment is rapidly advancing, placing increased emphasis on transparency, accountability, and corporate governance.
   - **Caremark Doctrine:** The principle of "reasonable" oversight for mission-critical risks, such as cybersecurity, has extended fiduciary responsibilities to directors and officers, including CISOs.
   - **Landmark Cases:** Cases like *Marchand v. Barnhill* and the SolarWinds enforcement highlight the need for rigorous oversight of cybersecurity risks at the board and executive levels.
2. **Key Regulatory Developments**
   - **SEC Cybersecurity Rule:** Requires public companies to disclose material cybersecurity incidents promptly and provide annual overviews of cybersecurity governance, strategy, and risk management.
   - **FTC Safeguards Rule:** Mandates robust cybersecurity programs for financial institutions, including regular risk assessments, data security controls, and incident response plans.
   - **NYDFS Cybersecurity Rule:** Implements prescriptive requirements, such as risk-based cybersecurity programs, third-party service provider management, and independent penetration testing, for financial services operating in New York.
3. **Enhanced Accountability for CISOs**
   - **CISO Liability:** CISOs face personal liability for failures in cybersecurity governance, risk disclosure, and incident reporting.
   - **Collaboration with Legal Teams:** Effective compliance requires close coordination between CISOs and legal professionals to ensure timely and accurate disclosures.
   - **Proactive Risk Management:** Emphasizes quantifying risks, predetermining materiality thresholds, and aligning cybersecurity efforts with regulatory expectations.
4. **Lessons from Enforcement Actions**
   - **Transparency and Timeliness:** Cases like SolarWinds and Uber underscore the importance of transparent and timely reporting of cybersecurity incidents.
   - **Risk Quantification:** Establishing data-driven approaches to assess risks and materiality thresholds strengthens cybersecurity programs and demonstrates due diligence.
   - **Best Practices for Compliance:** Adopting regulatory standards and leveraging consent orders as benchmarks ensure a defensible cybersecurity posture.

5. **Practical Implications for Organizations**
   - **Strategic Integration:** Cybersecurity is now a front-office function, integral to business strategy and risk management.
   - **Investor Protection:** Enhanced disclosure rules provide investors with greater insights into a company's cybersecurity posture and risk profile.
   - **Cultural Shift:** Organizations must foster a culture of accountability, transparency, and proactive governance to navigate the evolving regulatory landscape.

3.8.2 Reflection and Exploration Questions

Understanding the Concepts

1. How does the shifting regulatory landscape redefine the responsibilities of CISOs and corporate boards?
2. How does the concept of "mission-critical risks" influence corporate responsibilities for cybersecurity?

Connecting the Dots

3. How does this chapter build on earlier discussions of risk management and business alignment?
4. In what ways do the regulatory developments outlined here connect to broader trends in corporate governance?
5. How does the concept of materiality introduced in this chapter connect to earlier discussions about risk quantification?
6. In what ways do the SEC, FTC, and NYDFS regulations complement each other?

Real-World Application

7. Can you identify organizations or incidents that highlight the consequences of noncompliance with the discussed regulations?
8. How might your organization implement practices to align with the SEC, FTC, or NYDFS requirements?

Personal and Professional Growth

9. How can the principles of risk quantification and materiality enhance your approach to cybersecurity?
10. What steps can you take to strengthen collaboration between security and legal teams in your organization?

Further Exploration

11. What additional questions do you have about regulatory compliance and its implications for cybersecurity leadership?
12. How might future regulatory changes shape the role of the CISO and the broader cybersecurity landscape?
13. What skills and knowledge are essential for navigating the complexities of the current and future regulatory environment?

Part 2

Business and Risk Alignment

4 The Language of Business

4.1 INTRODUCTION TO THE LANGUAGE OF BUSINESS

I'll never forget the first time I sat in on a board meeting and felt completely lost in a sea of financial jargon and corporat-ese. It was like listening to a foreign language. That's when I realized that to be an effective CISO, I needed to learn the language of business. It's not just about understanding the numbers: it's about understanding what those numbers mean for the company and how cybersecurity can impact them.

Bridging the gap between technical expertise and strategic business acumen is what truly sets the CISO 3.0 apart. They grasp that cybersecurity isn't merely about safeguarding data; it's about empowering business growth and innovation. By speaking the language of business, CISO 3.0s become persuasive advocates for security investments, demonstrating the return on investment (ROI) of their initiatives and earning the trust and support of key stakeholders.

The call for CISOs to "speak the language of business" and "align their strategy to the business" has echoed through countless books, podcasts, and cybersecurity conferences, but there has been little guidance on how to actually do that. This chapter is the first step in that journey, spelling out what constitutes the language of business and how you can learn it. Hopefully, this will serve as a guide for CISOs seeking to elevate their profiles and business acumen. We will explore best practices for achieving cybersecurity/business alignment and provide an overview of the financial metrics that support the language of business.

4.1.1 The Language of Business

Sticking with the language analogy, consider the four primary "languages" to understand accounting (historical financial data), finance (predictive data based on future assumptions), economics (external data regarding the general economy at macro and micro levels), and risk (likely loss exposure). Even though there are other critical areas in business, such as sales, marketing, human resources, and operations, the fundamental principles and vocabulary of these four languages are useful guides to understanding the information context used in organizational decision-making processes.

> *The fundamental principles and vocabulary of these four languages [accounting, finance, economics, and risk] are useful guides to understand the information context used in organizational decision-making processes.*

Commercial organizations operate to create products or services that meet consumer needs, ultimately generating revenue. For a CISO to truly grasp the business, they must understand not only what drives cash inflows but also what resources are consumed in the process, impacting cash outflows. This understanding forms the basis of comprehending the business operating model.

DOI: 10.1201/9781003510789-6

Since industries vary, it is imperative for CISOs to understand the specific industry and the operating model of their organization. This includes recognizing whether the company follows a centralized and standardized approach or a decentralized and non-standardized model where business units may compete for resources and market share. Each approach brings unique challenges and opportunities for cybersecurity strategy, making this insight foundational to aligning security initiatives with business goals.

4.2 THE LANGUAGE OF ACCOUNTING

Once the operating model is understood, the next step is to gain an understanding of the accounting language. Accounting focuses on historical financial information presented in accordance with Generally Accepted Accounting Principles (GAAP). This provides a standardized and reliable way to assess a company's financial performance and position.

4.2.1 Three Core Financial Statements

A CISO 3.0 should have a solid understanding of the three primary financial statements, as they offer valuable insights into the organization's financial health and can guide cybersecurity decisions:

1. **The Balance Sheet:** This provides a snapshot of the company's financial position at a specific point in time.
   Assets: The resources the company owns, like cash, inventory, equipment, or intellectual property. Understanding these helps a CISO prioritize security investments based on which assets are most critical.
   Liabilities: What the company owes, such as loans or accounts payable. Knowing the company's financial obligations helps identify areas of vulnerability that could impact security decisions.
   Shareholders' Equity: This represents the company's net worth after liabilities are deducted from assets. It offers a glimpse into the company's overall financial stability, which is important for long-term planning.
2. **The Income Statement:** This shows the company's revenues and expenses over a period, resulting in either a net income or loss.
   Revenues: The income generated from selling goods or services. Understanding where revenue comes from can help CISOs align their security efforts with key revenue-generating activities.
   Expenses: The costs incurred in earning revenue, including operational costs, interest, and taxes. This helps CISOs identify where security measures could reduce costs or improve efficiency.
3. **The Cash Flow Statement:** This details the movement of cash in and out of the company over a period, broken down into operating, investing, and financing activities.
   Operating Activities: Cash flow from day-to-day business operations, like sales and supplier payments. It shows how well the company generates cash from its core business, which is essential for funding security initiatives.

Investing Activities: Cash flow related to buying or selling long-term assets like property or equipment. This reveals where the company is focusing its growth efforts and asset management, which can influence security priorities.

Financing Activities: Cash flow from financial transactions, like issuing stock or borrowing money. This highlights how the company is funded and its reliance on external financing, which may impact long-term security planning.

A fundamental principle in financial reporting is the assumption that a company will continue to operate on what's known as the "going concern" basis, which is an accounting term that means the company will be financially stable and able to meet its obligations for the foreseeable future. However, when bankruptcy becomes a real possibility due to shrinking cash flow or strained lender relationships, the notes in the financial statements offer terrific insights. These notes shed light on the company's financial struggles and the strategies it's considering to stay afloat.

For CISO 3.0, understanding these financial challenges is more than just a theoretical exercise: it's the key to aligning the cybersecurity strategy with the company's goals. In times of financial uncertainty, the CISO 3.0 plays a key role in prioritizing security investments that support the company's most urgent needs, becoming a strategic partner in its fight for survival.

If you're looking to better understand financial statements and how to interpret them, John A. Tracy's *How to Read a Financial Report: Wringing Vital Signs Out of the Numbers* is an excellent resource for uncovering the stories behind the numbers.

4.2.2 Financial Statement Trends to Watch for CISOs

4.2.2.1 The Balance Sheet

1. Working Capital (Current Assets – Current Liabilities): Is working capital increasing or decreasing?
2. Capital Structure: Is debt increasing or decreasing, and when is it due for repayment (i.e., current vs. long-term classification)?

4.2.2.2 The Income Statement

1. Top-Line Revenue Growth: Is there an increase in revenue period over period, and what are the drivers of this trend (e.g., new customers, higher prices, and acquisition of another firm)?
2. Net Income: Is there an increase or decrease period over period, and what are the drivers of this trend (sales growing faster than expenses to result in higher net income, interest expense growing faster than revenue because debt is being used to finance growth via acquisition, etc.)?

4.2.2.3 The Cash Flow Statement

1. Cash Balance Changes: What category (i.e., operating, investing, and financing) is generating or using the most cash?
2. Is cash from operating activities increasing or decreasing?

Seeing these trends often can be a catalyst for further discussion and insights from the folks in the finance group. This understanding can help a CISO align their security program by understanding the financial goals and priorities of the organization. For example, if the focus is on increasing revenue, the CISO may prioritize security measures that support business development initiatives. If the focus is on reducing expenses, the CISO may prioritize security measures that help streamline and automate processes by driving toward a centralized and standardized control model. Understanding the cash flow statement can help the CISO identify areas where security investments can have a positive impact on the company's liquidity. The accounting language and GAAP financial statements serve as a means of establishing a common and consistent mechanism used to report financial performance, which includes tracking revenues, expenses, capital, and other items. It also allows companies to track and communicate these results to the interested parties.

How familiar are you with your organization's financial statements? Have you even bothered to take a look? If so, how did you use this financial knowledge to prioritize security investments and demonstrate their impact on the bottom line?

4.3 THE LANGUAGE OF FINANCE

After you understand the basic accounting principles, it's important to understand the language of finance. Finance isn't just about spreadsheets and formulas; it's about understanding the value drivers of the business and how to align security investments with those drivers. It's about speaking the language of the CFO and the board, demonstrating how cybersecurity can contribute to the bottom line. Finance is used to predict what might happen in the future when organizations estimate future outcomes that involve projecting revenues, expenditures, and associated cash flows. Oftentimes, organizations will set internal target rates of return, and projects that do not meet such rates are not approved. As organizations mature, this type of financial discipline becomes a common practice and is used to evaluate capital projects.

For a CISO 3.0, understanding key finance concepts is essential to harmonize the security program with the organization's broader financial goals. *Value assessments* are pivotal, requiring CISOs to go beyond technical evaluations and assign a tangible or relative worth to information assets, systems, and processes. This involves factoring in the potential financial fallout from events like data breaches, downtime, and brand damage, aiding in the prioritization of security investments. For instance, a quantitative risk assessment could estimate potential financial losses from a ransomware attack, considering recovery costs, lost revenue, and regulatory fines. This data supports investment in preventive measures like security awareness training or threat detection tools.

Future projections, another core financial concept, involve forecasting and modeling the organization's financial trajectory. CISOs can provide insights into potential cyber risks and their financial ramifications, ensuring proactive risk mitigation and budget allocation for security measures.

For instance, a CISO might project the potential cost of a data breach based on industry trends and the organization's vulnerabilities, justifying investments in data loss prevention or cyber insurance.

Return rates, often measured as ROI, are key to financial decision-making. CISOs must articulate the ROI of security initiatives, demonstrating their bottom-line impact through cost savings, loss avoidance, or new revenue streams. Calculating the ROI of a multifactor authentication solution and comparing its cost against potential savings from preventing breaches exemplify this.

Finally, *capital allocation*, the strategic distribution of finite financial resources, demands CISOs to compete effectively for funding. They must present compelling business cases for security investments, highlighting their alignment with the organization's strategic goals and quantifying potential financial benefits. A detailed business case for a new security project would outline risks, their financial impact, the proposed solution's benefits, and the estimated ROI, aiding decision-makers in understanding and prioritizing its value.

In essence, grasping the language of finance and its decision-making processes allows the CISO 3.0 to become a powerful advocate for cybersecurity. They can translate technical concerns into financial terms, positioning security investments as strategic assets that safeguard and enable business growth.

4.4 THE LANGUAGE OF ECONOMICS

Unlike accounting and finance, which mainly focus on the internal workings of an organization, economics looks at the external factors that affect the organization, such as macro- and micro-markets and regulatory expectations. The world of economics might seem far removed from the day-to-day realities of cybersecurity, but the truth is that they're deeply intertwined. Market forces, supply and demand, as well as consumer preferences can all impact our security strategies. The CISO 3.0 needs to be aware of these broader economic trends and how they might affect the organization's risk profile. Economics informs market pricing that reacts to external factors such as inflation and rising and falling interest rates.

Some of the key concepts in economics that managers should understand include the following:

Supply and Demand (and Equilibrium)

Supply and demand explain how prices for goods and services are influenced by their availability and the level of consumer interest. For CISOs, understanding these market forces is critical, especially when navigating the costs of security products and services. For instance, when demand for cybersecurity professionals outstrips supply, wages increase, and hiring becomes more competitive. This is why skill shortages in cybersecurity drive up salaries and increase recruitment challenges. Grasping supply and demand dynamics allows CISOs to anticipate pricing changes and better plan for resource allocation.

Substitutes

Substitutes are products or services that can replace one another. In the world of cybersecurity, this means alternative solutions that provide similar functions. Being aware of available substitutes allows CISOs to negotiate better terms with vendors. For instance, if multiple vendors offer comparable security products, the CISO can leverage this competition to secure more favorable pricing or flexible contract terms. Understanding the availability of substitutes enhances bargaining power and maximizes value from security investments.

Price Elasticity

Price elasticity measures how sensitive demand is to changes in price. If demand is elastic, a small price increase can lead to a significant drop in demand, whereas inelastic demand means pricing has less impact on usage. For CISOs, understanding price elasticity helps anticipate how changes in the cost of security tools may affect their adoption within the organization. If the demand for a particular security tool is highly elastic, a price hike could result in reduced usage, which may leave the organization exposed to vulnerabilities.

Indifference Curves

Although more theoretical, indifference curves provide insight into how consumers make trade-offs between different products based on preferences and budget constraints. This concept translates to cybersecurity when organizations weigh investments in security against other business priorities, such as cost savings or operational convenience. CISOs who understand these trade-offs can better communicate the value of security investments to business leaders, making the case for balancing security needs with other organizational objectives.

Consumer Preferences

Consumer preferences, which include factors like price, quality, reputation, and perceived value, help shape purchasing decisions. For security leaders, understanding these preferences means recognizing what stakeholders value most. If customers place a high priority on data privacy, for example, the CISO should focus on investing in technologies that enhance privacy protection. By aligning cybersecurity efforts with what matters to customers and stakeholders, CISOs can strengthen the organization's competitive position and foster trust.

Beyond these concepts, broader economic trends play a role in shaping cybersecurity strategies. During economic downturns, companies may reduce budgets, limiting what can be spent on security initiatives. Conversely, during periods of economic growth, there are often more opportunities for investment in cybersecurity programs. CISOs must be attuned to these broader trends to adjust their strategies accordingly.

By applying these economic principles, CISO 3.0s can take a more strategic, forward-thinking approach to their role. This understanding not only helps them navigate market shifts but also ensures that cybersecurity investments align with the organization's overall financial goals and priorities, leading to more effective and sustainable security practices.

4.5 THE LANGUAGE OF RISK

The last language is risk. It should be top of mind for the CISO and complement the three languages of accounting, finance, and economics. In today's dynamic business environment, risk management is an increasingly vital part of the decision-making processes for businesses. It is important for CISOs to clearly articulate the differences and pervasiveness between cyber risk and other business risks when thinking about the "language of business". Recognizing that business risks include all types of risks that can impact the business, including cybersecurity risk, we will separate these two types of risks for the purposes of this chapter. Business risks include financial risks, such as market fluctuations or economic downturns, and operational risks, such as supply chain disruptions or natural disasters. Understanding an organization's business risks, which are often a function of a greater strategy, is necessary for CISOs to ensure that their security programs are aligned with the organization's goals and objectives. Cyber risks are specific to the information technology sector and can include risks such as unauthorized access, internal and/or integrated third-party supplier data breaches, business interruption due to a cyberattack, and related system failures. Improperly addressed cyber risks can result in a catastrophic impact on an organization's operations, finances, and reputation, making it critical for CISOs to have a deep understanding of these risks and the specific organizational responses. Because cyber risks are pervasive and fast-moving based on technology, they require ongoing and diligent efforts to manage and mitigate, often requiring specialized knowledge and expertise. Cyber risks can have a cascading impact on an organization's operations and reputation, requiring specific considerations when developing risk management strategies. Consequently, CISOs need to understand and be able to articulate how cyber and other business risks differentiate, complement, interlock, and/or influence each other. We will dig deeper into this idea and discuss risk at length in Chapter 6.

4.6 DETERMINING VALUE AND RESOURCE ALLOCATION

Once you have a solid understanding of the language of business, you can start to ask questions about how your organization determines value. Understanding value drivers is fundamental to the alignment of cybersecurity strategies with business goals. It helps CISOs identify and prioritize assets that need the most protection. To understand value, CISOs should consider what factors add value and what factors detract from value, along with the timeframes involved. This can create tension, and the savvier CISOs who understand the different languages of business are best equipped to navigate these common challenges. For example, some stakeholders (like certain executives) are very short-term focused. And if incentive compensation is based on minimizing quarterly spend to achieve an accounting metric, such as earnings before interest, taxes, depreciation, and amortization (EBITDA), compound annual growth rate (CAGR), gross profit (GP), or earnings per share (EPS), then cybersecurity resources and related investment could be constrained, resulting in excessive risk and horizon "technical debt" over a longer time. Conversely, owners and customers may have a long-term time horizon and are vested in seeing revenue growth due to a strong market presence and high customer trust that is reinforced by strong, sustainable cybersecurity reputation. This can result in a rising stock price, increasing dividends, and significant wealth creation. CISOs who understand the most important short- and long-term value drivers can factor in that information when making resource allocation decisions for cybersecurity resourcing. Decision factors can include trade-offs impacting operational efficiency, brand reputation, customer satisfaction, and innovation. These trade-offs will vary depending on the industry and the company's business model.

For example, a software company may create value through the development of innovative new products where security by design is a critical element of the DEV/SEC/Ops process. A manufacturing company that is not capturing personal information but has risk around an integrated supply chain may create value through operational efficiency, inventory velocity, and supply chain digital resiliency, resulting in better cost control. In the next chapter, we'll explore how and when value is created for different types of businesses.

4.6.1 Common Financial Metrics

Understanding financial metrics is important for any CISO in order to align a cybersecurity program with an organization's financial goals and, more importantly, ensure that the cybersecurity resource allocation decision delivers value that aligns with financial targets while not taking on undue risk. The CISO can achieve a win–win when working with the chief financial officer (CFO) by modeling different cybersecurity spend scenarios and understanding the impact on the organization's most relevant financial metrics.

Here are some of the most common financial metrics that every CISO should be familiar with:

CAGR (Compound Annual Growth Rate): This metric shows the annual growth rate of a company over a specific period. It's particularly useful for fast-growing startups and growth-focused companies as it reflects steady, long-term growth.

EBITDA (Earnings Before Interest, Taxes, Depreciation, and Amortization): EBITDA measures a company's operating performance and profitability, serving as a proxy for cash flow. It's a widely used metric, and it's important for CISOs to understand how cybersecurity spending – both operational expenses (OPEX) and capital expenditures (CAPEX) – affects EBITDA.

COGS (Cost of Goods Sold): This metric tracks the direct costs involved in producing goods or services. It's especially relevant for manufacturers and retailers as it reflects production efficiency and profitability.

Gross Profit (GP): Gross profit represents the revenue left after subtracting the COGS. It's a key indicator of a company's profitability before other expenses like operating costs are factored in.

SG&A (Selling, General, and Administrative) Costs: These are the operating expenses not directly tied to producing goods or services. This metric is crucial for companies investing heavily in growth and market expansion, such as new software firms or startups.

Earnings per Share (EPS): EPS shows how much profit a company generates for each outstanding share of its common stock. It's an important metric for publicly traded companies, as it gives shareholders an idea of the company's profitability on a per-share basis.

4.7 OPEX VERSUS CAPEX AND BUDGETING

CISOs should be aware of the difference between CAPEX and OPEX and how they impact financial metrics like EBITDA. CAPEX are typically organizational spend made on assets that generate use or benefits for periods longer than a year, such as computer hardware or certain types of intangible assets. OPEX are typically organizational spend made in ongoing costs, like salaries or software subscriptions, which are necessary to keep the business running and are not eligible to be capitalized based on company policy and Generally Accepted Accounting Principles (GAAP). To put it another way, buying a capital good is like adopting a pet – it's cute and cuddly at first, but then you realize it needs constant attention and costs a fortune. Renting is more like visiting the zoo, you get to see all the cool animals without having to clean up after them. Traditionally, we thought of a piece of hardware (e.g., a firewall) as CAPEX and the software-as-a-service (SaaS) version of that control (e.g., a SASE) as OPEX. However, that traditional view is changing. The old equation of "on-premises equals capital expenditure, cloud equals operating expenditure" is no longer valid. Businesses now have the flexibility to choose OPEX funding models across their entire IT infrastructure, whether it's on-premises, hybrid, or public cloud. Leasing on-premises or private cloud infrastructure, combined with managed services, offers a predictable monthly expense and frees up resources for core business activities. Even hybrid solutions, leveraging on-premises infrastructure with public cloud for specific workloads, can be delivered through a true OPEX model with a single monthly bill. It's important to note that the recent lease accounting rules may change where this example applies, so confirm the accounting treatment with the CFO. The general industry trend for IT and cybersecurity applications and services has been moving towards OPEX and away from CAPEX, motivated in large part because there is a high level of stickiness with subscription-type services, and investors value recurring revenue streams. In addition, switching costs can be high, so when evaluating where to allocate cybersecurity resources, understanding the total cost of ownership, including switching costs, requires a longer-term view and a more strategic understanding of the organization. For example, if the organization was interested in being acquired, it may not make sense to invest a large amount of capital in building up a security operations center (SOC) infrastructure compared to using an outsourced provider with lower upfront costs offering a subscription-type service.

4.7.1 Lease versus Buy Scenario

Using a simple example of a lease versus buy scenario, imagine you wanted to start a delivery business that would make $4,000 per month, but you first needed to buy a delivery truck. You could buy the truck up front for $100,000 cash, or you could lease it for $1,800 a month for 60 months, which would result in a total cost of

$108,000. Saving $8,000 might seem like a better decision at first, but consider that the business would only have to pay $1,800 for the first month, and then the truck would make $4,000 per month. That income would more than cover the truck's payment, and since you only spent $1,800 instead of $100,000 for the first month, you would have $98,200 to invest in other parts of your business. For IT programs, other benefits of "renting" capacity can include quicker time to value, immediately engaging experts, not incurring an internal learning curve, and allowing the focus to be on the core competencies of the organization. For instance, it takes much less time to consume the outputs of a managed service provider (MSP) than to build and develop the expertise to run an entire in-house SOC. Although the accounting treatment of cybersecurity budgets should not solely determine cybersecurity spend rationale, it is useful to understand the businesses' budget buckets for CAPEX and OPEX and ways to adjust security projects to meet those budgets. For example, CAPEX is often favorable for firms that want higher EBITDA since capital costs are depreciated/amortized. This means they do not directly impact EBITDA. Conversely, if CAPEX commitments are accounted for based on estimated useful lives (and a recent trend we have observed is that CFOs are extending the initial useful life estimates of technology), then CISOs may have to wait longer than initially anticipated to replace certain technologies. For example, if a piece of on-prem hardware is acquired and expected to be depreciated over 3 years, but the CFO elects to revise that estimate to 4 years based on "facts and circumstances" (including if the on-prem is still fully functional after 3 years), then the CISO may be in for a surprise if there was an initial plan to replace the device in 3 years. Keep in mind that some CFOs are reluctant to approve CAPEX for items that are not fully depreciated.

4.7.2 SG&A and COGS for CISOs

Two financial statement components that influence EBITDA are selling, general, and administrative (SG&A) costs and cost of goods sold (COGS). Each component can be influenced by CISOs. Typically, direct product costs, labor, and overhead are the largest elements of COGS, and such costs are directly associated with the sales of goods or services a company makes (e.g., think of the direct cost to NVIDIA for each chip it sells). Costs that cannot be directly associated with sales, such as rent, office supplies, insurance, and the salaries of the IT and security teams, are classified as SG&A. SG&A expenses are broad expenses that the entire company incurs rather than being associated with a specific good or service and help determine a company's profitability and operating margins. Consequently, SG&A costs often become a prime target for cost-cutting measures when companies seek to boost their bottom line.

In some organizations that are not viewed as industry leaders, IT and cybersecurity are often viewed as support functions and classified as "cost centers", their value overshadowed by their perceived expense. This can make securing budget and buy-in a constant struggle for CISOs of those organizations.

An astute CISO 3.0 recognizes an opportunity within this challenge. The opportunity to continue expanding the focus on cybersecurity externally by focusing not only on peer executives but also on how cybersecurity practices will impact an organization's customers and create enterprise value. A savvy CISO 3.0 knows that aligning security costs with revenue-generating activities can be a game-changer. It shifts the narrative from "cost center" to "value driver", making it easier to secure budget and demonstrate the strategic importance of cybersecurity.

One often overlooked opportunity is to take credit for what may be happening or should be happening in product development. Since many products firms sell today involve some type of technology, there is a growing market expectation that those products are secure from bad actors and ransomware. To achieve these market expectations, privacy and security by design practices are being pushed by regulators and being adopted by leading firms. These design practices involve cybersecurity resources, and it is common to seek input from the organization's CISO and their security teams during various elements of the product development lifecycle. However, it is far less common for the CISO and their security teams to monitor time spent with the goal of capturing their costs so such amounts can be capitalized. Capitalizing expenses under the accounting rules let organizations record expenses as assets on the balance sheet rather than expensing them immediately on the income statement, provided such expenses meet capitalization criteria. This approach acknowledges that certain expenses, such as those related to software product development, create future benefit and can be capitalized, and then the capitalized costs will be amortized over the expected useful life. In the context of new product development, companies can capitalize the costs of software development, including salaries of developers, software licenses, and other related expenses. These capitalized costs are then gradually expensed over the estimated useful life of the software, reflecting their contribution to revenue generation over time.

Thus, under accounting standards known as GAAP, it is possible to capitalize the development of security features and controls for new product lines, and perhaps most importantly, this approach makes cybersecurity an integral part of product development and highlights the value of the CISO and their cybersecurity teams.

Examples of cybersecurity costs that might be eligible for capitalization include the following:

- **Development of Security Features:** Costs associated with designing and implementing security features within a new product or service.
- **Security Testing and Vulnerability Assessments:** Costs incurred to identify and address security weaknesses in a product before its release.
- **Security Training for Development Teams:** Costs associated with educating developers on secure coding practices and security best practices.
- **Security Certifications and Compliance:** Costs incurred to obtain security certifications or comply with industry-specific regulations.

It's important to note that not all cybersecurity costs can be capitalized. Ongoing maintenance and operational expenses, such as security monitoring or incident response, are typically expensed as incurred. The decision to capitalize or expense cybersecurity costs should be made in consultation with the finance team and in accordance with GAAP guidelines.

I've seen this work best at high-tech firms where SaaS or other technology-driven products are the primary revenue generators. However, any organization that capitalizes development costs can potentially capitalize the associated security development costs, thereby reducing SG&A expenses and improving profitability.

By strategically capitalizing eligible cybersecurity costs, CISOs can demonstrate the direct link between security investments and revenue generation. This can help shift the perception of security from an expense center to an integral part of driving revenue, which will make it much easier to secure budget and buy-in for security initiatives and position the CISO as a strategic partner who contributes to the organization's financial success.

4.8 ALIGNING CYBERSECURITY WITH FINANCIAL BUSINESS GOALS

There are a number of ways to align cybersecurity budgets with your organization's financial goals. For example, cybersecurity can serve as a competitive advantage or even a revenue driver by building trust. When customers feel confident that their data is safe, they're more likely to choose your business, especially if they place a high value on security and privacy.

Here are a few creative ways to align cybersecurity spending with broader business objectives:

Turn Cybersecurity Expenditures into Cost Reductions: CISOs can identify areas where cybersecurity spending can lead to cost savings. For example, sun-setting legacy systems reduces licensing and maintenance costs while simultaneously improving the risk posture.

Turn Cybersecurity Expenditures into Cost Avoidance: CISOs can demonstrate how investing in cybersecurity can help the company avoid costs associated with data breaches or cyberattacks. For example, investing in employee training and security awareness programs can reduce the likelihood of successful phishing attacks or other social engineering tactics that could lead to a data breach.

Turn Cybersecurity Expenditures into Competitive Differentiators: CISOs can help the company differentiate itself from competitors by investing in cybersecurity that addresses customer concerns. For example, a company that invests in secure mobile applications or a secure e-commerce platform can attract more customers who value security and privacy.

Turn Cybersecurity Expenditures into Revenue Generation: CISOs can help the company generate revenue by leveraging cybersecurity capabilities to create new products or services, for example, offering managed services, incident response, or penetration testing to their customers to generate additional revenue streams.

Leverage Cybersecurity to Improve Customer Retention: CISOs can show how investing in cybersecurity can improve customer retention by reducing the risk of data breaches or cyberattacks that could compromise customer data. For example, a company that implements multifactor authentication

(MFA) for its online services can increase customer confidence in its security and reduce the likelihood of customer churn. Additional examples could include complementary service-level agreements (SLAs) for customer incident response, security updates, reports, threat intel, education, or awareness programs.

Align Cybersecurity Spending with Business Units: CISOs can work with business unit leaders to identify cybersecurity risks and priorities that align with the financial goals of each unit. For example, a sales team may prioritize secure collaboration tools to close deals faster, while a finance team may prioritize secure payment processing to reduce the risk of fraud.

Invest in Cybersecurity to Enable New Business Models: CISOs can demonstrate how investing in cybersecurity can enable new business models or revenue streams. For example, a company that invests in blockchain technology and smart contracts can create a secure platform for conducting digital transactions, which could enable new business models and generate additional revenue streams.

4.9 CONCLUSION

Let's face it, talking about cybersecurity in the boardroom can feel like you're speaking a different language. But here's the secret: If you want your security program to get the attention and funding it deserves, you need to learn to speak the language of business. It's not just about understanding the financial lingo: it's about connecting the dots between cybersecurity and the company's bottom line. When you can clearly demonstrate how your security efforts positively impact revenue, control costs, and mitigate risks, you're not just a CISO – you're a strategic partner.

Think of it this way: Every dollar you spend on cybersecurity isn't just an expense – it's an investment. And like any good investment, it needs to show a return. Whether it's by preventing costly data breaches, streamlining operations, or even opening up new revenue streams, your security program can directly contribute to the company's financial success. When you can articulate this value in a way that resonates with the CFO and the board, you'll find it much easier to get the buy-in and budget you need.

But it's not just about the numbers. It's also about translating those complex technical risks into something everyone can understand – the potential impact on the business. Can a cyberattack disrupt operations and cost us customers? Could it damage our brand value and tank our stock price? By painting a clear picture of these potential consequences, you're not only scaring people, but you're also empowering them to make informed decisions.

The CISO 3.0 is a translator, a bridge between the IT world and the boardroom. They can speak both languages fluently, making sure everyone understands the stakes and the value of a strong security program. And that's what sets them apart – the ability to drive innovation, protect critical assets, and build a culture of security, all while keeping the company's financial goals in sight.

4.9.1 Key Takeaways

1. **The Strategic Role of the CISO 3.0**
    - Modern CISOs must bridge technical expertise with business acumen, positioning themselves as strategic partners who align cybersecurity with financial and operational goals.
    - By learning the language of business, CISOs can effectively communicate the ROI of cybersecurity initiatives and secure stakeholder trust.
2. **The Four Languages of Business**
    - **Accounting:** Understanding historical financial data through core financial statements (balance sheet, income statement, and cash flow statement) to guide security investments.
    - **Finance:** Leveraging future projections, ROI assessments, and value drivers to align cybersecurity strategies with financial goals.
    - **Economics:** Interpreting external market forces like supply and demand, substitutes, and consumer preferences to adapt security measures.
    - **Risk:** Differentiating between cyber and business risks while integrating risk management into organizational decision-making processes.
3. **Financial Statements and Metrics**
    - **Balance Sheet:** Provides a snapshot of assets, liabilities, and shareholder equity, highlighting areas critical for security prioritization.
    - **Income Statement:** Tracks revenues and expenses, enabling alignment of security measures with key revenue drivers and cost efficiencies.
    - **Cash Flow Statement:** Details cash movements, emphasizing the importance of liquidity in funding security initiatives.
    - Key financial metrics like EBITDA, COGS, and EPS help CISOs demonstrate how security investments impact profitability and operational efficiency.
4. **Value Assessments and ROI**
    - Quantifying the financial impact of cybersecurity initiatives aids in prioritizing investments and securing executive buy-in.
    - Demonstrating value through cost reductions, cost avoidance, and revenue generation establishes cybersecurity as a value driver.
5. **Capital Allocation and Budgeting**
    - Differentiating between CAPEX and OPEX allows CISOs to align security initiatives with organizational budgeting practices.
    - Strategic allocation of resources ensures that cybersecurity contributes to long-term growth and resilience.
6. **Aligning Cybersecurity with Business Goals**
    - Collaborating with business units to identify and address cybersecurity risks enhances alignment with organizational objectives.
    - Leveraging cybersecurity to build trust, improve customer retention, and enable new business models positions the CISO as a proactive leader.

4.9.2 Reflection and Exploration Questions

Understanding the Concepts

1. In what ways can the language of business enhance the role of the CISO in aligning cybersecurity with business goals?
2. Why is it important for CISOs to understand the four languages of business?

Connecting the Dots

3. How does knowledge of financial statements support better decision-making for cybersecurity investments?
4. In what ways do risk management principles discussed here connect with earlier chapters on risk quantification?

Real-World Application

5. Can you think of examples where cybersecurity investments led to measurable business outcomes like cost savings or revenue growth?
6. How might understanding the economics of supply and demand influence decisions on hiring cybersecurity talent or purchasing tools?

Personal and Professional Growth

7. How can you apply the principles of ROI and value assessments to your own role or organization?
8. What steps can you take to improve your fluency in the languages of accounting, finance, economics, and risk?

Further Exploration

9. What additional financial metrics or economic concepts could enhance a CISO's ability to align with business objectives?
10. How might emerging trends like digital transformation or regulatory shifts impact the intersection of cybersecurity and business strategy?
11. Can you brainstorm creative ways to align security initiatives to business goals and help IT security drive revenue and not just cost?

5 Ownership and Boards of Directors

5.1 INTRODUCTION

The world of corporate governance can sometimes feel like a secret society with its own unspoken rules and power dynamics. However, for a cybersecurity leader today, it is imperative to understand how the boardroom operates. It's not only about knowing who's who, but it's also about understanding the motivations and priorities that drive decision-making at the highest level.

Think of it as learning a new language, the language of the boardroom. Once you're fluent, you can tailor your cybersecurity strategies to resonate with everyone from the CEO to the newest board member. You'll be able to address their concerns head-on and show them that you are the right person to manage their cyber risk.

In this chapter, we're going to pull back the curtain on board governance. We'll explore the different types of ownership structures, break down what board members are actually responsible for, and shed light on the sometimes blurry line between board oversight and day-to-day management. Consider this your crash course in boardroom navigation. By the end, you'll have the confidence and clarity to craft a cybersecurity message that not only informs but also persuades.

But before we dive in, let's consider a few fundamental questions about value within an organization:

What provides value to your organization?

Who determines that value? Is it stockholders, partners, owners, VC/PE firms, hedge funds, or someone else?

What delivers value? Is it share price, dividends, capital appreciation, or a combination of factors?

Where is value created? Does it reside in assets, market-based mechanisms, discounted cash flow, or other sources?

When is value created? Is it through quarterly or annual dividends, the sale of equity, or other events?

Can you answer these questions about your organization? These are not just theoretical questions. They have real-world implications for how you communicate the importance of cybersecurity to the decision-makers in your organization. The answers will vary depending on the specific ownership structure and strategic goals of your company.

This exploration begins with a discussion of how value is determined, followed by an overview of common ownership structures, ranging from publicly traded companies to privately held entities, each with its unique set of governance dynamics.

DOI: 10.1201/9781003510789-7

Next, we'll explore the structure and duties of boards of directors, emphasizing their essential role in protecting shareholder interests, promoting ethical practices, and guiding the company toward long-term success.

This chapter will further examine various board governance models, each with its own approach to decision-making and oversight. We will explore the fiduciary duties that bind board members, emphasizing their legal and ethical obligations to act in the best interests of the company and its stakeholders. In contrast, we will also define the role and fiduciary duties of executive management, elucidating the differences between board oversight and the day-to-day management of the company.

Before you jump into the complexities of value creation and the metrics that drive different ownership structures, you need to make a key ally: Your chief financial officer (CFO). The CFO is not only a treasure trove of information on these topics, but they also have a personal stake in the game. CFOs are often compensated based on metrics like earnings per share (EPS), earnings before interest, taxes, depreciation, and amortization (EBITDA), and other financial performance indicators, making them natural allies in your quest to align security initiatives with the company's overarching goals.

Think of it this way: If you can demonstrate how your cybersecurity program directly contributes to the metrics that determine your CFO's bonus, you've essentially unlocked the path to budget and buy-in. It's a win–win scenario where security and finance work in concert to drive the company's success. So, don't hesitate to tap into your CFO's expertise and insights as you navigate the world of board communications and strategic decision-making. Their perspective can be invaluable in translating the language of cybersecurity into the language of business, ensuring that your security initiatives are not only seen as necessary expenses but also as strategic investments that contribute to the bottom line.

With that said, here are some generalized answers to the questions of who, what, when, and how value is created for each of the different organizational types. Remember, these are just starting points, and the specifics will vary depending on the unique circumstances of your organization.

5.2 CORPORATE VALUATIONS

The concept of corporate valuation lies at the heart of investment decisions, mergers and acquisitions, and strategic planning. I've seen firsthand how a strong cybersecurity posture can impact a company's valuation, especially during mergers and acquisitions. When potential buyers or investors see that a company takes security seriously, it instills confidence and often leads to a higher valuation. It's a powerful reminder that cybersecurity is not just about protecting assets: it's about creating value.

The process of determining the economic value of a business is influenced by a myriad of factors, including financial performance, market conditions, and future growth prospects. The timing and methods of valuation can vary drastically depending on the company's ownership structure and stage of development. In this section, we will explore various valuation methods and how they address the fundamental questions of value creation, delivery, and timing, providing insights into how the CISO 3.0 can strategically influence these metrics to enhance the company's overall value proposition.

5.2.1 Methods for Determining Value

- **Asset-Based Valuation:** This method focuses on the company's net asset value, calculated by subtracting liabilities from assets.
  - *When it's used*: It's often used for companies with sizable tangible assets, those nearing liquidation, or in distressed situations where the focus is on the value of the underlying assets.
  - *Where value is created*: The value resides primarily in the company's assets and their potential resale value.
  - *What delivers value*: The realization of value typically occurs through the sale or liquidation of assets.
  - *Who provides value*: The value is primarily determined by the company's assets and their potential buyers or liquidators.
- **Market-Based Valuation:** This approach compares a publicly traded company to similar companies, using metrics like price-to-sales (P/S) ratio or price-to-earnings (P/E) ratio.
  - *When it's used*: It's commonly used for publicly held companies where there's an active market for their shares.
  - *Where value is created*: Value is created through market-based mechanisms, reflecting investor sentiment and the company's perceived growth potential.
  - *What delivers value*: Share price appreciation is the primary driver of value for shareholders.
  - *Who provides value*: The value is determined by the collective actions of shareholders in the stock market.
- **Income-Based Valuation:** This method estimates the company's value based on its future income streams, typically using discounted cash flow (DCF) analysis.
  - *When it's used*: It's suitable for companies with predictable cash flows and growth potential, particularly in mature industries or for established businesses.
  - *Where value is created*: Value is created through the company's ability to generate future cash flows.
  - *What delivers value*: The present value of those future cash flows, discounted back to today, represents the company's value.
  - *Who provides value*: The company's management and operations, which generate the income streams, are the primary providers of value.

5.2.2 Specific Valuation Methods

The methods for determining value provide a broad framework for understanding the underlying principles and approaches to valuation. The specific valuation methods, on the other hand, are the practical tools and techniques used to implement those approaches and arrive at a concrete estimate of the company's value. These methods involve specific calculations and analyses based on financial data, market information, and other relevant factors.

- **Multiples Valuation:** This approach involves applying a multiple (e.g., P/E ratio and EBITDA multiple) to a financial metric (e.g., earnings and EBITDA) to estimate the company's value. A common example is the P/E ratio, which is determined by dividing the company's market price per share by its EPS. The P/E ratio offers a snapshot of how much investors are willing to pay for every dollar of the company's earnings. Typically, a higher P/E ratio suggests that investors anticipate stronger future growth.
  - *When it's used*: It's a common method used in various contexts, including private equity and venture capital investments, and for comparing companies within the same industry.
  - *Where value is created*: Value is derived from the company's financial performance relative to its peers.
  - *What delivers value*: The multiple applied to the financial metric reflects market sentiment and growth expectations.
  - *Who provides value*: The company's financial performance and the market's perception of its potential drive value.
- **Discounted Cash Flow (DCF):** Calculates the present value of a company's projected future *free cash flows*, essentially the cash generated by operations that can be distributed to investors or reinvested into the business. This approach considers several key factors, including the company's expected growth rate; the discount rate, which accounts for the investment's associated risks; and the terminal value, which is an estimate of the company's value at the end of the forecast period. Together, these elements help determine the company's overall value based on its future potential.
  - *When it's used*: It's often used for mature companies with stable cash flows or those with high growth potential, particularly when future cash flows can be reasonably forecasted.
  - *Where value is created*: Value is created through the company's ability to generate future cash flows.
  - *What delivers value*: The present value of those future cash flows, discounted back to today, represents the company's value.
  - *Who provides value*: The company's management and operations, which generate the future cash flows, are the primary providers of value.
- **Comparable Transactions Method:** This approach analyzes the prices paid for similar companies in recent mergers and acquisitions.
  - *When it's used*: It provides a market-based benchmark for valuing the company, especially in private equity transactions or when considering potential acquisitions.
  - *Where value is created*: Value is derived from the prices paid for similar companies in the market.
  - *What delivers value*: The transaction prices of comparable companies serve as a reference point for valuing the target company.
  - *Who provides value*: The market and the buyers in those comparable transactions determine the value.

- **Precedent Transactions Method:** Similar to the comparable transactions method, this approach looks at past transactions involving the same company, such as previous funding rounds or acquisitions of its divisions.
    - *When it's used*: It can be useful for valuing companies with a history of transactions, providing insights into how the market has valued the company or its parts in the past.
    - *Where value is created*: Value is derived from the company's historical transaction values.
    - *What delivers value*: Past transaction prices serve as a basis for estimating the company's current value.
    - *Who provides value*: The market and the buyers/investors in those past transactions determine the value.
- **Liquidation Value:** This method estimates the value of the company's assets if they were sold off individually.
    - *When it's used*: It's typically used in bankruptcy or liquidation scenarios and provides a floor value for the company.
    - *Where value is created*: The value resides in the company's assets and their potential resale value in a liquidation scenario.
    - *What delivers value*: The sale of assets in a liquidation event realizes the value.
    - *Who provides value*: The value is determined by the potential buyers of the assets in a liquidation scenario.
- **Replacement Cost:** This approach estimates the cost of replicating the company's assets and operations.
    - *When it's used*: It's often used for companies with unique assets or in industries with high barriers to entry, where replicating the business would be costly.
    - *Where value is created*: Value is derived from the cost and difficulty of replicating the company's assets and operations.
    - *What delivers value*: The high cost of replacement can contribute to a higher valuation for the company.
    - *Who provides value*: The company's unique assets, intellectual property, and established operations contribute to the value.

To be a CISO 3.0, your strategic influence should extend beyond the immediate purview of cybersecurity. Gaining a deep understanding of what drives your company's value and aligning security initiatives with those specific value drivers make the CISO 3.0 a major player in the organization's financial success and long-term growth. Whether it's protecting revenue streams, streamlining operations, facilitating compliance, or building trust, the CISO 3.0's actions can have a tangible impact on the company's bottom line and its overall value proposition.

5.3 COMMON OWNERSHIP STRUCTURES

The ownership structure of a company influences its priorities, strategic direction, and the metrics that drive its decision-making. It's important to understand these structures to truly achieve the elusive goal of "aligning your security program to the business".

5.3.1 Publicly Held Companies

Publicly held companies are those whose shares are traded on stock exchanges, allowing a wide range of investors to buy and sell them. Examples include major companies like Apple, Microsoft, and Amazon. These businesses are closely monitored by regulatory bodies such as the Securities and Exchange Commission (SEC), which means they must follow strict reporting rules and maintain transparency in their financial dealings. This level of scrutiny, along with the pressure from a diverse group of investors, places significant emphasis on financial performance.

One key aspect of creating shareholder value for public companies is the payment of dividends, which are regular distributions of profits to shareholders. Dividends provide a direct return on investment (ROI), attracting investors looking for steady income and signaling the company's financial health. Companies choose to offer dividends for various reasons: To show confidence in their future earnings, to attract a broader range of investors, and to build a reputation as a financially stable and reliable investment.

In the world of publicly traded companies, two important metrics stand out: EPS and share price. EPS measures how much profit is attributed to each outstanding share, giving insight into the company's profitability on a per-share basis. Meanwhile, the share price represents the current market value of a single share, determined by supply and demand in the stock market. While both metrics offer a snapshot of financial performance, they serve different purposes. EPS helps investors gauge profitability and compare it with other companies, while dividends provide direct financial returns to shareholders, shaping their view of the company's overall financial health.

Beyond EPS and dividends, other key financial metrics like Earnings Before Interest and Taxes (EBIT) and EBITDA (Earnings Before Interest, Taxes, Depreciation, and Amortization) play a crucial role in assessing a company's operational profitability. These figures offer a look at how well a company is performing before accounting for financial costs like interest and taxes, as well as noncash expenses such as depreciation and amortization. Strong EBIT or EBITDA figures generally translate to higher net income, leading to improved EPS. This relationship highlights the importance of operational efficiency in boosting shareholder value and driving the company's overall financial success.

What Provides Value: Share price appreciation and dividends are the primary sources of value for shareholders in publicly held companies.

Who Determines Value: Shareholders, through their buying and selling activities in the stock market, ultimately determine the share price and, indirectly, the value of the company.

What Delivers Value: Share price appreciation is driven by the company's growth prospects and financial performance, while dividends offer a regular income stream to investors.

Where Is Value Created: Value is created through a combination of market-based mechanisms (supply and demand for shares) and the company's ability to generate consistent earnings and cash flow.

When Is Value Created: Value is created continuously as the company's performance and market sentiment fluctuate. However, key value realization events for shareholders include receiving dividends and selling their shares at a higher price than they purchased them.

How Is It Valued: Market-based valuation and multiples valuation are commonly used, with a focus on metrics like P/E ratio, EPS, and dividends.

5.3.1.1 How Security Can Align

For publicly held companies, where shares are traded on a stock exchange and a wide range of investors hold a stake, the CISO 3.0's role extends beyond mere protection. They must actively contribute to enhancing shareholder value, which is primarily driven by share price appreciation and dividends. To achieve this, the CISO 3.0 can implement security measures that directly impact the company's financial performance. For instance, robust data protection initiatives can prevent costly data breaches, safeguard revenue streams, and bolster investor confidence, ultimately leading to share price appreciation. Similarly, by optimizing security operations and reducing downtime, the CISO can contribute to improved operational efficiency, positively impacting EBIT and EBITDA, which in turn can drive EPS growth and increase shareholder value.

Focus on Financial Performance and Shareholder Value: The CISO 3.0 can demonstrate how cybersecurity investments protect revenue streams, reduce costs, and address risks that could have a negative impact on the company's financial performance and share price. For example, implementing robust security measures to prevent data breaches can safeguard customer trust and loyalty, leading to increased revenue and share price appreciation.

Prioritize Transparency and Disclosure: Public companies operate under intense scrutiny from regulators and investors. The CISO 3.0 ensures that cybersecurity risks and incidents are disclosed in a timely and transparent manner, adhering to regulatory requirements and maintaining investor confidence. This could involve working closely with the legal and finance teams to prepare accurate and informative disclosures for SEC filings. We'll discuss this more in Chapter 17.

Align with Investor Expectations: Public companies are often evaluated based on their environmental, social, and governance (ESG) performance. The CISO 3.0 can demonstrate how cybersecurity contributes to the "G" (governance) aspect of ESG, showcasing the company's commitment to responsible and ethical data management practices.

The CISO 3.0 can strategically choose security investments with a higher CAPEX component (e.g., infrastructure upgrades and security software development) to defer expenses and improve short-term profitability, positively impacting EPS. For example, investing in the development of a proprietary security analytics platform could involve significant upfront CAPEX costs for research, development, and testing. However, once operational, this platform could lead to long-term cost savings by automating threat detection and response, reducing the need for manual intervention, and improving overall security efficiency. These savings would positively impact the bottom line and EPS, showcasing the strategic value of the investment.

5.3.2 Private Equity-Held Firms

For private equity (PE) firms, the primary value lies in the potential for sizable financial returns upon exiting their investment in the company. The PE firm, as the majority owner, ultimately determines the value of the company based on its assessment of the company's potential for profitability and growth. PE firms acquire companies using a combination of debt and equity financing, aiming to improve their performance and sell them at a profit within a few years. PE firms are highly focused on maximizing returns on their investment, often employing a "buy-and-build" strategy that involves acquiring a platform company and then expanding that business through the acquisition of complementary organizations to increase revenue and profitability. Value is delivered through improving the company's financial performance, often by implementing operational efficiencies and strategic acquisitions, and ultimately selling the company at a higher valuation. Value is created through the PE firm's active involvement in the company's operations, leveraging their expertise and resources to enhance its value proposition. Value is realized primarily during the exit phase, when the PE firm sells its stake in the company, typically through a sale to another company or an initial public offering (IPO).

The financial goals of a PE-held company evolve over time, shifting from investment to exit.

Investment Phase: During the initial investment phase, the focus is often on injecting capital into the company to fuel growth initiatives, like creating new products, expanding into different markets, or acquiring competitors. Key metrics in this phase include revenue growth, market share expansion, and customer acquisition.

Restructuring Phase: In some cases, a PE firm may initiate a restructuring phase to improve the company's financial performance and streamline its operations. This often requires cost-cutting measures, such as reducing selling, general, and administrative (SG&A) expenses, streamlining processes, or divesting noncore assets.

Divestment Phase: Once the company has reached a certain level of maturity and profitability, the PE firm will start to prepare for an exit. This involves maximizing the company's valuation and preparing it for a sale or IPO. Key metrics in this phase include enterprise value, return on invested capital, and EBITDA multiples.

Exit Phase: The final phase of the PE lifecycle involves selling the company, through either a strategic sale to another company or an IPO. The primary metric in this phase is the final valuation of the company, which determines the PE firm's ROI.

A PE typically holds assets for a limited time, usually 3–5 years. This timeframe creates a sense of urgency to achieve major growth and profitability, influencing the metrics that are prioritized over time.

What Provides Value: The primary value for PE firms lies in the potential for sizable financial returns upon exiting their investment in the company.

Who Determines Value: The PE firm, as the majority owner, ultimately determines the value of the company based on its assessment of the company's potential for growth and profitability.

What Delivers Value: Value is delivered through improving the company's financial performance, often by implementing operational efficiencies, strategic acquisitions, and ultimately selling the company at a higher valuation.

Where Is Value Created: Value is created through the PE firm's active involvement in the company's operations, leveraging their expertise and resources to enhance its value proposition.

When Is Value Created: Value is realized primarily during the exit phase, either through an IPO or when the PE firm sells its stake in the company to another company or PE firm.

How Is It Valued: Multiples valuation and DCF analysis are often used, with an emphasis on EBITDA, cash flow, and enterprise value.

5.3.2.1 How Security Can Align

In the world of private equity, where the focus is on maximizing returns within a relatively short timeframe, the CISO 3.0's role is to align security investments with the PE firm's financial goals. This could involve optimizing CAPEX and OPEX spending to enhance profitability and improve the company's valuation. For example, by strategically leveraging cloud-based security solutions (OPEX) instead of investing heavily in on-premises infrastructure (CAPEX), the CISO can reduce upfront costs and improve cash flow, making the company more attractive to potential buyers during the exit phase. Additionally, the CISO can contribute to SG&A reduction by streamlining security processes and automating routine tasks, freeing up resources and improving operational efficiency, further enhancing the company's financial performance and its appeal to investors.

Focus on Financial Returns and Exit Strategy: PE firms are laser-focused on maximizing returns on their investments. The CISO 3.0 can demonstrate how cybersecurity investments contribute to the company's financial performance and enhance its valuation, making it more attractive to potential buyers or investors during the exit phase.

Example: Implementing security measures that streamline operations, reduce costs, and protect intellectual property can positively impact the company's bottom line and its attractiveness to potential buyers. A CISO might demonstrate how automating patch management reduces IT overhead and minimizes the risk of costly breaches, directly contributing to a higher EBITDA.

Align with the PE Firm's Investment Thesis: Understand the PE firm's investment thesis and strategic goals for the company. Tailor your security program to support these goals, whether it's rapid growth, operational efficiency, or preparing for an IPO. This alignment demonstrates the CISO's understanding of the PE firm's priorities and their ability to contribute to their success.

Example: If the PE firm's goal is rapid growth through acquisitions, the CISO 3.0 can focus on developing a scalable security architecture and robust due diligence processes to ensure that acquired companies don't introduce undue risk. This proactive approach can help avoid costly security incidents that could derail the growth strategy and negatively impact the company's valuation.

Prioritize Due Diligence and Risk Mitigation: PE firms conduct extensive due diligence before investing in a company, and cybersecurity is a critical component of this process. The CISO 3.0 ensures that the company's security posture is robust and that potential risks are identified and addressed proactively, facilitating a smooth due diligence process and increasing the likelihood of a successful investment.

Example: During the exit phase, conducting a thorough security assessment and addressing any identified vulnerabilities can instill confidence in potential buyers or investors, maximizing the company's valuation and ensuring a successful exit for the PE firm.

Other Measures

Cybersecurity Insurance: While technically a form of risk transfer, strategically utilizing cyber insurance can also demonstrate to PE firms that the company is proactively managing its risk exposure. This can be particularly valuable during due diligence, showcasing the company's commitment to financial prudence and risk mitigation.

Security Metrics and Reporting: The CISO 3.0 can develop a set of key security metrics and provide regular reports to the PE firm, demonstrating the effectiveness of the security program and its contribution to the company's overall success. This level of transparency can build confidence in the CISO's leadership and the company's security posture.

The CISO 3.0 should strategically balance CAPEX and OPEX to maximize the company's financial performance. This directly influences key metrics such as return on invested capital (ROIC) and EBITDA multiples by ensuring that security investments contribute to the company's overall profitability and valuation. For example, during the *investment phase*, opting for cloud-based security solutions (OPEX) instead of investing heavily in on-premises infrastructure (CAPEX) can reduce upfront costs and improve cash flow, allowing the company to allocate more resources to growth initiatives and improve its financial metrics.

While private equity firms are known for their active involvement in mergers and acquisitions (M&A), they're certainly not the only ones pursuing this growth strategy. Public companies often turn to acquisitions as a swift way to boost EPS, making it essential for the CISO 3.0 to be well-versed in M&A and the pivotal role security can play within it. A CISO can assess the target company's cybersecurity posture, even if they can't conduct a full assessment of the target company before the deal closes.

The importance of thorough cybersecurity due diligence during M&A cannot be overstated. The Yahoo breach of 2014, which led to a $350 million reduction in

Verizon's acquisition price, and Marriott's acquisition of Starwood, resulting in a hefty General Data Protection Regulation (GDPR) fine due to a preexisting breach, serve as stark reminders of the financial repercussions of overlooking cybersecurity risks.

Despite the potential value CISOs can bring to the due diligence process, they are often excluded. This exclusion often stems not from concerns about insider trading or regulatory violations but from a lack of awareness or understanding of cybersecurity risks. Some organizations, especially those in industries not traditionally targeted by cyberattacks, may underestimate the significance of these risks. The focus on short-term financial gains in M&A transactions can also overshadow cybersecurity concerns. There's also a misconception that involving CISOs early on can slow down the deal, even though addressing these concerns upfront can prevent costly surprises later.

The CISO 3.0 needs a strong grasp of the M&A process, so they can explain how the security team can add value by addressing cybersecurity concerns early on, how they can help mitigate risks, and how they can avoid costly surprises that could impact the deal's valuation or even jeopardize its completion.

5.3.3 Venture Capital-Held Firms

Venture capital (VC) firms target early-stage companies, particularly in technology and innovation, that show strong potential for rapid growth, for example, Uber, Airbnb, and SpaceX. VC firms take on higher risks than PE firms in exchange for the potential for substantial returns. Their primary focus is on rapid growth and achieving a successful exit, through an acquisition or IPO. Key metrics for VC-held firms include user growth, market share, and revenue multiples.

5.3.3.1 Difference between PE and VC

While both PE and VC firms aim to generate returns through their investments, their strategies and target companies differ significantly. PE firms usually invest in established companies with a proven track record and steady cash flow, often using a mix of debt and equity to finance the deal. Their focus is typically on improving operations, cutting costs, and boosting profitability. In contrast, VC firms invest in early-stage companies that have high growth potential but unproven business models. They rely mostly on equity financing and concentrate on helping these startups scale, grow their customer base, and eventually achieve a successful exit.

VC firms also tend to hold on to their investments longer, often for 5–7 years. During this time, they support the company through different stages of growth, such as the seed stage; series A, B, and C funding rounds; and eventually the growth, maturity, and exit stages. The financial metrics that matter change throughout this journey: In the early stages, the focus is on growth and revenue multiples, while profitability and cash flow take center stage as the company matures.

VCs help guide companies through the following key stages:

Seed Stage: For example, a VC firm might invest in a promising tech startup with a unique product idea. The key metric here would be the early adoption rate, indicating the product's potential appeal.

Series A, B, C . . . Funding: As the startup gains traction, the VC firm might provide additional funding to scale operations, hire more employees, and expand marketing efforts. The key metric here could be the compound annual growth rate (CAGR) of user acquisition, demonstrating the company's growth trajectory.

Growth Stage: The startup might focus on expanding into new markets or launching new product features to further accelerate growth. The key metric here could be the market share growth rate, which would highlight the company's ability to capture more market share.

Maturity Stage: As the company matures, it might start generating revenue and achieving profitability. The key metric here could be the gross margin, indicating the company's ability to generate profit after accounting for the cost of goods sold (COGS).

Exit Stage: The VC firm might exit its investment through an IPO or acquisition. The final valuation of the company in this phase determines the VC firm's ROI.

What Provides Value: The primary value for VC firms is the potential for exponential returns on their investment in high-growth companies.

Who Determines Value: The VC firm, along with other investors and market participants, collectively determines the value of the company based on its growth potential, market traction, and future prospects.

What Delivers Value: Value is delivered through rapid growth in user base, market share, and revenue, leading to a successful exit, usually through an IPO or acquisition.

Where Is Value Created: Value is created through the company's innovation, disruption of existing markets, and ability to capture a greater share of a growing market.

When Is Value Created: Value is realized primarily during the exit phase, similar to PE-held firms. There is an IPO, or the VC firm sells its stake in the company to another company or investment firm.

How Is It Valued: Multiples valuation is prevalent in early stages, while DCF gains importance as the company matures.

5.3.3.2 How Security Can Align

For VC-held firms, where the emphasis is on rapid growth and innovation, the CISO 3.0's role is to enable these objectives while managing risk effectively. This might involve prioritizing investments in security measures that support scalability and agility, such as cloud-native security solutions that can adapt to the company's rapid growth. The CISO can also contribute to the company's valuation by demonstrating a proactive and mature security posture, assuring investors of the company's commitment to protecting its intellectual property and customer data. This can be achieved through regular security assessments, vulnerability management programs, and incident response preparedness, showcasing the company's ability to manage cyber risks effectively and maintain investor confidence.

Prioritize Scalable Security Solutions: The CISO 3.0 should focus on security measures that support scalability and agility.

Example: Implement cloud-native security solutions that can adapt to the company's rapid growth. This directly influences the key metrics of user growth and market share by ensuring that the company's security infrastructure can handle increased demand and expansion into new markets without compromising security.

Demonstrate a Mature Security Posture: The CISO 3.0 should regularly communicate the company's security posture to investors, highlighting security certifications, awards, and successful risk mitigation efforts. This directly impacts the company's overall valuation by assuring investors that the company is taking security seriously and is committed to protecting its assets and data.

Example: Obtaining industry-recognized security certifications such as ISO 27001 or SOC 2 to demonstrate the company's adherence to security best practices and commitment to data protection, enhancing its credibility and attractiveness to investors.

Seed Stage: Implementing a cloud-based security solution that can scale with the company's growth. This aligns with the key metric of *early adoption rate* by ensuring that the company's security infrastructure can support rapid user acquisition and product adoption.

Series A, B, C . . . Funding: Investing in security automation and orchestration tools to streamline security operations and reduce manual effort. This aligns with the key metric of *CAGR of user acquisition* by enabling the security team to keep pace with the company's rapid growth and focus on strategic initiatives.

Growth Stage: Expanding the security team and implementing advanced threat detection and response capabilities to protect the company's expanding attack surface. This aligns with the key metric of *market share growth rate* by ensuring that the company can securely expand into new markets and acquire new customers.

Maturity Stage: Implementing a robust data protection solution to protect sensitive customer and intellectual property data. This aligns with the key metric of *gross margin* by reducing the risk of data breaches that could result in financial losses and damage the company's reputation.

Exit Stage: Conducting a thorough security assessment and obtaining relevant security certifications to demonstrate a mature security posture to potential acquirers or investors. This directly impacts the company's *final valuation* by increasing its attractiveness and minimizing any security-related concerns during the due diligence process.

The CISO 3.0 can strategically choose security investments with a higher CAPEX component (e.g., infrastructure upgrades and security software development) to defer expenses and improve short-term profitability, positively impacting *revenue multiples*. For example, investing in a new security information and event management (SIEM) system might have a higher upfront CAPEX cost, but it can lead to long-term operational efficiencies and cost savings, ultimately benefiting the bottom line and improving revenue multiples.

5.3.4 Limited Liability Companies

For limited liability company (LLC) owners (or members), the value of their business comes from its profitability and long-term sustainability, offering both income and the potential for capital growth. As the owners, members gauge the company's value based on its financial performance, assets, and future growth opportunities. This value is built by generating profits, increasing the value of assets, and potentially selling the business down the line.

Value creation happens when the LLC generates revenue, manages costs efficiently, and builds a solid reputation in the marketplace. While value is continually built as profits and assets grow, it's often realized during key events, such as selling the business or distributing profits to the members.

What Provides Value: The value for LLC owners (members) lies in the profitability and sustainability of the business, providing income and potential for capital appreciation.

Who Determines Value: The members of the LLC, as the owners, determine the value of the company based on its financial performance, assets, and future growth prospects.

What Delivers Value: Value is delivered through generating profits, increasing the value of the company's assets, and potentially selling the business in the future.

Where Is Value Created: Value is created through the LLC's ability to generate revenue, manage costs effectively, and build a strong reputation in the market. (The Big 3 Revenue, Cost, and Risk. This will come up again)

When Is Value Created: Value is created continuously as the LLC generates profits and increases its assets, but value realization events may include the sale of the business or the distribution of profits to members.

5.3.4.1 How Security Can Align

In LLCs, where profitability and sustainability are key, the CISO 3.0's focus is on safeguarding revenue and reducing risks that could harm the company's financial well-being. This might include implementing cost-effective security measures to lower the chances of costly incidents like data breaches or ransomware attacks. Beyond protecting the bottom line, the CISO can add value by strengthening the company's reputation and building trust with customers and partners through strong security practices and clear communication about how data is protected.

Implement Cost-Effective Security Measures: The CISO 3.0 should prioritize security investments that offer a high ROI and minimize the risk of costly security incidents. This directly influences the key metrics of *profitability* and *sustainability* by reducing the financial impact of security breaches and ensuring the company's long-term viability.

Example: Implementing multifactor authentication (MFA) for all employees to reduce the risk of unauthorized access and data breaches, protecting sensitive customer and financial information at a relatively low cost.

Protect Revenue Streams: The CISO 3.0 should focus on securing critical systems and processes that directly contribute to revenue generation. This directly impacts the *profitability* metric by preventing disruptions to business operations and ensuring the company can continue generating income.

Example: Implementing robust security measures for e-commerce platforms and payment systems to protect against fraud and ensure uninterrupted online sales, safeguarding a critical revenue stream for the LLC.

Mitigate Risks to Financial Health: The CISO 3.0 should proactively identify and address any risks that could impact the company's financial stability. This directly influences the *sustainability* metric by reducing the likelihood of financial losses due to security incidents and ensuring the company's long-term viability.

Example: Regularly perform risk assessments to identify and prioritize potential threats, and then implement the necessary security controls to reduce risks and limit their potential financial impact.

Enhance Reputation and Trust: The CISO 3.0 can foster trust with customers and partners by demonstrating a clear dedication to security and data protection. This indirectly impacts the *profitability* metric by attracting and retaining customers who value security and are willing to pay a premium for products or services from a trusted provider.

Example: Publishing a clear and concise privacy policy that outlines how the company collects, uses, and protects customer data, demonstrating transparency and building trust with customers.

The CISO 3.0 can strategically choose security investments with a higher CAPEX component (e.g., on-premises security appliances) to improve short-term cash flow and profitability. This directly impacts the *profitability* metric by reducing immediate expenses and allowing the company to reinvest those savings in other areas of the business.

5.3.5 Family-Owned Companies

Value for family-owned businesses can be complicated, encompassing financial returns, social impact, employee well-being, and most importantly preservation of family control and legacy. Family members who own and control the company ultimately determine the value, often considering both financial and nonfinancial factors. Value is delivered through a combination of financial performance, maintaining family control, upholding the company's values and traditions, and contributing to the community. Value is created through the company's ability to generate profits, sustain long-term growth, and maintain a strong brand reputation. Value creation in family-owned businesses is often a continuous process, with a focus on long-term sustainability and intergenerational wealth transfer.

What Provides Value: Value for family-owned businesses can be complicated, encompassing financial returns, preservation of family legacy, social impact, and employee well-being.

Who Determines Value: Family members who own and control the company ultimately determine the value, often considering both financial and nonfinancial factors.

What Delivers Value: Value is delivered through a combination of financial performance, maintaining family control, upholding the company's values and traditions, and contributing to the community.

Where Is Value Created: Value is created through the company's ability to generate profits, sustain long-term growth, and maintain a strong brand reputation.

When Is Value Created: Value creation in family-owned businesses is often a continuous process, with a focus on long-term sustainability and intergenerational wealth transfer.

5.3.5.1 How Security Can Align

In family-owned businesses, where preserving legacy and maintaining family control are often priorities, the CISO 3.0's responsibilities go beyond just financial protection. They must also safeguard sensitive family information and protect the company's reputation. This might involve enforcing strict access controls, using data encryption, and implementing other security measures to keep confidential information secure. Additionally, the CISO can play a key role in the business's long-term sustainability by developing a solid succession plan for the security team, ensuring that the company's expertise and security practices are passed on to future generations.

Protect Sensitive Family Information: The CISO 3.0 should implement strict access controls, data encryption, and other security measures to safeguard confidential family and business data. This directly influences the key metrics of *family legacy* and *maintaining family control* by keeping sensitive information secure and protected from unauthorized access or exposure.

Example: Implementing a robust data loss prevention (DLP) solution to prevent sensitive family and business information from being leaked or shared inappropriately, safeguarding the family's privacy, and maintaining control over critical data.

Safeguard Company Brand Value: The CISO 3.0 should focus on protecting the company's brand and reputation by preventing security incidents that could damage public trust. This directly impacts the key metric of *brand value* by ensuring that the company is perceived as trustworthy and reliable, which is the key to long-term success and intergenerational wealth transfer.

Example: Establish a security awareness training program to educate employees on common cyber threats and best practices for data protection, helping minimize the risk of human error that could harm the company's reputation.

Develop a Succession Plan: The CISO 3.0 should create a robust succession plan for the security team to ensure continuity and knowledge transfer across generations. This directly influences the key metric of *long-term sustainability* by ensuring that the company's security expertise and capabilities are preserved and passed on to future generations, enabling the company to continue thriving in the long run.

Example: Establishing a mentorship program within the security team to facilitate knowledge transfer from experienced professionals to younger generations, ensuring that the company's security expertise is preserved and passed on.

Align Security with Family Values: The CISO 3.0 should ensure that the company's security practices align with the family's values and ethical principles. This directly influences the key metric of *upholding company values and traditions* by demonstrating that the company's commitment to security is rooted in its core values, fostering a sense of pride and continuity within the family.

Example: If the family values environmental sustainability, the CISO 3.0 could prioritize security solutions that minimize environmental impact, such as energy-efficient hardware and cloud-based services.

The CISO 3.0 can strategically balance CAPEX and OPEX security investments to align with the family's financial goals and risk tolerance. For instance, investing in long-term security infrastructure (CAPEX) can provide stability and cost savings over time, while OPEX investments can offer flexibility and adaptability to modern threats. This directly impacts the *financial performance* metric by optimizing security spending and ensuring that it aligns with the family's long-term financial objectives.

5.3.6 Sole Proprietorships

The value for sole proprietors is primarily derived from the income generated by the business, which supports their livelihood and financial goals. The sole proprietor, as the sole owner and operator, determines the value of the business based on its profitability, assets, and future prospects. Value is delivered through generating profits, managing expenses effectively, and potentially selling the business in the future. Value is created through the sole proprietor's skills, expertise, and effort in running the business. Value is created continuously as the business generates income and builds a customer base. Value realization events may include selling the business or transferring ownership.

What Provides Value: The value for sole proprietors is primarily derived from the income generated by the business, which supports their livelihood and financial goals.

Who Determines Value: The sole proprietor, as the sole owner and operator, determines the value of the business based on its profitability, assets, and future prospects.

What Delivers Value: Value is delivered through generating profits, managing expenses effectively, and potentially selling the business in the future.

Where Is Value Created: Value is created through the sole proprietor's skills, expertise, and effort in running the business.

When Is Value Created: Value is created continuously as the business generates income and builds a customer base. Value realization events may include selling the business or transferring ownership.

5.3.6.1 How Security Can Align

In sole proprietorships, where the owner's livelihood is directly tied to the business's success, the CISO 3.0's focus is on protecting income streams and minimizing disruptions. This could involve implementing cost-effective security solutions that are easy to manage and maintain, such as cloud-based security services or open-source tools. The CISO can also educate the sole proprietor about cyber risks and best practices, empowering them to make data-driven decisions about security investments and prioritize measures that directly impact their bottom line.

Implement Cost-Effective Security Solutions: The CISO 3.0 should prioritize security solutions that are affordable and easy to manage for sole proprietors, such as cloud-based security services or open-source tools. This directly influences the key metric of *income generation* by minimizing the financial impact of security incidents.

Example: Implementing a cloud-based antivirus and anti-malware solution to protect the sole proprietor's computer and data from cyber threats, providing essential security without requiring extensive technical expertise or expensive on-premises infrastructure.

Minimize Disruptions: The CISO 3.0 should focus on preventing security incidents that could disrupt the sole proprietor's business operations and impact their income. This directly influences the key metric of *income generation* by ensuring that the business can continue serving customers and generating revenue without interruptions.

Example: Establishing a dependable backup and disaster recovery plan ensures that a sole proprietor's data is secure and can be swiftly restored after a security incident, reducing downtime and minimizing potential income loss.

Educate the Sole Proprietor: The CISO 3.0 should provide the sole proprietor with information and training on cyber risks and best practices, empowering them to make decisions based in data about security investments. This indirectly influences the key metric of *income generation* by enabling the sole proprietor to prioritize security measures that directly protect their business and minimize the risk of financial losses due to security incidents.

Example: Regularly providing security awareness training for the sole proprietor, focusing on topics like phishing, password safety, and secure browsing practices, helps minimize the risk of human error that could lead to security breaches.

The CISO 3.0 should prioritize security solutions with a lower upfront cost and predictable monthly or annual fees (OPEX) to minimize the financial burden on the sole proprietor. This directly impacts the *income generation* metric by avoiding large capital expenditures and ensuring that security investments are manageable and sustainable for the business, for example, opting for a cloud-based email security solution with a monthly subscription fee instead of investing in an expensive on-premises email security appliance, reducing upfront costs and providing predictable expenses for the sole proprietor.

5.3.7 Public Organizations

Public organizations create value by delivering essential services to the public, improving quality of life, and promoting social welfare. The government or other stakeholders, such as taxpayers and service users, ultimately determine the value of public organizations. Value is delivered through efficient and effective service delivery, meeting the needs of the community, and maintaining public trust. Value is created through the organization's ability to utilize resources effectively, deliver high-quality services, and achieve its mission. Value is created continuously as the organization fulfills its mandate and serves the public good. Many people find it had to articulate IT value to Federal and State, Local, and Education (SLED) organizations because revenue is not the primary driver for public organizations. You'll find it's actually not all that different. You just have to replace revenue with service delivery in the big three triad – revenue, cost, and risk. As long as you can articulate how your IT or security program, or initiative impacts service delivery, cost, and risk, you will be able to earn budget and buy-in.

What Provides Value: Public organizations create value by delivering essential services to the public, improving quality of life, and promoting social welfare.

Who Determines Value: The government or other stakeholders, such as taxpayers and service users, ultimately determine the value of public organizations.

What Delivers Value: Value is delivered through efficient and effective service delivery, meeting the needs of the community, and maintaining public trust.

Where Is Value Created: Value is created through the organization's ability to utilize resources effectively, deliver high-quality services, and achieve its mission.

When Is Value Created: Value is created continuously as the organization fulfills its mandate and serves the public good.

Understanding the diverse ways in which value is created and delivered across different ownership structures is essential for crafting targeted communication strategies. If you recognize the unique motivations and priorities of each group, you can effectively articulate the value of cybersecurity initiatives in a way that resonates with their specific concerns and interests.

5.3.7.1 How Security Can Align

In the public sector, where service delivery and public trust are paramount, the CISO 3.0's role is to ensure that cybersecurity enables efficient and effective service delivery while safeguarding sensitive citizen data. This could involve prioritizing investments in security measures that protect critical infrastructure, prevent data breaches, and maintain the confidentiality of citizen information. The CISO can also demonstrate the value of their program by quantifying the potential cost savings from preventing cyber incidents or the improved efficiency resulting from streamlined processes.

Protect Critical Infrastructure: The CISO 3.0 should prioritize security investments that protect critical infrastructure and ensure the continuity of essential services. This directly influences the key metric of *service delivery* by preventing disruptions and ensuring that citizens can access the services they rely on.

Example: Implementing robust security measures for power grids, water treatment plants, and transportation systems to protect against cyberattacks that could disrupt essential services and impact public safety.

Prevent Data Breaches: The CISO 3.0 should focus on safeguarding sensitive citizen data and preventing data breaches that could compromise privacy and erode public trust. This directly influences the key metric of *maintaining public trust* by demonstrating the organization's commitment to protecting citizen information and upholding its responsibility to the public.

Example: Implementing strong encryption and access controls for sensitive data, conducting regular security audits, and providing ongoing security awareness training to employees to minimize the risk of data breaches.

Maintain Confidentiality of Citizen Information: The CISO 3.0 should ensure that citizen information is handled in accordance with privacy regulations and ethical guidelines. This directly influences the key metric of *maintaining public trust* by demonstrating respect for citizen privacy and upholding the organization's commitment to ethical data handling practices.

Example: Implementing privacy-enhancing technologies (PETs) to enable data analysis and sharing while protecting individual privacy, demonstrating the organization's commitment to responsible data use.

Quantify Cost Savings: The CISO 3.0 should demonstrate the value of cybersecurity investments by quantifying the potential cost savings from preventing cyber incidents. This directly influences the key metric of *efficient resource utilization* by highlighting the financial benefits of proactive security measures and justifying budget allocations for cybersecurity initiatives.

Example: Conducting a cost–benefit analysis to estimate the potential financial impact of a data breach, including costs associated with incident response, legal fees, and brand damage, and comparing those costs to the investment required for preventive security measures.

Demonstrate Improved Efficiency: The CISO 3.0 should showcase how cybersecurity initiatives can streamline processes and improve operational efficiency. This directly influences the key metric of *effective service delivery* by highlighting how security measures can enable faster and more efficient service provision, benefiting both the organization and the citizens it serves.

Example: Implementing automation and orchestration tools to streamline security operations and incident response, freeing up staff to focus on other critical tasks and enabling faster resolution of security issues, ultimately improving service delivery.

The CISO 3.0 can strategically balance CAPEX and OPEX security investments to align with the organization's budgetary constraints and long-term goals. For instance, investing in long-term security infrastructure (CAPEX) can provide stability and cost savings over time, while OPEX investments can offer flexibility and adaptability to new threats. This directly impacts the **efficient resource utilization** metric by optimizing security spending and ensuring that it aligns with the organization's financial objectives.

By understanding the unique value drivers and priorities of each ownership structure, the CISO 3.0 can tailor their cybersecurity program and procurement activities to align with the organization's overarching goals. This strategic alignment not only enhances the effectiveness of the security program but also positions the CISO as a valuable business partner who contributes to the organization's overall success.

5.4 ROLE OF THE BOARD OF DIRECTORS

5.4.1 Understanding the Board's Role

The vast majority of the company types we've explored will have some kind of board of directors (BOD). Communicating effectively with this board is vital for any CISO or executive in a technology-focused role. To achieve this, it's important to have a clear understanding of the board's specific function within the organization.

The corporate BOD serves as a key decision-making body, elected by shareholders to represent their interests and ensure the company is on the right track. Its role extends beyond mere oversight – it guides the company's overall direction, weighing in on significant decisions like approving strategic initiatives, appointing top executives, and closely monitoring financial health. In essence, the board helps shape the long-term vision of the organization while keeping management accountable for delivering results. Think of them as the guardians of the company's long-term success.

Board interactions can be intimidating, especially for CISOs who are used to dealing with technical matters. But remember, the board is there to support you, not grill you. They want to understand the risks and how you're addressing them. Be confident, be transparent, and, most importantly, speak their language. We'll talk a lot more about how to do that in Chapter 16.

The makeup and dynamics of a board can vary depending on whether a company is publicly traded or privately held. Public companies usually have larger, more diverse boards, often including a mix of independent directors and individuals with connections to the company. These boards face stricter regulations and greater scrutiny since they're accountable to a wide range of shareholders.

Privately held companies, on the other hand, tend to have smaller boards with fewer independent members. Their focus may be less about maximizing shareholder value and more about long-term growth or preserving family values, depending on how the company is owned.

In PE or VC-backed firms, board composition is often shaped by the investors. Investors typically negotiate for board seats as part of their investment deal, giving them influence over the company's strategic direction and ensuring their investment is safeguarded. These boards often include a mix of investor representatives, independent directors, and members of the management team.

Even sole proprietorships, LLCs, or family-owned businesses can benefit from having a board, although it's not legally required. These boards can offer valuable guidance and oversight and may consist of family members, trusted advisors, or industry experts. Their role may differ from public or investor-driven boards, focusing more on issues like succession planning, family dynamics, and upholding the company's values and legacy.

5.4.2 The Board's Primary Responsibilities

The board's role is primarily one of oversight, not day-to-day management. While they provide guidance and set direction for the company's executives, they typically don't get involved in the daily operations. The board focuses on setting the company's long-term strategy, and the executive team is responsible for carrying it out. This includes approving key strategic decisions, such as mergers, acquisitions, and major investments.

Another important role of the board is overseeing the company's risk management. They help identify potential risks, evaluate them, and ensure that proper controls are in place to mitigate those risks. Financial oversight is also critical: they review financial statements, approve budgets, and set policies on dividends.

One of the board's key responsibilities is hiring and, when necessary, replacing the company's CEO. They may also be involved in selecting other top executives. In addition, the board determines executive compensation, including salaries, bonuses, and stock options.

The board is also responsible for ensuring strong corporate governance. This means setting ethical standards, maintaining transparency, and protecting the rights of shareholders.

One of the most significant responsibilities a board director has is the fiduciary duty owed to the company and its shareholders – a topic we'll explore in detail in the next section.

5.4.3 Key Focus Areas for Boards

Now that you understand the board's role, let's consider their top priorities. Take a moment to think about what boards of directors care about most. What keeps them up at night? Jot down your top three guesses.

Have you written them down? Good. Now, let's reveal the answers:

1. **Revenue:** Revenue encompasses inflows of funds from business activities, essential for fueling growth and maximizing shareholder returns.
2. **Cost:** Revenue together with cost creates margin, which drives shareholder value. Boards want to ensure that the company is generating sufficient revenue to cover its costs and generate profits. They are also interested in ensuring that costs are being managed effectively.
3. **Risk:** From a corporate board point of view, risk is the potential for events or circumstances to negatively impact the company's objectives. There are two types of risk: Positive risk, which is the kind of risk from innovation and investments, and negative risk, which is loss exposure. Risk involves identifying, evaluating, and mitigating uncertainties in the business environment to safeguard the company's assets and ensure long-term success. We'll take a closer look at the concept of risk and how to manage it in a later chapter.

Later in this chapter, we will discuss how to craft messages that speak to the roles and needs of both boards and executive management. CISOs that know the board's priorities and how they view risk can tailor your communications to resonate with their concerns and demonstrate the value of cybersecurity to the organization.

5.4.4 Board Governance Models

The effectiveness of a BOD often depends on its governance model, which defines the board's structure, roles, and responsibilities. While the specific model varies based on the company's size, industry, and ownership, understanding common governance frameworks can help CISOs and executives navigate how their board operates.

5.4.4.1 Common Board Governance Models

Carver Model: This model focuses on policy-making and strategic oversight as the board's primary roles. The board sets the company's direction, while the CEO and executive team handle day-to-day operations. There's a clear division between governance and management, promoting accountability and transparency. This model is popular among larger, publicly traded companies where there's a distinct separation between ownership and management.

Traditional Model: In this model, the board takes a more hands-on approach, often engaging in the company's operations. Board members might provide specific expertise and guidance to management. This approach is commonly used in smaller companies, privately held firms, or industries with complex regulations, where closer board involvement is beneficial.

Advisory Board Model: Often used by startups and smaller companies, this model relies on a group of experienced professionals who offer guidance to the CEO and management. However, an advisory board doesn't have the legal responsibilities of a formal BOD. It's more informal and focused on advice rather than oversight.

Two-Tier Board Model: Common in some European countries, particularly Germany, this model divides the board into two tiers: A supervisory board that oversees strategic direction and a management board that handles daily operations. While this structure provides strong checks and balances, it can also slow decision-making.

Lead Director Model: In this model, an independent director is appointed as the lead director, working closely with the CEO and leading executive sessions of the board in this model discussions are held without the CEO present. This model strengthens board independence and is especially useful when the CEO is also the chairman of the board, a structure common in US companies. It's often chosen by organizations looking to enhance oversight and promote transparency.

Understanding the governance model of your board can help you tailor your communication. When your cybersecurity insights are aligned with the board's structure and responsibilities, they're more likely to be understood and acted upon in a way that fits the company's overall governance approach.

5.5 FIDUCIARY DUTIES

One of the core responsibilities of a board director is their fiduciary duty, which is a legal and ethical obligation to act in the best interests of the company and its shareholders. This duty is at the heart of corporate governance, ensuring that directors make decisions that support the organization's long-term success.

Fiduciary duties cover a range of responsibilities, requiring directors to act with integrity, diligence, and loyalty. These aren't just formalities: they are legally enforceable obligations, and failure to uphold them can result in serious consequences, including personal liability for damages.

At its core, fiduciary duty acts as a moral guide for corporate leaders, ensuring their actions are focused on protecting and growing shareholder value. It fosters trust and confidence in the company's leadership by putting the interests of the organization and its shareholders first.

Let's take a closer look at the specific fiduciary duties that board members and officers owe to the company and its shareholders.

5.5.1 Duty of Loyalty

The duty of loyalty mandates that directors act in good faith, putting the interests of the company and its shareholders above their own. This means avoiding conflicts of interest, refraining from self-dealing, and making decisions that benefit the company as a whole, rather than personal gain. For instance, a director cannot use their position to secure a lucrative contract for a company they own or have a financial stake in.

5.5.2 Duty of Obedience

The duty of obedience requires board members to follow all relevant laws, regulations, and the company's bylaws. This means directors cannot make decisions that break the law, even if they think it might benefit the company financially. For instance, a trucking company can't instruct its drivers to speed in order to make deliveries faster, even if it could boost revenue. Legal compliance always comes first.

5.5.3 Duty of Care

The duty of care obliges directors to exercise a standard of reasonable care in their decision-making. This involves being informed about the company's operations, risks, and opportunities, and making decisions based on careful consideration and due diligence. Failure to exercise reasonable care could result in a breach of fiduciary duty and potential legal liability. The standard of care is determined by what a "reasonable person" would do in similar circumstances, considering the specific context and industry norms.

5.5.4 Duty of Prudence

The duty of prudence is a key component of the duty of care. It emphasizes the importance of being aware of risks and exercising caution in decision-making. Directors must consider the potential consequences of their decisions and take steps to mitigate risks where possible. This involves staying informed about industry trends, emerging threats, and potential vulnerabilities that could impact the company's operations.

Understanding and upholding fiduciary duties is not only a legal requirement: it is also a fundamental ethical obligation for board directors. Directors can ensure that their decisions contribute to the long-term success and sustainability of the organization by prioritizing the interests of the company and its shareholders.

These duties are enshrined in legal precedent and ethical standards, and are owed by directors and officers to the corporation and its shareholders. The origins of these duties can be traced back to landmark legal cases like *Caremark*, which established the board's obligation to act in good faith and make decisions in the best interests of the company.

When shareholders believe that directors have violated their fiduciary duties, they can sue them. We talked about those types of Caremark lawsuits in Chapter 3 while exploring changes to the regulatory landscape. While it is possible to sue corporate directors for breach of duty, these cases are notoriously difficult to win because they require proving that the directors acted in bad faith or with intentional dereliction of duty, a high standard to meet. This difficulty stems from the business judgment rule, which presumes that directors acted in good faith, with reasonable prudence, and in the best interests of the company unless proven otherwise. To overcome this presumption, shareholders must prove that the directors acted recklessly, with gross negligence, or in bad faith.

5.6 CAREMARK AND LEGAL PRECEDENTS

In Chapter 3 we looked at recent Caremark claims through the lens of their impact on cybersecurity governance and the standard for creating a "reasonable" security program. We are going to look at those cases again, but this time we'll focus on what the cases mean for the fiduciary duties and liabilities faced by directors and executives.

5.6.1 Caremark Case

The Caremark case set a significant precedent by ruling that corporate directors may be personally liable if they fail to make a good faith effort to establish an effective compliance program. Essentially, directors have a duty of care that requires them to implement and maintain systems designed to reasonably detect and prevent legal violations. Ignoring this responsibility could be seen as a breach of their fiduciary duty, with serious personal consequences. This ruling carries substantial weight in the realm of cybersecurity, as it underscores the board's obligation to oversee and mitigate cyber risks, making it clear that cybersecurity is no longer just a technical issue but a governance concern as well.

5.6.2 Marchand (Blue Bell Ice Cream) Case

The *Marchand* case, involving Blue Bell Ice Cream, further emphasized the board's duty of loyalty in overseeing mission-critical operations. The court held that the board's failure to implement a "*reasonable*" system of monitoring and reporting for essential activities, such as food safety in this case, could constitute a breach of their duty of loyalty. This case expanded the scope of the board's oversight responsibilities, emphasizing the need for directors to understand and monitor the key operational aspects of the business that are essential to its success.

5.6.3 Boeing Co. Case

The *Boeing* case extended the definition of "*mission-critical*" activities to include large corporations with diverse product lines, not just companies with a single product focus. The court held that the board's duty of oversight extends to any activity that is essential to the company's mission and operations, regardless of the company's size or complexity. This case further broadened the scope of the board's responsibilities, requiring them to have a thorough understanding of the company's operations and the risks associated with them.

5.6.4 McDonald's Corp. Case

In the *McDonald's Corp.* case, the Delaware Court of Chancery *extended the fiduciary duty from just directors to include officers* of the company. The court emphasized that officers play a central role in identifying and reporting potential risks or "red flags" to the board, and they are responsible for addressing these issues if they fall within their area of expertise. This decision highlights the need for strong collaboration and open communication between the board and executive management because both groups play a vital role in steering the company toward long-term success and protecting the interests of its shareholders

The case was dismissed on appeal, but the Delaware Supreme Court did not explicitly reject the Chancery Court's holding that officers owe fiduciary duties. This leaves the question of officers' fiduciary duties somewhat unsettled, but it is clear that officers also play a part in corporate governance and risk management.

5.6.5 Sorenson Case

In the *Sorenson* case, the court explored the applicability of the *Caremark precedent to cybersecurity oversight failures.* While the case was ultimately dismissed, it signaled a potential shift in corporate governance, suggesting that a lack of cybersecurity oversight could be grounds for a Caremark claim, especially if it involves a violation of laws or regulations. This case stresses the growing importance of cybersecurity in the boardroom and the need for directors to proactively address cyber risks.

The legal precedents set by the *Caremark*, *Marchand*, *Boeing*, and *McDonald's* cases have interesting implications for corporate governance and cybersecurity. These cases collectively establish that directors owe a fiduciary duty to the company and its shareholders, to which they can be held personally liable. This duty encompasses the obligation to implement and maintain "reasonable" systems to oversee "mission-critical" functions, which could potentially extend to cybersecurity. Also, the *McDonald's* case suggests that not only board directors but also executive management owe fiduciary duties, emphasizing the shared responsibility for risk management and oversight within the organization.

5.7 THE CISO'S FIDUCIARY DUTY

The CISO holds the primary responsibility for safeguarding a company's digital assets and ensuring the security of its operations. In this role, the CISO owes a fiduciary duty to both the shareholders and the BOD. This dual responsibility stems from the interconnected nature of corporate governance and the critical importance of cybersecurity to businesses today.

Shareholders, as the owners of the company, provide the capital that fuels its operations and growth. In return, they expect a duty of loyalty from both the BOD and the executives, including the CISO. This duty of loyalty entails acting in the best interests of the shareholders, protecting their investments, and maximizing shareholder value.

The BOD, in turn, communicates to the shareholders their oversight of cyber risk, a critical component of the company's overall risk profile. The CISO is expected to provide the board with relevant information about "material" cyber risks. The CISO is responsible for communicating key information about significant cyber risks to the board, ensuring that even nontechnical members can understand the potential impacts of cyber threats on the company's financial performance, reputation, and strategy. We'll explore this in greater detail in Chapter 6.

The CISO's fiduciary duty to the board involves providing regular updates on the company's cyber risk posture, covering emerging threats, vulnerabilities, and the effectiveness of existing security controls. The board, in turn, provides oversight of the CISO's activities, ensuring that the company's cybersecurity strategy aligns with its overall risk tolerance and strategic goals. This collaborative relationship between the CISO and the board is essential for effective cyber risk management and fulfilling the fiduciary duty owed to shareholders (Figure 5.1).

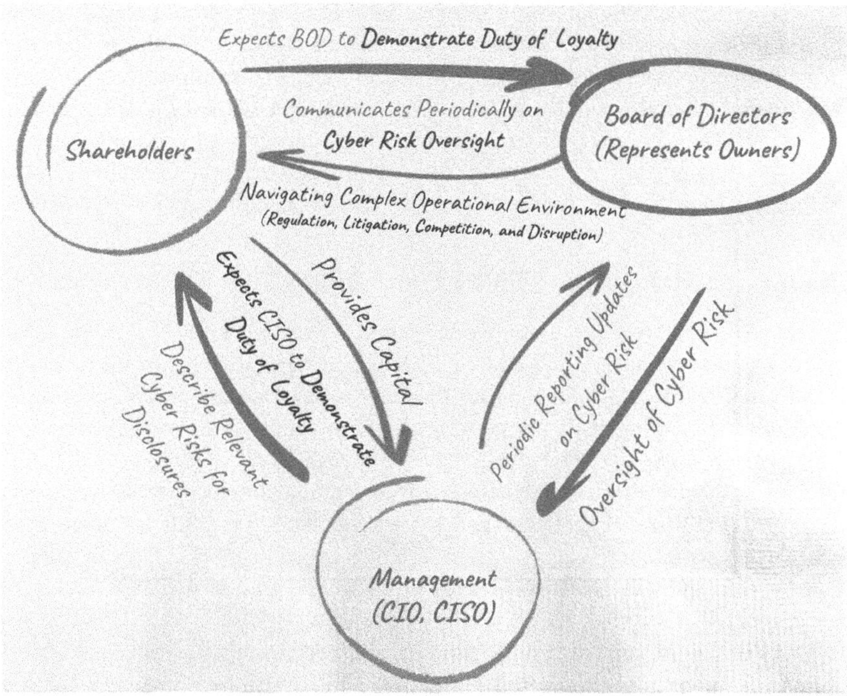

FIGURE 5.1 CISO Duties Napkin Drawing.

5.8 THE DYNAMICS OF GOVERNANCE: BOARD DIRECTORS VERSUS EXECUTIVES

I like to compare the relationship between the BOD and executive management to the legislative and executive branches of government. This analogy highlights the distinct yet complementary roles each group plays in steering the company toward success.

5.8.1 THE BOD: THE LEGISLATIVE BRANCH

The BOD functions much like the legislative branch of a government, responsible for setting the overall direction and framework for the company. They define the corporate vision and mission, which guide the organization's activities. The board oversees the development and execution of corporate strategy, ensuring it aligns with the company's long-term goals and shareholder interests. They also set performance targets, providing benchmarks for success and holding the executive team accountable for achieving them.

One of the board's most important tasks is hiring, evaluating, and, if necessary, replacing the CEO. This decision significantly shapes the company's future, as the CEO is charged with turning the board's vision into reality. The board also ensures strong corporate governance, promoting ethical standards, maintaining transparency, and protecting shareholder rights. They act as stewards for the shareholders, ensuring the company's actions align with shareholder expectations.

The board also ensures that the company complies with legal and regulatory requirements. They review policies and procedures, monitor adherence to laws, and take action when needed. Additionally, they review and approve executive compensation, making sure it is fair, competitive, and tied to the company's performance.

5.8.2 Senior Management: The Executive Branch

Senior management, like the executive branch of government, is responsible for carrying out the board's vision and strategy. They take the board's directives and turn them into actionable plans, managing the day-to-day operations of the company. This includes managing budgets, making decisions about resource allocation, and reporting progress back to the board.

Executives also oversee the development and management of employees, creating a culture that fosters excellence and innovation. They set goals for departments, ensuring every team contributes to the company's overall success. They develop and execute marketing and sales strategies to drive revenue growth and expand the company's reach.

In short, the board provides strategic direction and oversight, while the executive team focuses on execution and implementation. This division of responsibilities creates a system of checks and balances, promoting accountability and transparency while keeping the company on track to meet its goals (Figure 5.2).

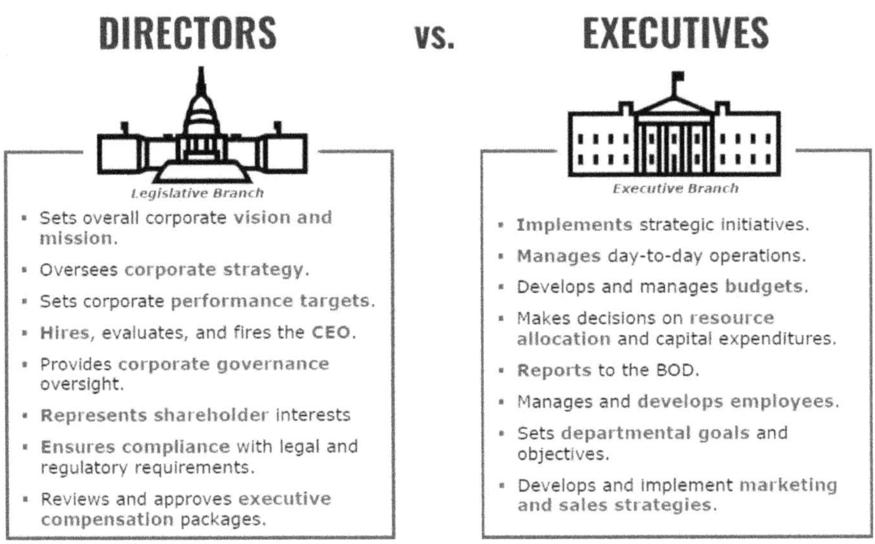

FIGURE 5.2 Directors versus Executives.

The key to successful communication with both the BOD and executive leadership teams (ELTs) lies in understanding their distinct roles, priorities, and information needs. While both groups are involved in driving the company's success, their perspectives and decision-making processes differ. Therefore, tailoring your message to resonate with each audience is essential for achieving your desired outcomes.

5.8.3 Communicating with the BOD

When presenting to the board, your primary goal is to earn their buy-in and instill confidence in your cybersecurity program. The board is less interested in technical details and more concerned with the big picture: How cybersecurity impacts the company's overall strategy, financial performance, and risk profile.

Your presentation should be a tightly woven story, ideally clocking in under 15 minutes. Highlight how you've leveraged company resources, showcasing tangible wins in language that resonates with the board – revenue protected, downtime minimized, and brand value safeguarded. Paint a vivid picture of the current risks, outlining the potential financial fallout of a cyber incident. Then, unveil your strategic plan: How you intend to fortify the company's defenses and the anticipated return on this investment. Be prepared to defend your choices, demonstrating that your approach isn't just sound but the most cost-effective path to resilience.

Directors often come to meetings with their own questions, influenced by various sources such as friends, board groups like the National Association of Corporate Directors (NACD), or even casual observations from magazines or advertisements. These questions might include the following:

What is our current relevant threat landscape?
How do we compare to our peers?
What are the emerging trends?
Are we compliant with a particular framework?
Should we be using a specific cybersecurity tool?

While these questions indicate engagement and a desire to understand, they can easily derail your presentation if not handled strategically. If you find yourself fielding such questions frequently, it might be a sign that you haven't effectively framed the cybersecurity challenges, set the right tone, or conveyed what matters most to the business.

The key is to tell a compelling risk story that articulates the potential financial impact of cyber threats in clear, quantifiable terms. We'll cover risk and quantification in depth in the next chapter. When you focus on what matters most to *your* business and how cybersecurity investments directly impact the bottom line, questions about peer comparisons or the latest trendy tool become less relevant. While a conversation about these topics is always welcome, ensure it doesn't overshadow the core message of your presentation: How your cybersecurity strategy protects and enhances shareholder value.

5.8.4 Communicating with ELTs

When interacting with executive leadership, your primary goal is to secure their buy-in and obtain the necessary budget for your cybersecurity initiatives. The ELT is more involved in the day-to-day operations of the company and will be interested in understanding the specific risks and threats facing the organization.

Your presentation to the ELT can be more detailed and technical than your board presentation, typically lasting up to 60 minutes. You should clearly articulate the risks and threats in business terms, explaining their potential impact on the company's operations, financial performance, and brand value. Outline your proposed plan for addressing these risks, including the resources required and the expected outcomes. Be prepared to present multiple options, each with its own cost–benefit analysis, and justify your recommendations based on their alignment with the company's strategic goals and risk tolerance (Figure 5.3).

In both cases, effective communication requires a clear understanding of your audience's priorities and concerns. You can build trust, secure buy-in, and ensure that your cybersecurity initiatives receive the necessary budget considerations simply by tailoring your message to resonate with your audience's specific needs.

We will cover board and executive communications extensively in Chapter 16.

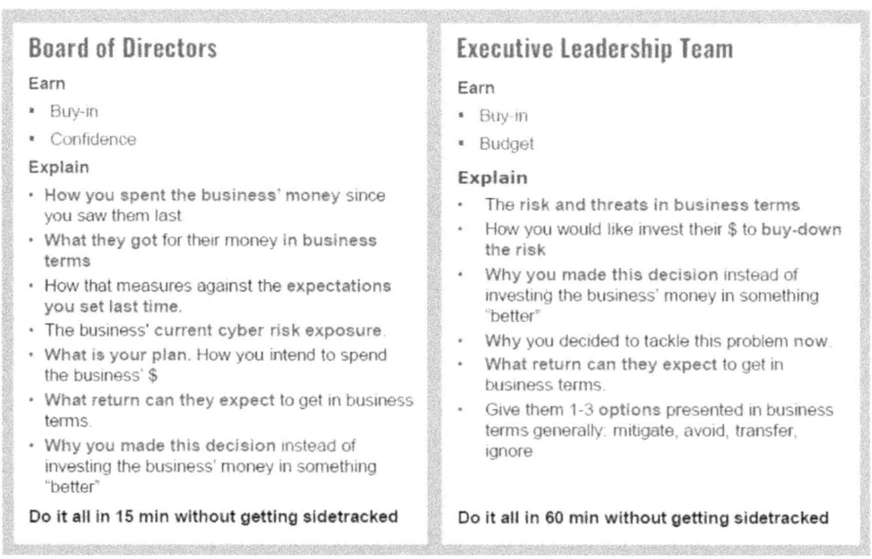

FIGURE 5.3 Board Directors versus ELT.

5.9 CONCLUSION

In this chapter, we've explored the relationship between ownership structures, board governance, and effective communication. We've seen how different ownership models, from publicly traded giants to tight-knit family businesses, each have their own unique motivations and priorities. For a CISO aiming to be a true strategic leader, understanding these nuances is key. It's about more than just knowing the org chart: it's about getting into the heads of your stakeholders and tailoring your security approach to fit their specific needs.

We underscored the BOD's role in safeguarding the company's long-term success and their fiduciary duty, encompassing loyalty, obedience, care, and prudence, ensuring that their decisions align with the company's best interests. But remember, the board sets the direction, while management steers the ship day-to-day. As a CISO, you need to adapt your communication style to each group, speaking their language and addressing their concerns.

By understanding the various ownership structures and what makes them tick, you can align your IT and security programs with the bigger picture. Effective communication with the board and executives is your golden ticket to securing the budget and support you need for your cybersecurity initiatives. And when you can clearly articulate how your program impacts the company's revenue, costs, and risks, you're not solely a tech person: you're a strategic asset.

In the end, it all boils down to communication and understanding. The keys are tailoring your message, navigating the complexities of corporate governance, and aligning your security efforts with the company's core values. With those keys you can create a security culture that doesn't just protect the company but actively contributes to its success. That's the essence of a CISO 3.0, a leader who understands the business, speaks the board's language, and builds a security program that's not only effective but also strategically aligned and valued by the entire organization.

5.9.1 Key Takeaways

1. **Ownership Structure and Its Impact on Value**
   - Different ownership structures, including publicly traded companies, PE firms, VC-backed firms, family-owned businesses, LLCs, and sole proprietorships, have unique value drivers and priorities.
   - Understanding these structures allows CISOs to align cybersecurity strategies with organizational goals effectively.
2. **Value Drivers across Ownership Models**
   - **Public Companies:** Value is driven by share price appreciation and dividends, influenced by financial performance and market sentiment.
   - **PE Firms:** Value stems from operational efficiency, profitability, and preparation for a profitable exit.
   - **VC Firms:** Focused on rapid growth, innovation, and successful exits through IPOs or acquisitions.
   - **Family-Owned Businesses:** Value includes financial returns, preservation of legacy, and alignment with family values.

- **LLCs and Sole Proprietorships:** Profitability, sustainability, and risk management are key drivers.

3. **Board Governance and Fiduciary Duties**
   - BOD oversee long-term success, manage risk, and ensure ethical corporate governance.
   - Directors owe fiduciary duties, including loyalty, care, prudence, and obedience, to the company and its shareholders.
   - The board's governance models influence its structure and priorities, requiring tailored communication strategies.
4. **The Evolving Role of the CISO**
   - CISOs must navigate boardroom dynamics and align cybersecurity initiatives with organizational value drivers.
   - They hold fiduciary responsibilities to shareholders and the board, emphasizing communication of material cyber risks and proactive risk management.
   - CISO 3.0 leaders focus on translating cybersecurity into strategic business outcomes.
5. **Communication with Boards and Executives**
   - Tailoring messages to board priorities (revenue, cost, and risk) enhances effectiveness.
   - Boards prioritize high-level impacts, while executive teams need detailed plans and cost–benefit analyses.
   - Successful communication involves demonstrating how cybersecurity aligns with and enhances the company's strategic goals.
6. **Lessons from Legal Precedents**
   - Cases like Caremark, Marchand, and Boeing emphasize the critical role of boards in managing mission-critical risks, including cybersecurity.
   - Directors and officers can face personal liability for failing to meet oversight responsibilities, underscoring the importance of proactive governance.

5.9.2 Reflection and Exploration Questions

Understanding the Concepts

1. In what ways do ownership structure impact value?
2. How do fiduciary duties influence board decision-making?

Connecting the Dots

3. How does the role of a CISO differ across various ownership models?
4. In what ways do the board's responsibilities connect to earlier discussions about aligning cybersecurity with business objectives?

Real-World Application

5. Can you identify examples of how cybersecurity initiatives enhanced organizational value?

6. How might understanding fiduciary duties improve your approach to communicating with the board?

Personal and Professional Growth

7. How can you apply the concept of tailoring communication strategies to your current role?
8. What steps can you take to better understand the value drivers specific to your organization?

Further Exploration

9. What additional skills or knowledge might you need to effectively communicate with boards and executives?
10. How can you determine what creates value for your organization and align your program and initiatives to that value creation?

6 Risk

6.1 INTRODUCTION

In cybersecurity, understanding risk is essential. It's not only about identifying threats and vulnerabilities but also about comprehending the potential impact they can have on your organization. In this chapter, we're going to dive deep into the concept of risk. We'll break down what it really means, explore the different flavors it comes in, and see how it all ties back to those threats and vulnerabilities. We'll also talk about your organization's risk appetite, which is how much risk you're willing to stomach, and the basics of putting numbers on those risks. Then, we'll lay out your options for dealing with risk, whether it's avoiding it altogether, transferring it to someone else, or facing it head-on.

By the time we're done, you'll have a solid grasp of modern risk management. You'll be able to make smarter decisions that not only protect your company but also support its overall strategy. Because let's face it, in today's world, a CISO needs to be more than just a tech expert. They need to be a risk whisperer, able to translate the complex language of cybersecurity into the language of business. That's what sets a CISO 3.0 apart, the ability to see risk not only as a technical problem but also as a core part of the business landscape.

6.2 RISK STRATEGY

Risk management is an essential function of the modern CISO, but it is distinct from security strategy. While security strategy focuses on the implementation of controls, processes, and technologies to mitigate threats, *risk strategy* is a higher-order discipline concerned with quantifying, prioritizing, and financially structuring risk to ensure organizational resilience. A well-defined risk strategy enables informed decision-making at the executive level, aligning cybersecurity risk with broader enterprise risk management.

6.2.1 Risk Strategy versus Security Strategy

Risk strategy and security strategy are interdependent but serve distinct roles within an organization. *Security tactics* – as discussed in Chapter 12 – both inform and are informed by the overarching *security strategy,* detailed in Chapter 9. However, *risk strategy is separate* and should be defined *in conjunction with the chief information security officer (CISO), chief financial officer (CFO), and Chief Risk Officer (CRO).* This cross-functional approach ensures that cybersecurity risks are assessed alongside financial and operational risks, allowing the organization to optimize risk transfer, mitigation, and acceptance strategies.

At its core, the *goal of risk strategy* is to achieve resilience to *material cyber losses* by implementing a structured and financially informed approach to managing risk. To achieve this, organizations should adopt the following *CISO 3.0 Risk Strategy Framework.*

DOI: 10.1201/9781003510789-8

6.2.2 The CISO 3.0 Risk Strategy Framework

Objective: Build a risk-informed cybersecurity strategy that aligns with business objectives and ensures resilience to material losses.

1. **Establish a Risk Baseline**
    - **Define Risk in Business Terms:** Translate cybersecurity risks into financial and operational impact.
    - **Determine Risk Appetite and Tolerance:** Align with executive leadership and the board on acceptable levels of risk.
    - **Categorize Risk Types:** Distinguish among IT risk, enterprise risk, regulatory risk, financial risk, etc.
2. **Quantify and Prioritize Risk Exposure**
    - **Utilize Risk Quantification Models:** Implement FAIR, Monte Carlo simulations, or other data-driven approaches.
    - **Assess Business Impact:** Identify potential financial, reputational, and operational consequences.
    - **Develop Risk Registers:** Maintain an updated risk register that reflects evolving threats and organizational changes.
3. **Determine Materiality and Disclosure Requirements**
    - **Define Material Cyber Risk Events:** Establish the criteria for what constitutes a material risk event.
    - **Align with Regulatory Expectations:** Incorporate Securities and Exchange Commission (SEC), Federal Trade Commission (FTC), and Caremark fiduciary duty considerations.
    - **Prepare Board-Level Risk Reporting:** Develop structured reports for leadership and governance bodies.
4. **Develop a Risk Treatment Strategy**
    - **Evaluate Risk Response Options:**
        - **Mitigate:** Deploy controls and security measures to reduce risk.
        - **Transfer:** Leverage cyber liability insurance and third-party agreements.
        - **Avoid:** Exit or alter high-risk business activities.
        - **Accept:** Determine how to finance and prepare for residual risks.
    - **Prioritize Risk Reduction Initiatives:** Align security investments with business objectives.
5. **Implement Risk Transfer Mechanisms**
    - **Assess Cyber Liability Insurance Coverage:** Evaluate policy terms, exclusions, and coverage limits.
    - **Align Incident Response with Insurance:** Ensure response plans support claim readiness.
    - **Explore Alternative Risk Financing:** Consider self-insurance, captives, or structured reserves.
6. **Allocate Resources for Accepted Risk**
    - **Adjust Financial Reserves:** Set aside capital to cover anticipated losses.
    - **Develop Self-Insurance Models:** Explore internal funding options for cyber-related losses.

- **Align Risk Financing with Business Strategy:** Ensure risk management decisions support long-term growth.

7. **Define and Monitor Key Risk Indicators (KRIs)**
   - **Establish KRIs Aligned with Business Goals:** Measure risk trends and security effectiveness.
   - **Develop Actionable Metrics:** Distinguish between leading (predictive) and lagging (historical) indicators.
   - **Embed Risk Monitoring into Business Processes:** Integrate real-time risk insights into decision-making.
8. **Maintain Continuous Risk Governance and Compliance**
   - **Embed Cybersecurity into Governance Frameworks:** Ensure risk oversight at the executive and board levels.
   - **Conduct Regular Risk Reviews:** Reassess risk exposure and update treatment strategies as necessary.
   - **Align Risk Management with Regulatory Compliance:** Ensure adherence to SEC, FTC, and other relevant regulations.
9. **Enhance Cyber Resilience and Response Preparedness**
   - **Develop a Cyber Recovery Strategy:** Ensure alignment with business continuity and disaster recovery planning.
   - **Test and Refine Incident Response Plans:** Conduct tabletop exercises and simulations.
   - **Foster a Resilience Culture:** Educate leadership and stakeholders on risk management best practices.

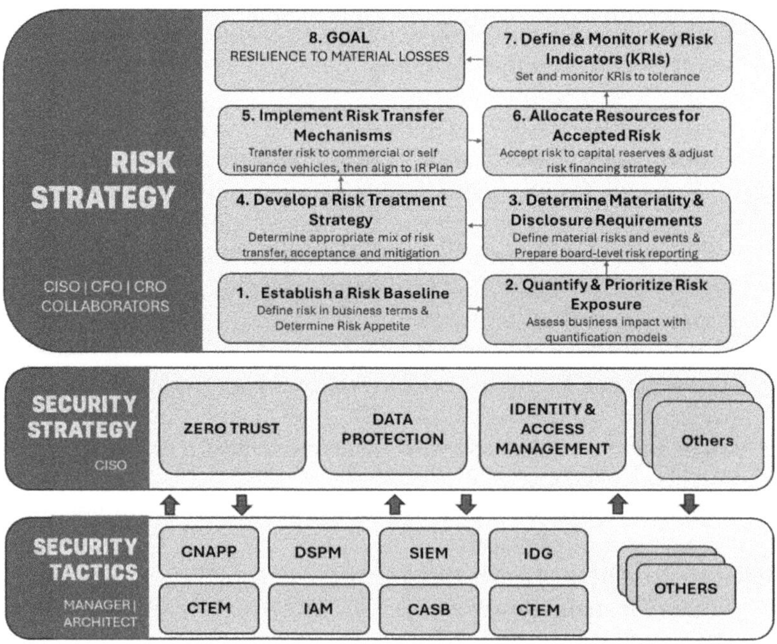

FIGURE 6.1 Risk Strategy.

A well-structured *risk strategy* provides the foundation for *cyber resilience*, ensuring that an organization is not just defending against cyber threats but is financially and operationally prepared for inevitable incidents. *By integrating cybersecurity risk into the broader enterprise risk framework, CISOs can shift cybersecurity from a cost center to a strategic enabler* – one that protects business value while supporting long-term growth (Figure 6.1).

By adopting the *CISO 3.0 Risk Strategy Framework*, organizations can move beyond compliance-driven approaches and *establish a proactive, business-aligned risk management culture.* This ensures that cybersecurity risks are *quantified, prioritized, and mitigated* in a manner that supports *strategic objectives and operational resilience.*

We will discuss over the course of this chapter and the rest of the book how to perform each of these steps so you can develop your own risk strategy.

6.3 RISK OVERVIEW AND DEFINITION

Risk, in essence, is the potential for loss or harm due to exposure to a threat. It's a complex concept that encompasses both the likelihood of an event occurring and the magnitude of its impact. The fundamental formula for risk is

$$\text{Risk} = \text{Probability} \times \text{Impact},$$

where

- probability is the likelihood of a specific event or incident occurring. It's usually expressed as a percentage or a frequency (e.g., once a year or twice a month);
- impact is the magnitude of the loss or harm that would result if the event occurred. This can be measured in financial terms, operational disruptions, brand damage, or other relevant metrics.

It's important to note that it's very common to see the formula written as likelihood x impact, but that isn't exactly correct. While likelihood is often used to express the probability of an event occurring, it's technically a more qualitative assessment, whereas probability is a more precise, quantitative measure.

6.3.1 Likelihood versus Probability

While sometimes used interchangeably, likelihood and probability have nuanced distinctions in the context of risk assessment:

- **Likelihood:** Often used to describe a qualitative assessment of how likely an event is to occur. It is a subjective measure based on expert judgment, historical data, and other relevant factors. Likelihood can be expressed in terms like "rare", "unlikely", "possible", "likely", or "almost certain".

- **Probability:** A more quantitative measure of the chance of an event happening, typically expressed as a numerical value between 0 (impossible) and 1 (certain). Probability can be calculated based on statistical models, historical data analysis, or simulations.

6.3.1.1 Coin Toss Analogy

Consider the analogy of a coin toss:

- **Likelihood:** Before flipping a coin, we might say the likelihood of getting heads is "equally likely" as getting tails. This is a qualitative assessment based on our understanding of how coins work.
- **Probability:** However, the probability of getting heads on a single coin toss is precisely 0.5 or 50%. This is a quantitative measure based on the possible outcomes of the event.

Now, let's extend this to multiple coin tosses:

- **Likelihood:** If we flip a coin 100 times, we might say the likelihood of getting exactly 50 heads and 50 tails is "unlikely", This is because while it's the most probable outcome, there are many other possible combinations.
- **Probability:** We can calculate the probability of getting exactly 50 heads and 50 tails using statistical methods. This probability will be lower than 50%, but it can be precisely calculated.

Likelihood is calculated by starting with a numerical possibility and gathering evidence from outcomes to support or refute it. We can use likelihood to make judgments. For example, if a coin is tossed 100 times and lands on heads only four times, the likelihood of the coin being unbiased is very low.

This analogy illustrates the difference between likelihood and probability. Likelihood is a broader, subjective assessment of an event's chances, while probability is a specific, numerical measure based on the underlying factors and possible outcomes.

Consequently, a risk analyst might say that a data breach is "likely" to occur within the next year (likelihood) based on the company's current security posture and industry trends. They might then use a quantitative model to estimate that the probability of a breach is 0.3 or 30% (probability). In risk management, we need both measures, and it's important to understand the difference.

6.4 RISK VERSUS THREAT

One of the most basic parts of risk management is understanding the difference between risk and threat, as well as other related terms like vulnerability and exploit. These terms are often conflated, leading to confusion and misaligned risk management efforts. In fact, misunderstanding the difference between risk and threat is one of the most common problems I see among CISOs, and it undermines effective board-level communications (Figure 6.2). Let's clarify their definitions and provide examples of how they are sometimes misused:

Risk: The potential for loss or harm due to exposure. It is quantified by assessing the likelihood (or probability) and impact of potential events. Risk supports informed decision-making based on cost–benefit analysis and can be mitigated through various risk management strategies, such as risk avoidance, risk transfer (e.g., insurance), risk mitigation, or risk acceptance.

Threat: A specific event or actor who could potentially cause harm or disruption. Threats can be intentional or unintentional, external or internal, and can be caused by human, environmental, or technological factors. Threats are identified through threat intelligence and other security measures. They are mitigated through the implementation of security controls and countermeasures.

Vulnerability: A weakness or flaw in a system, process, or asset that can be exploited by a threat to cause harm or disruption. Vulnerabilities can be technical (e.g., software bugs or misconfigurations) or nontechnical (e.g., lack of employee training and inadequate policies). Vulnerabilities are identified through vulnerability scanning and assessments.

Exploit: A specific method or tool used to take advantage of a vulnerability. Exploits can be developed by threat actors or discovered through security research.

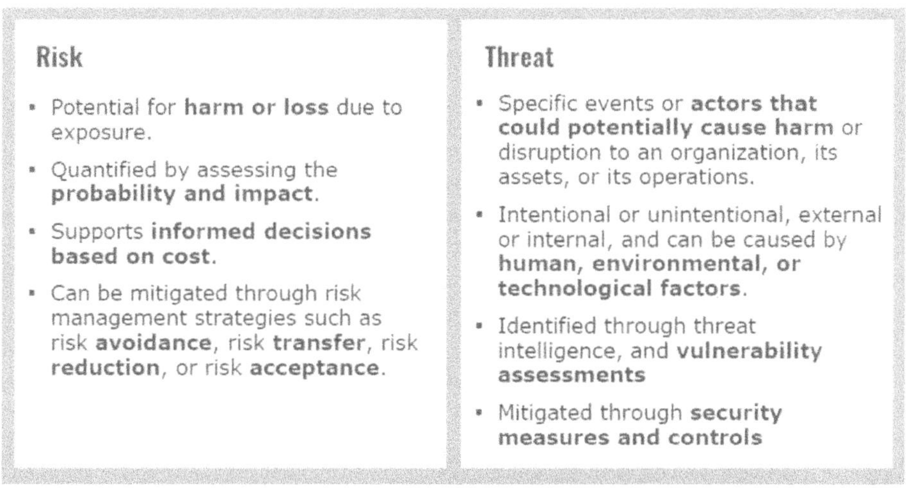

FIGURE 6.2 Risk versus Threat.

6.4.1 Common Misconceptions in Risk Registers

It's not uncommon to see risk registers cluttered with entries that are not actually risks but rather threats or vulnerabilities. This misalignment can hinder effective risk management by focusing on the wrong elements. Here are some examples:

- **"Ransomware" Listed as a Risk:** Ransomware is a threat, not a risk. The risk would be the potential financial loss, operational disruption, and opportunity costs that could result from a ransomware attack.
- **"Legacy Unpatchable OS" Listed as a Risk:** This is a vulnerability, not a risk. The risk would be the potential for a threat actor to exploit this vulnerability to gain unauthorized access, cause disruptions, or steal data.
- **"Phishing Attacks" Listed as a Risk:** Phishing is a threat, not a risk. The risk would be the potential for employees to fall victim to a phishing attack, leading to unauthorized access, data breaches, or financial loss.

To ensure your risk register is accurate and actionable, it's important to clearly define and differentiate between risks, threats, vulnerabilities, and exploits. Focus on identifying the potential impact of threats exploiting vulnerabilities and use this information to prioritize and implement effective risk mitigation strategies.

Communicating threats to the board lacks the all important "so what?" from a business perspective and can often be interpreted as fear, uncertainty, and doubt (FUD). It can also lead to presenting the board with technical performance metrics like the number of blocked emails, which is something you should never do. We'll discuss effective board-level communication in more detail in Chapter 16.

6.5 IT RISK VERSUS ENTERPRISE RISK

IT risk is a subset of enterprise risk. While IT risk focuses on the potential for loss or disruption due to failures or breaches in IT systems, enterprise risk encompasses a broader range of risks. From a board's perspective, IT risk is just one of many risks the organization faces. These risks include liquidity risk (the risk of not having enough cash on hand to meet short-term obligations), health and safety risk (the risk of injury or illness to employees or customers), contract risk (the risk of financial loss due to a breach of contract), access to credit risk (the risk of being unable to obtain loans or credit at favorable terms), land management risk (the risk of environmental damage or legal disputes related to land ownership or use), and many others.

If you read a 10-K financial statement, you'll generally find that IT or cyber risk is just one of many disclosed risks, usually somewhere in the middle of the list. Consequently, *it's important to present risks in "business terms" when communicating with the board.* This means focusing on the potential impact of IT risks on the organization's overall objectives rather than on technical details. This is the very first step in the risk strategy framework.

The key business risks associated with cyber incidents include the following:

- Financial:
  - Response costs
- Operational:
  - Productivity loss
- Regulatory compliance:
  - Restrictions
  - Fines
- Legal:
  - Judgments
- Reputational:
  - Stock value
  - Customer churn
- Strategic:
  - Loss of competitive advantage
- Health and safety:
  - Loss of life

In addition to those categories of business risks, most organizations face many of the same challenges related to IT and security. Turning the security teams' challenges into these more universal business challenges is a great way to start translating security into the language of business.

The most common business challenges include the following:

- Production uptime/business continuity
- Support innovation for growth and scale
- Customer experience (CX)
- Operational efficiency, cost control, and reduction
- Cyber risk reduction
- Regulatory changes and compliance risk reduction
- Supply chain risk reduction
- Resource and talent limitations
- Environmental, social, and governance (ESG)

CISOs can also integrate their challenges and initiatives into these broader business challenges. For instance, concerns like business process continuity, cyber resilience, and cyber liability all directly contribute to the overarching goal of production uptime. Similarly, digital transformation, AI security, and cloud security are all part of the effort to support innovation for growth and scale, as well as CX. Beyond the CISO's challenges, the security teams' challenges and even individual security initiatives can be rolled up into higher-level challenges.

Think of the challenges as nested, where technical controls are a part of security team challenges, which contribute to the CISO's challenges, which are a segment of the broader business challenges. To illustrate this further, consider how a managed detection and response (MDR) initiative can be seen as part of the security team's Security Operations Center (SOC) modernization efforts, which, in turn, supports the

CISO's breach risk reduction goals, ultimately feeding into the business's overarching cyber risk reduction objective. This nested structure allows for a clear translation of technical security initiatives into tangible business risks, facilitating effective communication and motivating investment in cybersecurity.

When we consider business-level risks and challenges, we often think of financial loss due to breach response as the biggest risk. The direct financial cost of a breach is often not a major concern from a board's perspective. Consider the size and scope of the problem at an enterprise scale. It's a rare cyber event that would rise to the level of 5–10% of annual revenue, which is the usual threshold for financial materiality. Many organizations can chalk up tens of millions of dollars in losses to operational expenses. If you have a company that does $20 billion in annual revenue, a $10 million breach would represent only 0.0005% of annual revenue; as you can imagine, that is not going to be a huge concern for most board members.

The risks that truly concern boards are opportunity costs and market costs. It's not so much that $10 million was lost, but where it will come from. Does the breach response come from cash flow? If so, what was that cash going to be used for? To buy a piece of revenue-generating equipment? That type of opportunity cost is the real killer. I'm not a big believer in long-term reputational harm due to breaches. Virtually no one stopped shopping at Target or Home Depot because they had massive credit card breaches. They did a hit to brand value in the form of a short-term hit to stock value, but by the 12-month mark, both companies were back up to pre-breach values. So, the question becomes what was planned during the short-term effect. Maybe the breach happens during a mergers and acquisitions (M&A) cycle and affects the valuation. As we saw with Yahoo, when they had to announce their breach in the middle of being acquired by Verizon and lost hundreds of millions of dollars, by framing IT risks in terms of their potential impact on these key business areas, you can effectively communicate their significance to the board and ensure that cybersecurity is viewed as a strategic imperative rather than just a technical issue.

6.6 QUALITATIVE RISK ANALYSIS

The second step in the risk strategy framework is to assess and prioritize your risk exposure. Qualitative risk analysis is a common approach used to assess and prioritize risks based on subjective judgments and descriptive scales. It often involves assigning ratings or labels to risk likelihood (e.g., low, medium, or high) and impact (e.g., minor, moderate, or severe) to categorize and compare different risks.

I used to rely heavily on qualitative risk assessments, but I quickly realized their limitations. It's like trying to navigate a ship with a compass that's constantly spinning. You might have a general sense of direction, but you'll never reach your destination with precision. That's why I made the shift to a more quantitative approach, leveraging data and analytics to gain a clearer and more actionable understanding of our risks.

6.6.1 Challenges of Traditional Qualitative Approaches

One of the main challenges of traditional qualitative risk analysis is its inherent *subjectivity*. Assessments rely heavily on personal interpretations and expert judgments, which can vary between individuals or groups. This subjectivity often leads to inconsistencies and makes it challenging to compare risks across different projects or business units, hindering the development of a unified risk management approach.

Another limitation of qualitative analysis is the *lack of precision*. Descriptive scales, while easy to understand, fail to provide the numerical accuracy necessary for objective risk comparisons. This ambiguity sometimes makes it harder to accurately assess the relative severity of different risks, potentially leading to misallocation of resources and ineffective prioritization of mitigation efforts. For example, if you have 5 high-risk items and 15 medium-risk items, is it better to address one high or three mediums? You can't calculate that answer based on this scale. Qualitative approaches often conflate the concepts of likelihood and probability, further blurring the lines between subjective estimations and objective, quantifiable measures of risk.

Also, qualitative assessments often suffer from *limited actionability*. The descriptive nature of these assessments may not provide the level of detail and specificity required to guide concrete risk mitigation actions. This can result in generic or ineffective responses that fall short of addressing the root causes of risks, leaving organizations vulnerable to potential threats.

Communicating risk effectively to stakeholders, particularly those who are not well-versed in risk management terminology or the specific context of the analysis, presents yet another challenge. Qualitative assessments, with their reliance on subjective language and lack of concrete data, can be difficult to interpret and can lead to misunderstandings about the true nature and severity of risks.

6.6.2 Risk Matrix: A Common Tool with Shortcomings

A risk matrix, shown in Figure 6.3, is a widely used tool for visually mapping and comparing risks in qualitative risk analysis. It typically arranges risks on a grid, with one axis representing likelihood and the other impact. Each risk is placed on the matrix based on how likely it is to occur and its potential impact, providing a quick visual summary of current risks.

While risk matrices can be useful for making comparisons, they have limitations. One major drawback is the reliance on ordinal scales (such as low, medium, and high) for both likelihood and impact. These scales are not as precise as numerical values, which can lead to misinterpretation and make it difficult to prioritize risks. As previously mentioned, risk matrices and ordinal scales alone cannot be used to make calculations, making it impossible to prioritize risks effectively. For instance, a risk matrix with an ordinal scale cannot tell us whether it's more beneficial to address one "high" impact risk or another.

The subjective nature of placing risks on the matrix can also introduce bias, leading to inconsistencies in how risks are evaluated. Additionally, risk matrices tend to focus on just likelihood and impact, often leaving out other critical factors such as how quickly a risk could materialize (velocity) or how volatile the risk environment is. Furthermore, when risks are aggregated across different departments or projects, using a matrix can become cumbersome, making it harder to see the organization's overall risk landscape clearly.

Risk Rating

Standard Risk Matrix for any Business

Impact						
	5	Medium / High	Medium / High	High	High	High
	4	Low / Medium	Medium / High	Medium / High	High	High
Impact	3	Low / Medium	Low / Medium	Medium / High	Medium / High	High
	2	Low	Low	Low / Medium	Low / Medium	Medium / High
	1	Low	Low	Low	Low / Medium	Low / Medium
		1	2	3	4	5
		Likelihood				

FIGURE 6.3 Risk Matrix.

Inconsistent risk assessments often undermine traditional risk management efforts. Different teams or individuals within the same organization may assess the same risks differently, leading to misaligned priorities and incomplete responses. This lack of consistency can create gaps in the organization's defenses, leaving it vulnerable to threats that may go unaddressed.

Another common issue is the reliance on historical data to predict future risks. While past incidents can offer useful lessons, they may not accurately represent today's rapidly evolving threat landscape. Overdependence on this outdated data can create a false sense of security, making organizations ill-prepared for emerging threats.

Traditional approaches also tend to evaluate risks in isolation, overlooking how risks are interconnected and how one can trigger others in a cascading effect. This siloed approach can lead to an underestimation of overall risk. One event could trigger a chain reaction, amplifying the impact of other risks. Mitigation strategies built on this isolated view might fall short when it comes to addressing the full scope of potential consequences. Focusing on compliance rather than true risk management is another common pitfall. Meeting regulatory standards is important, but it should not overshadow the need to build a robust security strategy based on actual risk.

6.6.3 Moving to a More Quantitative Approach

To overcome the limitations of traditional qualitative risk analysis, organizations are increasingly adopting more quantitative approaches, such as the FAIR framework. These approaches leverage data and statistical models to provide more precise and objective risk assessments, facilitating informed decision-making and effective risk mitigation.

Several factors are driving this shift. First, quantitative analysis reduces subjective biases by offering a data-driven, consistent approach for comparing risks across departments or business units. This leads to more reliable risk prioritization and better resource allocation.

Second, quantifying risks in financial terms makes it easier to communicate with senior leadership and stakeholders who may not be versed in technical cybersecurity details. Framing risks in terms of their potential financial impact helps justify security investments and fosters more informed decision-making at the executive level.

A quantitative approach also provides a clearer foundation for action. By precisely measuring risk factors, organizations can develop cost-effective strategies to address the most significant threats, optimizing their security budgets and maximizing the return on investment (ROI). Regularly tracking these risks allows organizations to see how effective their mitigation strategies are and make data-backed adjustments when needed.

The move toward quantitative risk analysis represents a new and exciting (at least I find it exciting) advancement in the field of cybersecurity risk management. Leveraging data and statistical models allows organizations to achieve greater objectivity, precision, and actionability in their risk assessments, ultimately leading to more informed decision-making, optimized resource allocation, and enhanced security.

6.7 RISK QUANTIFICATION BASICS

Risk quantification involves assigning numerical values to factors like probability and impact, making it easier to assess risks objectively and make more informed, data-driven decisions. It can feel overwhelming at first, but think of it like learning a new language. Once you understand the basics, it opens up a whole new perspective. You'll start to see risks more clearly, make better decisions, and communicate your findings in ways that resonate with the business.

If you're looking to dive deeper into this topic, I highly recommend *How to Measure Anything in Cybersecurity Risk* by Douglas Hubbard and Richard Seiersen. This book changed my entire way of thinking, and it's a great resource for exploring this complex subject in more detail. In this chapter, we will merely scratch the surface of cyber risk quantification (CRQ) to provide a foundational understanding of how it works.

One approach to risk quantification is *qualitative analysis*. This method relies on expert judgment and experience to assign subjective ratings to risk factors. While it can be useful for initial risk identification and prioritization, it lacks the precision and objectivity of quantitative methods, making it less suitable for complex risk assessments and decision-making.

In contrast, *quantitative analysis* uses mathematical models and statistics to estimate the probability and impact of risk events. This approach offers a more data-driven and objective assessment of risk. Still, it can be challenging to implement due to the complexity of the models and the need for accurate data inputs.

Semi-quantitative analysis attempts to bridge the gap between qualitative and quantitative approaches. It combines elements of both, using qualitative ratings as inputs to quantitative models. This can offer a balance between the ease of use of qualitative methods and the precision of quantitative analysis, but it still relies on subjective judgments to some extent.

Scenario analysis entails creating hypothetical situations and assessing how they might affect an organization. This approach can be helpful for understanding the potential consequences of different risk events, but it requires careful scenario design and can be resource-intensive.

The domain of cyber risk quantification also offers a variety of methodologies, each with its own strengths and complexities. Among the notable approaches are the FAIR framework, the insights of Hubbard and Seiersen, and the concept of cyber value at risk (CVaR). For those seeking a more streamlined approach, a simple yet effective formula can be employed to score and prioritize specific risks:

$$(\text{Threat} + \text{Vulnerability} + \text{Impact}) \times (\text{Probability} + \text{Velocity}) = \text{Risk}$$

This formula encapsulates the core elements of risk assessment, allowing for a quick evaluation and comparison of different risks.

The method you choose for risk quantification will depend on factors like the quality of your data, the complexity of the risks you're dealing with, and how advanced your organization's risk management processes are.

6.7.1 FAIR Framework

The Factor Analysis of Information Risk (FAIR) framework is a widely recognized model for quantifying and analyzing risk. It breaks down both probability and impact into subcategories that mostly follow the risk formula, providing a more granular understanding of risk factors.

- **Probability:** FAIR divides probability into two components:
  - **Threat Event Frequency (TEF):** How often a threat actor is likely to attempt to exploit a vulnerability. The subcategories for TEF include
    - **Contact Frequency:** How often a threat actor comes into contact with an asset.
    - **Action Frequency:** How often a threat actor, having made contact, takes action to exploit a vulnerability.
  - **Vulnerability (Vuln):** The probability that a threat event will be successful in exploiting the vulnerability. The subcategories for vulnerability include
    - **Threat Capability:** The level of skill, resources, and motivation possessed by the threat actor.
    - **Control Strength:** The effectiveness of the security controls in place to protect the asset.
- **Impact:** FAIR categorizes impact into six types:
  - **Productivity Loss:** Lost time and efficiency due to the incident.
  - **Response Costs:** Expenses incurred in responding to and recovering from the incident.

- **Replacement Costs:** Costs associated with replacing damaged or lost assets.
- **Competitive Advantage Loss:** Loss of market share or business opportunities.
- **Fines and Judgments:** Legal and regulatory penalties.
- **Reputation Damage:** Loss of customer trust and brand value.

By breaking down risk into these subcategories, the FAIR framework allows organizations to identify and prioritize the biggest risk factors, enabling more targeted risk mitigation strategies.

In addition to FAIR, there are several other common approaches to risk quantification, each with its strengths and weaknesses:

- **Monte Carlo Simulations:** This method involves running thousands of simulations to generate a range of possible outcomes for a given risk scenario. This can be helpful for understanding the potential variability of risk, but it requires complex modeling and can be computationally intensive. In terms of the risk formula, Monte Carlo simulations primarily focus on estimating the *impact* of a risk event by simulating various scenarios and their potential consequences.
- **Certainty Rating Experts:** In *How to Measure Anything in Cybersecurity Risk*, Hubbard and Seiersen suggest a method where experts are trained to provide estimates with a 90% confidence interval. This means that they are 90% certain that the true value falls within their given range. This approach aims to improve the accuracy and consistency of expert judgments, acknowledging the inherent subjectivity in qualitative assessments. It still relies on expert opinions, which can be influenced by biases and limited knowledge. It requires access to a team of qualified experts, which might not be feasible for all organizations. Even with 90% certainty, there's still a 5% chance the estimate could be significantly low, potentially leading to unforeseen "black swan" events, which are highly improbable events with severe consequences, as described in Nassim Nicholas Taleb's book *The Black Swan*. This method primarily focuses on estimating the *probability* or *likelihood* of a risk event, although experts might also be asked to provide estimations of potential impact.
- **Actuarial Methods and Tables:** This method utilizes statistical analysis of historical data to estimate the probability and impact of risk events. It is commonly used in the insurance industry, but its reliance on past data can be a limitation when dealing with emerging risks or rapidly changing threat landscape. Actuarial methods primarily focus on estimating the *probability or frequency* of risk events based on past occurrences.

Note that most of these methods require you to inventory all your assets and determine their values to ascertain the potential impact. This process can be a major undertaking, particularly for large organizations with complex IT environments.

No matter which method you choose, it's important to keep in mind that risk quantification is never perfect. There will always be some level of uncertainty and subjectivity. However, by taking a structured approach and making the most of the data you have, organizations can gain deeper insights into their risk exposure and make smarter decisions about how to protect their assets and operations.

I often advise CISOs who are new to risk quantification that risk management is all about trying to predict the future. Since perfect foresight is impossible, they should only take risk quantification as far as necessary to obtain sufficient data for informed decision-making. CISOs are often tempted to overcomplicate quantification by trying to clarify, justify, and explain every detail of their risk data and calculations. It's like drawing a picture with crayons: One crayon can sketch a recognizable car, while ten can create a highly detailed image. But was it necessary to invest the time and use ten crayons if the goal was simply to convey the concept of a car? Of course not. Similarly, don't expend more time and effort than needed to draw your conclusions and make sound decisions in risk quantification.

While risk quantification offers numerous benefits, we should acknowledge the challenges associated with its implementation. Many organizations have attempted to adopt CRQ but have struggled to achieve meaningful results. In fact, research suggests that nearly half of cybersecurity leaders will struggle to use CRQ effectively for decision-making. While improvements in credibility and risk awareness are common, achieving tangible, action-based results remains a challenge. However, the tides are shifting. In recent years, several companies, including X-Analytics® and Kovrr®, have successfully productized CRQ, offering solutions that address the challenges of traditional risk quantification methods. These solutions leverage the actuarial model and have established Application Programming Interface (API) relationships with cyber insurers, enabling them to access near real-time claims data. This data-driven approach allows them to determine the probability of cyber incidents based on the frequency of claims paid and the impact based on the payout amounts. They can tailor these assessments for organizations of similar size, geography, and/or industry vertical, providing more relevant and actionable insights.

The emergence of productized CRQ solutions offers a major advantage: There's no longer a need to rely solely on expert guesses or conduct extensive asset inventories and valuations. This streamlines the risk quantification process, making it more accessible and efficient for organizations of all sizes and maturity levels.

6.8 RISK QUANTIFICATION IN PRACTICE

Now that we have a grasp of the fundamental concepts of risk quantification, let's explore how to leverage it to gain a deeper understanding of our IT risks and translate them into the language of business, which is primarily dollars. This translation is the key to effective communication and decision-making at the board and executive levels.

In this section, we'll take a closer look at loss exceedance curves (LECs), which are visual tools that show the likelihood of losses surpassing certain thresholds, helping illustrate the potential financial impact. We'll also discuss inherent risk, which is the level of risk tied to an asset or process before any security controls are in place.

From there, we'll move to residual risk, which is the risk that remains even after implementing security measures, highlighting the exposure that still exists despite protective efforts.

We'll also touch on risk appetite, which is the amount of risk an organization is willing to tolerate while pursuing its goals. By understanding these concepts and using risk quantification techniques, organizations can make smarter decisions about how to manage risk, allocate resources, and invest in cybersecurity. This knowledge allows them to prioritize actions, strengthen their security posture, and ultimately safeguard their financial health.

6.8.1 LECs: Visualizing Financial Risk

A loss exceedance curve (LEC) is a powerful tool in risk quantification that visually depicts the probability of experiencing losses that surpass specific financial thresholds. It essentially shows the likelihood of different levels of financial impact occurring within a given timeframe, usually annually.

The horizontal axis of an LEC represents the potential financial loss amounts, often on a logarithmic scale to accommodate a wide range of values. The vertical axis represents the exceedance probability, indicating the likelihood of a loss exceeding the corresponding amount on the horizontal axis.

For instance, a point on the curve at $1 million and 80% signifies that there is an 80% chance of experiencing a loss of $1 million or more in a year. The shape of the curve provides valuable insights into the risk profile. A steep curve indicates a higher probability of large losses, whereas a flatter curve suggests a lower likelihood of significant financial impact.

6.8.2 Inherent Risk on the LEC

Inherent risk refers to the level of risk an organization faces before any security measures are put in place, often shown as the "baseline" curve on an LEC. This curve helps illustrate the potential financial losses the company could face without any protections. It serves as a starting point to gauge how well your risk mitigation efforts are working and the extent to which security investments are reducing potential financial exposure (Figure 6.4).

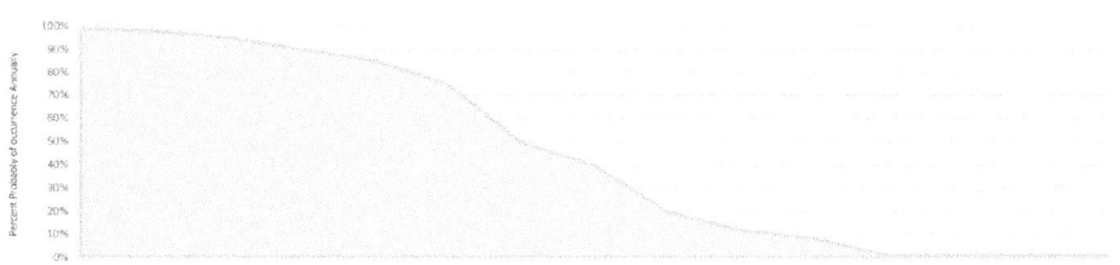

FIGURE 6.4 Inherent Risk Curve.

By comparing the inherent risk curve to the residual risk curve, which reflects the risk after controls are implemented, organizations can get a clear picture of how their security measures are paying off. This comparison helps guide decisions about whether to invest further or adjust their risk management strategy.

6.8.3 Risk Appetite: Balancing Risk and Reward

Risk appetite is the amount of risk an organization is willing to accept to achieve its business goals. It plays a key role in determining how much to invest in cybersecurity and which risk mitigation strategies to pursue. Every organization's risk appetite is different, shaped by factors like the industry it operates in, regulatory requirements, available resources, and strategic goals. A clear risk appetite statement should outline how much risk the organization is comfortable with across different areas and offer guidance on how to respond.

Within this framework, there are three important concepts to keep in mind. *Risk appetite* refers to the overall level of risk the company is prepared to take to reach its goals. *Risk tolerance* is the acceptable range or deviation from that appetite for specific risks. And *risk limit* is the maximum level of risk the organization can handle before it must take corrective action (Figure 6.5).

6.8.4 Risk Appetite, Tolerance, and Limit on a Loss Curve

Visualizing these concepts on an LEC provides further clarity. Risk appetite is depicted as a horizontal line, marking the maximum acceptable financial loss within a given probability. Risk tolerance can be represented as a band surrounding this risk appetite line, showing the acceptable range of variation in losses. The risk limit, however, is a definitive point on the curve beyond which the organization is unwilling to accept any further risk, necessitating immediate action to mitigate the exposure (Figure 6.6).

It is important to note that zero risk is not the goal. In fact, being overly conservative with risk can stifle an organization's growth and innovation. There are two types of risk: Positive risk, associated with pursuing opportunities and innovation, and negative risk, the risk of loss or harm due to threats and vulnerabilities. Organizations take on positive risks to achieve their strategic objectives and generate profits, while they seek to mitigate negative risks through security controls and risk

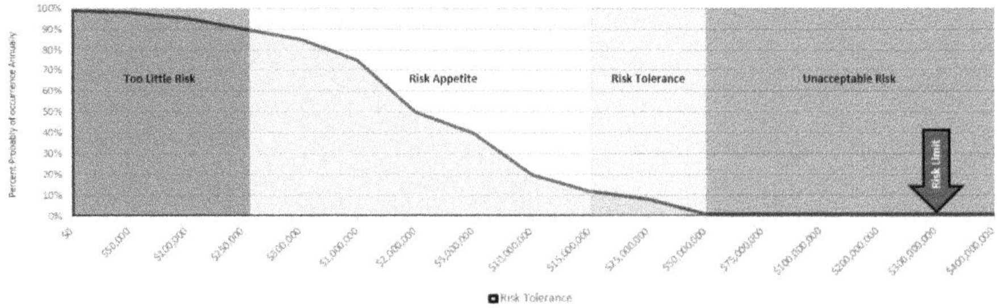

FIGURE 6.5 Risk Appetite Curve.

FIGURE 6.6 Risk Tolerance Speedometer.

management strategies. A certain amount of risk is both necessary and desirable for organizations to thrive. The challenge lies in finding the "right amount of risk", balancing the potential rewards of innovation and growth with the potential losses from cyber incidents.

Consider the analogy of a speed limit on a highway. Let's say the speed limit is 70 MPH. Many drivers have a tolerance for exceeding the speed limit slightly and are willing to go 80 MPH. Even so, there's a point where the risk becomes unacceptable, perhaps around 90 MPH. Beyond that, there's a limit where the consequences become severe, such as a fatal accident or losing your license. Driving too slowly can also be risky. Going 45 MPH in a 70-MPH zone can be just as dangerous as speeding.

6.8.5 Determining Risk Appetite

There are two main ways to determine your organization's risk appetite.

The first method is a straightforward approach: Estimating risk tolerance as a percentage of revenue. A good starting point might be 0.01% for low tolerance, 0.05% for medium tolerance, and 1% for high tolerance. These percentages can be adjusted based on your organization's specific needs and risk profile. This method offers a quick, simple way to establish an initial risk tolerance level.

The second approach is more in-depth and involves conducting a risk appetite exercise. This process includes interviewing directors, executives, and other key stakeholders to gather their views on risk. It helps you uncover your organization's underlying risk culture and spot any differences in how various groups perceive risk.

Typically, a risk appetite exercise involves a series of workshops or interviews where participants are asked to consider different risk scenarios and share their comfort levels with various outcomes. The aim is to better understand your organization's overall risk tolerance and to identify areas where further alignment or clarification may be needed. For those interested, there's a risk appetite worksheet available for download at CISO30.com that can help guide these interviews.

Sample questions for a risk appetite exercise might include the following:

- How much downtime could we tolerate in the event of a cyber incident?
- What resources can't we live without and for how long?
- What impact would a data breach have on our reputation and customer trust?
- How comfortable are we with investing in new technologies that might introduce additional risks?
- What level of financial losses would be catastrophic?
- What is our most sensitive information cannot be lost or made public?
- What could cause physical harm to our employees, customers, partners, or visitors?
- What level of financial loss would be considered unacceptable for our organization?
- Would you accept a 10% chance per year of losing $5 million dollars due to cyber security risk?
- Would you accept a 20% chance per year of losing $5 million dollars due to cyber security risk?
- Would you accept a much larger risk, say $20 million, at 1%, 5%, or 10% per year?

6.8.6 Risk Appetite Statements

After completing the risk appetite exercise, you can use the insights gathered to craft a clear risk appetite statement. This statement should outline the organization's tolerance levels for various types of risks and offer guidance on how to respond to them. It serves as a reference point for decision-making, ensuring that risk management efforts align with the organization's broader strategic goals.

Example Risk Appetite Statements

- "We are willing to accept a low level of financial risk in pursuit of innovation, but we will not tolerate any actions that could jeopardize the safety of our employees or customers".
- "We are comfortable with a moderate level of operational risk, but we expect to be informed of any potential disruptions to critical business processes".
- "We have a high tolerance for reputational risk, but we will take swift action to address any negative publicity that could damage our brand".

CISOs and CROs can use these statements as a practical tool for decision-making and communicating with stakeholders. For example, the CISO can rely on the risk appetite statement to prioritize security investments based on the organization's risk tolerance. The CRO can use it to communicate the company's risk profile and mitigation strategies to the board of directors. It can also guide the organization's incident response approach, ensuring that actions taken during an event align with the accepted level of risk.

By defining and sharing the organization's risk appetite clearly, you create a shared understanding of risk tolerance, enable informed decision-making, and foster a culture of risk awareness across the organization.

6.9 RISK TREATMENT OPTIONS

After identifying and quantifying risks, organizations need to decide how to address them. There are four main options: Avoidance, transfer, mitigation, and acceptance. The best choice depends on the specific risk, the organization's risk appetite, and available resources.

6.9.1 Risk Avoidance

Risk avoidance means eliminating a risk entirely by stopping the activity or process that causes it. While this approach completely removes the risk, it's not always practical or desirable, as it can limit business opportunities or stifle innovation.

In an IT/cyber context, risk avoidance could involve the following:

Not Adopting a New Technology or Service: If the risks associated with a new technology or service outweigh the potential benefits, the organization might choose not to adopt it.

Discontinuing a Legacy System: If a legacy system poses a big enough security risk and cannot be adequately secured, the organization might decide to retire it.

Exiting a High-Risk Market: If operating in a particular market or region exposes the organization to unacceptable levels of cyber risk, it might choose to exit that market.

6.9.2 Risk Transfer

Risk transfer means shifting the responsibility for a risk to a third party, often through mechanisms like insurance or outsourcing. This approach can be particularly useful for managing risks that the organization either doesn't want to fully absorb or isn't equipped to handle on its own. For more details on how cyber insurance can play a role in this process, we'll dive deeper into the topic in Chapter 7.

Examples of risk transfer in an IT/cyber context include the following:

Cyber Insurance: Purchasing insurance to cover the financial losses associated with cyber incidents.

Outsourcing: Contracting with a third-party provider to manage specific IT functions or processes, transferring some of the associated risks to the provider.

Service-Level Agreements (SLAs): Including clauses in contracts with vendors or service providers that hold them accountable for certain levels of security and performance, transferring some of the risk to them.

6.9.3 Risk Mitigation

Risk mitigation focuses on reducing either the likelihood or impact of a risk by putting security controls and countermeasures in place. It's the most common approach to managing risk and encompasses a variety of security practices and technologies. We'll explore how to use risk quantification to guide your mitigation strategies in more detail in Chapters 9 and 10.

In an IT/cyber context, risk mitigation might include the following:

Implementing Security Controls: Deploying firewalls, intrusion detection systems, antivirus software, and other security technologies to protect against threats.

Employee Training and Awareness: Educating employees about cybersecurity best practices and potential threats to reduce the risk of human error.

Incident Response Planning: Developing and testing plans for responding to and recovering from cyber incidents to minimize their impact.

Vulnerability Management: Regularly identifying and patching vulnerabilities to reduce the attack surface.

6.9.4 Risk Acceptance

Risk acceptance means acknowledging a risk and deciding to live with it often because the cost of mitigating it outweighs the potential benefits or because the risk falls within the organization's risk appetite.

It's important to note that only the risk owner should have the authority to accept a risk, not the line of business (LOB) owner. Often, stakeholders may casually say, "I accept the risk" when faced with vulnerabilities that need addressing. A common example is running unpatched or outdated operating systems on production machines that can't afford downtime. The LOB owner might claim this is acceptable, but they may not fully understand the potential financial impact of a breach compared to a scheduled outage. They are neither equipped nor the right person to make that call.

It's important to remember the distinction between vulnerability and a risk, and to only accept risks that have been properly quantified from a business perspective. Defining who actually "owns" the risk can be challenging. While the board might delegate risk ownership to the CISO, in a responsible, accountable, consulted, informed (RACI) framework, the board is ultimately accountable for material cyber risks and has a duty of oversight. Executive management, including the CEO, is responsible and accountable, meaning the CISO, CIO, or CRO might hold operational responsibility. However, the buck stops with the CEO, who might ultimately want to retain risk ownership. On the other hand, LOB owners may be consulted but are probably not responsible or accountable for cyber risk, so they should not own it.

Defining risk ownership can be part of the risk appetite exercise. Ensuring that the right stakeholders are making decisions about accepting risks improves an organization's risk management and helps guide decisions on cybersecurity investments and strategies.

Risk appetite plays a key role in determining how to treat risks. Risks that fall within the organization's appetite may be accepted, while those that exceed it could require mitigation or transfer. In some cases, risks that are too high to accept but can't be effectively mitigated or transferred might even require avoidance.

For example, an organization with a low-risk appetite might decide to transfer the risk of a major data breach through cyber insurance while accepting the lower risk of minor phishing attempts. Meanwhile, a company with a higher risk appetite might accept a broader range of risks, relying on its internal security controls and incident response capabilities to manage them.

The combination of investments across all four risk treatment strategies avoidance, transfer, mitigation, and acceptance, I refer to as *risk buy-down*. The ultimate goal of risk management is to balance your investments across these strategies to maximize risk reduction, or in other words, to "buy down" the inherent risk to an acceptable level within your organization's tolerance.

Risk treatment plans (RTPs) should be developed for each risk. That plan should include the specific mix of risk treatments that will be used to buy down that particular risk. Developing an RTP is the fourth step of the risk strategy framework, and we'll discuss how to determine the right mix of risk treatments in Chapter 8. For risks where their RTP includes some form of mitigation, you should also develop risk mitigation plans (RMPs). These plans should outline specific actions to be taken to mitigate the identified risks and enhance the effectiveness of existing controls.

Choosing the right risk treatment method involves carefully weighing the organization's risk appetite, the potential impact of the risk, and the cost-effectiveness of available mitigation or transfer options. Understanding these factors and incorporating them into a structured risk management process are key to making well-informed decisions.

6.10 TAIL RISK AND BLACK SWAN EVENTS

Tail risk in risk management refers to the chance of experiencing a significant loss from a rare event. These risks are defined by their low likelihood but potentially devastating impact, often sitting in the "tail" of a probability distribution curve. Technically, the tail stretches on indefinitely because it's impossible to fully eliminate the chance of any event, no matter how unlikely.

Examples of tail risks in a cybersecurity context could be as follows:

- A catastrophic data breach exposing highly sensitive customer information, leading to massive financial losses, regulatory fines, and severe opportunity costs.
- A successful cyberattack on critical infrastructure that causes extensive outages and potentially even loss of life.
- A novel and sophisticated cyberattack exploiting a previously unknown vulnerability, rendering existing security controls ineffective.

On an LEC, these tail risks appear on the far right, where the chances of occurrence are low, but the potential losses are enormous.

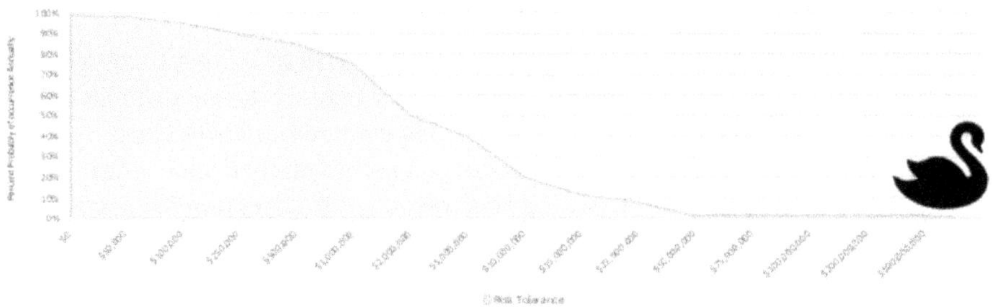

FIGURE 6.7 Black Swan Curve.

6.10.1 Black Swan Risks

Black swan risks, a concept popularized by Nassim Nicholas Taleb, are a specific type of tail risk characterized by their extreme rarity, severe impact, and retrospective predictability (i.e., after the event occurs, it seems obvious that it could have happened, even though it was not predicted beforehand). Black swan events are the nightmares that keep CISOs up at night. They're the unpredictable, high-impact events that can cripple an organization.

The global pandemic is a prime example of a black swan event. It forced a rapid migration to a decentralized, virtual cloud workspace, altering the way security is delivered and creating new attack vectors. Understanding that black swan events are inevitable is a fundamental principle of a robust cyber resiliency strategy. This approach acknowledges that cyber threats are inherently unpredictable, necessitating agility and adaptability (Figure 6.7).

Key takeaways regarding black swans:

Cyber Threats Are Not Predictable: Respond effectively to unexpected situations and "fail safely" when attacked.

Black Swan Events Depend on the Observer: Build in situational awareness to continually absorb and assess new information and understand core strengths and weaknesses.

Do Not Attempt to Predict Black Swan Events: Instead, build resilience to negative events with an agile and adaptable approach to business processes and operations.

We can't prevent every black swan event, and it's impossible to completely eliminate tail risks, but we can prepare for them by building a culture of resilience and investing in robust incident response capabilities. We'll discuss this in more depth in Chapter 13.

6.10.2 Cyber Value at Risk

In the scope of risk quantification, Cyber Value at Risk (CVaR) is a metric that gauges the potential financial loss an organization might face due to cyber threats within a specific timeframe and confidence level. It goes beyond merely identifying the average or expected loss and instead examines tail risk, where the potential for catastrophic financial impact resides.

In essence, CVaR answers the question: "What is the worst-case scenario we can expect, and how likely is it to happen?" It provides a more conservative estimate of potential losses compared to traditional metrics like annualized loss expectancy (ALE), which focuses on the average expected loss.

The calculation of CVaR involves several steps:

1. **Identify and Assess Cyber Risks:** The first step is to identify and assess the various cyber risks the organization faces, considering their likelihood and potential impact. This can be achieved through risk assessments, threat modeling, and vulnerability analysis.
2. **Estimate Loss Distribution:** The next step is to estimate the probability distribution of potential losses associated with each identified risk. This involves considering various scenarios and their potential financial impact, often using historical data, expert judgment, or simulation techniques.
3. **Determine Confidence Level:** Confidence level represents the desired level of certainty in the CVaR estimate. For instance, a 95% confidence level means that there is a 95% probability that the actual loss will not exceed the calculated CVaR value.
4. **Calculate CVaR:** Once the loss distribution and confidence level are determined, CVaR can be calculated as the average of the losses that exceed the threshold corresponding to the chosen confidence level. In simpler terms, it's the average of the worst-case scenarios within the specified confidence interval.

CVaR is particularly valuable in situations where the potential for catastrophic losses is a major concern. It helps organizations understand their exposure to tail risks, which allows them to make better decisions about risk mitigation strategies, insurance coverage, and capital allocation. To calculate CVaR, organizations require a few key elements. First and foremost, a thorough risk assessment is essential. This entails a thorough understanding of the organization's cyber risks, encompassing both their likelihood and potential impact. The second requirement is access to historical loss data or the capability to utilize simulation models. This data is needed for estimating the probability distribution of potential losses. Lastly, expertise in risk quantification is necessary. This involves applying statistical techniques and risk modeling to accurately calculate CVaR.

Take this example. Let's say a company estimates its potential losses from cyber incidents over the next year, and the results show a 5% chance of a $10 million loss, a 2% chance of a $20 million loss, and a 1% chance of a $50 million loss. To calculate the CVaR at a 95% confidence level, we focus on the losses that exceed the 95th percentile of the loss distribution. In this case, only the $50 million loss exceeds this threshold. Therefore, the CVaR at a 95% confidence level is $50 million. This means that there is a 95% probability that the company's losses from cyber incidents in the next year will not exceed $50 million.

CvaR can provide a valuable perspective on the potential financial impact of cyber risks, but be careful not to go too far trying to predict the unpredictable.

6.11 RESIDUAL RISK

Residual risk is the level of risk that remains after an organization has implemented security controls and mitigation measures. It represents the organization's exposure to potential losses or harm even with protective measures in place.

6.11.1 Calculating Residual Risk

Residual risk is calculated by subtracting the risk reduction achieved through security controls from the inherent risk. In simpler terms,

$$\text{Residual Risk} = \text{Inherent Risk} - \text{Reduction from Risk Treatments},$$

where

- **Inherent Risk:** The level of risk before any mitigation efforts.
- **Reduction from Risk Treatments:** The decrease in risk achieved through the implementation of security controls.

On an LEC, residual risk is typically depicted as a second curve that lies below the inherent risk curve. This illustrates the reduction in potential financial losses achieved through the implementation of security controls. The gap between the two curves represents the effectiveness of the risk mitigation efforts.

Ideally, the residual risk curve should fall below the organization's risk appetite line, indicating that the remaining risk is within an acceptable level. If the residual risk curve still intersects or exceeds the risk appetite line, it signals that further risk mitigation or transfer measures might be necessary to bring the risk exposure within the desired tolerance.

Understanding residual risk helps you make better decisions about your cybersecurity investments and risk management strategies. By regularly tracking and evaluating residual risk, organizations can spot where extra controls or changes are necessary to keep risk levels within acceptable limits.

6.12 LOSS EXCEEDANCE EXAMPLE: VANDELAY INDUSTRIES

Let's explore a practical example of how a fictional company, Vandelay Industries, utilizes an LEC to illustrate its risk buy-down strategy.

6.12.1 Scenario

Vandelay Industries has a published formal risk appetite statement that includes a loss (risk) tolerance statement. This statement outlines a low tolerance for risks that could result in losses exceeding $50 million or 15% of earnings before interest, taxes, depreciation, and amortization (EBITDA) in a single fiscal year. This risk appetite is visually represented in Figure 6.8 by a solid line, which steadily drops as it approaches the $50 million mark.

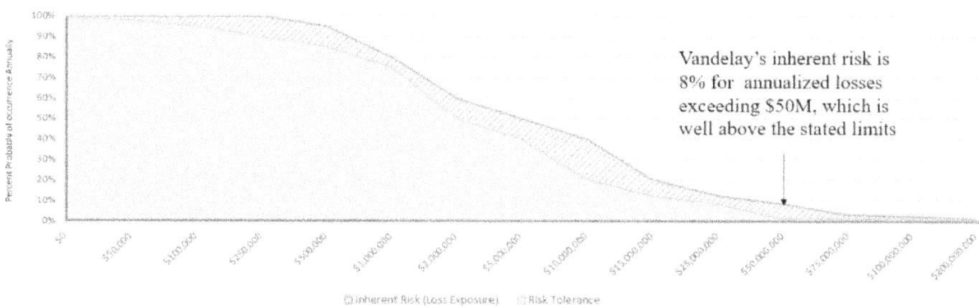

FIGURE 6.8 Risk Exceedance Curve.

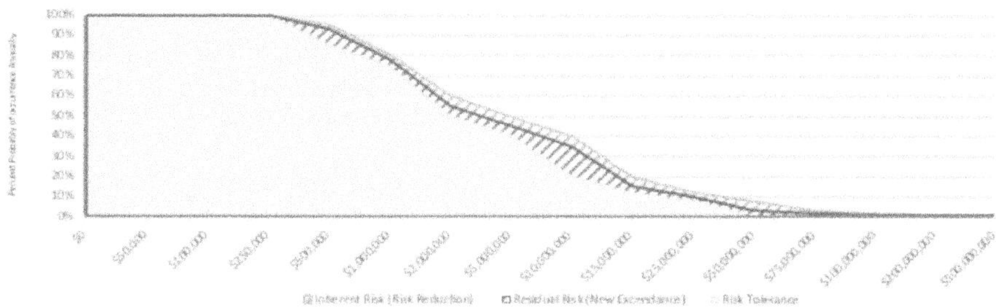

FIGURE 6.9 Residual Risk Curve.

The line in the graph in Figure 6.8 depicts Vandelay's inherent risk, which is the untreated, raw risk associated with its business activities. The inherent risk values were calculated using historical losses and loss simulations.

Observe where the hashed line breaks above the solid limit. This indicates that the inherent risk surpasses the stated loss tolerance, particularly in the $10 million to $50 million exposure range.

The subsequent graph, Figure 6.9, demonstrates how the proposed projects will effectively buy down Vandelay's IT security financial risk. Implementation of these projects will result in a set of residual exposures that fall comfortably within the stated limits.

The LEC incorporates the risk reductions expected from the improvement plan into the original loss tolerance graph. This projection brings the residual risk down below the loss tolerance limits across all financial impact levels, successfully addressing the excess exposure for risks above $10 million.

Focusing on the $10 million mark, we observe that the probability of a loss exceeding $10 million has been reduced from 20 to 10%. In simpler terms, this translates to a shift from experiencing such a loss once every 5 years to once every 10 years. A $10 million loss over 5 years averages to $2 million annually, while over 10 years, it's only $1 million per year, effectively halving the risk exposure at the $10 million loss range. Applying the same logic to the $50 million upper tolerance limit, a $5 million exposure is reduced to $500,000. When these calculations are performed for

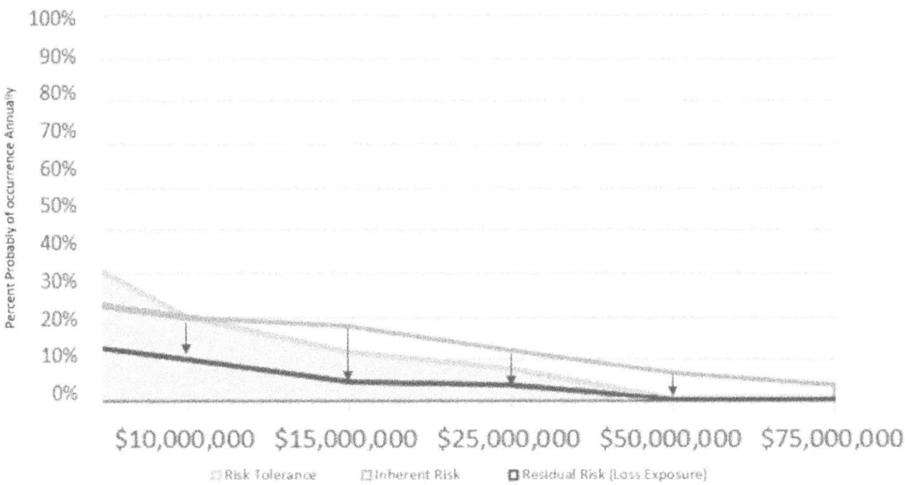

FIGURE 6.10 Risk Buy-Down Curve.

all values across the board, we see a 75–80% reduction in the financial risk exposure rate that surpasses Vandelay's stated tolerance limit. The proposed $1.3 million annual security investment buys down the annualized IT security financial risk. This translates to a reduction in exposure from $3 million to $5 million per year down to $500,000 to $1 million annually. The effectiveness of this risk buy-down is visually represented in Figure 6.10. This example demonstrates how risk quantification, particularly through the use of LECs, can help organizations understand their risk exposure to achieve their desired risk tolerance levels.

The Vandelay Industries scenario illustrates the power of LECs in visualizing and quantifying risk. Simply plotting inherent and residual risk against the organization's risk appetite allows decision-makers to clearly see where their risk exposure exceeds their tolerance levels. This visualization enables them to make informed choices about risk mitigation strategies and resource allocation, ensuring that their investments effectively reduce financial risk and align with the company's overall objectives.

While LECs are valuable tools for internal risk analysis, they might not be the most effective way to communicate risk information to others, particularly board members who might not be familiar with the technical nuances of risk quantification. In Chapter 15, we'll explore more impactful and accessible methods for sharing and articulating risk information, ensuring that it resonates with diverse audiences and drives informed decision-making all across the organization.

6.12.2 Defining Materiality in Risk

Understanding and defining material risk is at the core of every CISO's mission. Rick Howard states in his book *Cybersecurity First Principles: A Reboot of Strategy and Tactics* that reducing the probability of material impact is the absolute cybersecurity first principle. We will cover risk materiality and how to determine it in depth in Chapter 17, but for now, it's enough to understand the basic meaning of materiality.

In essence, *materiality* refers to the significance or importance of a particular piece of information or event. In the context of cybersecurity, a material risk is one that has the potential to impact the organization's operations, financial performance, or brand value.

Material risk is the fundamental foundational building block of your entire program. Determining what risks are material and how to reduce material risk is the entire game. Identifying, assessing, and communicating material risks effectively are key differentiators for the CISO 3.0 and comprise the third step in building a risk strategy. We'll dig deeper into how to determine materiality in Chapter 17.

6.13 THE RISK REGISTER 3.0

The risk register should be a centralized repository for documenting and tracking an organization's risks. It's fundamentally important for risk management, providing a structured framework for identifying, assessing, prioritizing, and monitoring risks. The register typically includes information about each risk, such as its description, likelihood, impact, potential consequences, and mitigation strategies.

The creation of a risk register often involves a collaborative effort among various stakeholders, including risk owners, subject matter experts, and risk management professionals. The process typically begins with identifying potential risks through brainstorming sessions, interviews, or surveys, or by reviewing historical data and industry reports. Once identified, each risk is assessed in terms of its likelihood and potential impact, often using qualitative or quantitative methods. The risks are then prioritized based on their severity and potential consequences, allowing the organization to focus its resources on addressing the most critical risks first. Finally, mitigation strategies are developed and documented in the risk register, along with responsible parties and timelines for implementation. The register is then regularly reviewed and updated to reflect changes in the stated risks or the effectiveness of mitigation efforts.

The traditional approach to risk registers, however, often falls short. One common pitfall I've observed is the tendency for CISOs to create risk registers filled with threats and vulnerabilities, mistaking them for actual risks. Remember that risk comprises both probability and impact. The risk formula further clarifies this by separating risk into threat + vulnerability + impact. So, to accurately understand our risks, we need to meticulously track both vulnerabilities and the threats that might exploit them.

National Institute of Standards and Technology (NIST) provides a structured approach to risk assessment through a seven-step process: Classification, inventory, threat analysis, vulnerability analysis, impact analysis, risk ranking, and the development of a risk treatment strategy (Figure 6.11).

You can also think of risk as a lifecycle or a series of relationship statements. It begins when threats exploit vulnerabilities, which creates risk that impacts assets, causing exposures that require controls to mitigate the threats. You can see these relationships in the cycle graphic (Figure 6.12).

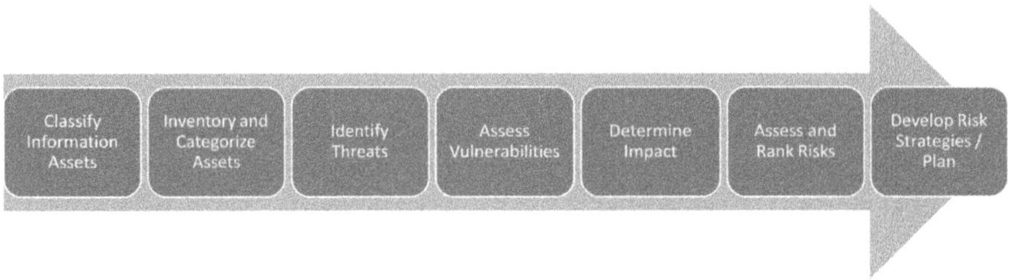

FIGURE 6.11 Risk Assessment Arrow.

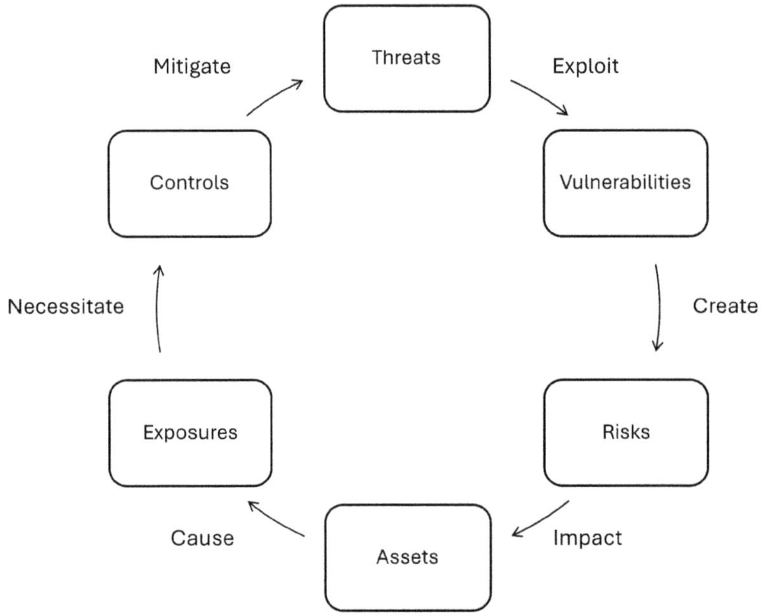

FIGURE 6.12 Risk Lifecycle.

With a clear understanding of the risk lifecycle, I propose creating a risk register that mirrors these models. This involves establishing separate registers for assets, threats, and vulnerabilities; ensuring accurate risk identification; and preventing the misclassification of threats as risks.

To facilitate this, I've provided a Risk Register 3.0 template you can find in the download section at CISO30.com. It features multiple tabs for creating distinct registers for assets, threats, vulnerabilities, and risks. If you already utilize a governance, risk, and compliance (GRC) tool like Archer or a risk platform like MetricStream or have a configuration management database (CMDB) in place, you can easily cross-reference your data and won't require the Excel template. However, you can still apply the Risk Register 3.0 concepts within those tools.

The first step is to create an asset register. Ideally, you should have an asset inventory or CMDB, but if not, you can begin by listing individual systems, grouping them by application, or using general categories like servers, applications, end-user systems, etc.

Possible asset categories include the following:

- Servers
- Applications
- End user systems
- Terminal and POS systems
- ICS and SCADA systems
- On-board and embedded systems
- IOT devices
- Mobile devices
- Storage and backup systems
- Offline media and data
- People.

Comparing your threats and vulnerabilities against these asset groups provides a preliminary risk understanding without a full inventory. Once you have your asset list, classify systems as critical (you can't function without them) or sensitive (they contain confidential data) and assign values to them, using either qualitative metrics (high, medium, or low) or, preferably, quantified values.

The next step is to establish a register for your threats. The template offers two distinct approaches to accomplish this. The first involves creating groups, similar to how we categorized assets, but this time focusing on threats and the actors behind them.

Threat categories may include the following:

- System and network intrusion
- Theft, loss
- Crimeware
- Web app attacks
- Distributed denial-of-service (DDoS) attacks
- Errors
- Misuse

Common threat actors include the following:

- Cyber crime
- Nation states
- Third-party partners
- Corporate espionage
- Insider – disgruntled
- Insider – mistake
- Terrorist
- Activist
- Vandals.

This grouping method aligns with how X-Analytics®, the CRQ platform I mentioned earlier, maps risk. You can see in Figure 6.13 that they've quantified the risk at the intersection of each asset group and threat category.

Threat Category (Low Threat – High Threat)

Asset Group	WebApp Attack	PoS Intrusion	Misuse	Error	Theft/Loss	Crimeware	Skimmers	Cyber Espionage	DoS Attack	Everything Else
Server/Apps	3.364	0	0.13	0.143	0.016	4.979	0	0.927	1.329	2.518
Network	0.124	0	0.003	0.159	0.005	0.941	0	0.329	0.785	0.313
End User Systems	0.519	0	0.036	0.046	0.19	3.358	0	0.845	0.243	1.021
Terminal	0.044	0	0.003	0.005	0.005	0.082	0	0.016	0.027	0.066
ICS/SCADA	3.364	0	0.13	0.143	0.016	4.979	0	0.927	1.329	2.518
Healthcare Devices	0	0	0	0	0	0	0	0	0	0
Onboard Systems	1.942	0	0.083	0.094	0.103	4.168	0	0.886	0.786	1.769
Critical IoT	0.519	0	0.036	0.046	0.19	3.358	0	0.845	0.243	0.608
Non-Critical IoT	0.519	0	0.036	0.046	0.19	3.358	0	0.845	0.243	0.608
Media & Offline Data	0.044	0	0.04	0.228	0.202	0.082	0	0.072	0.027	0.322
People	1.287	0	0.13	0.228	0.202	3.801	0	0.707	0.074	1.305

FIGURE 6.13 X-Analytics Asset/Threat Graph.

The second option is to utilize a preexisting threat catalog, such as the one provided by the HITRUST framework. This comprehensive catalog has already done the heavy lifting for you, categorizing threats into physical, logical, and organizational, with further subcategories like intentional, unintentional, and compliance, which are then broken down into groups like misappropriation, misuse, failure, and nefarious actions.

The template includes a list of possible threats as an example. However, the catalog cannot assign values or scores to these threats, as their impact will likely be unique to your organization. You'll need to add these scores yourself.

Next, we'll establish a register for your vulnerabilities. If you're already utilizing a vulnerability scanner or a GRC tool to manage vulnerabilities, you can just use that or you can import those vulns into the first tab of the template. If you lack these tools, don't worry. The second tab template provides an alternative method to populate your vulnerability register. Similar to assets and threats, we can categorize vulnerabilities, and thankfully, several organizations offer valuable resources to assist in this process. For web application vulnerabilities, the Open Web Application Security Project (OWASP) list, included in the template, is a great starting point. The Cloud Security Alliance's (CSA's) Global Security Database (GSD), accessible via APIs and GitHub tools, is another excellent resource for tracking cloud vulnerabilities. Plus, Cybersecurity and Infrastructure Security Agency's (CISA's) Known Exploited Vulnerabilities list, downloadable as a CSV, offers a focused approach by highlighting vulnerabilities with confirmed exploitation. Lastly, the MITRE ATT&CK framework can help categorize vulnerabilities. I included the MITRE Tactics in the template, but you could go further and drill down to the techniques level if you wanted. If you go down to the techniques level, be mindful of potential overlap with threats, especially if you're using a pre-made threat catalog. When it comes to scoring vulnerabilities, you have options. You can utilize common vulnerability scoring system (CVSS) scores or create your own rubric. The template demonstrates a simple approach using exposure and severity scales from 1 to 5 to generate a vulnerability score.

Now that you have meticulously curated registers of your assets, threats, and vulnerabilities, the next phase involves aligning and mapping your assets to their corresponding vulnerabilities. If you're fortunate enough to have a vulnerability management program in place, your existing tools are likely already performing this task, allowing you to seamlessly import the scores. In the absence of such tools, manual mapping is a viable alternative. You can use asset categories like systems and servers, or specify a system like your enterprise resource planning (ERP), as the asset, and then map its vulnerabilities. For instance, an ERP system might be susceptible to the MITRE tactic TA0009 – Collection, where adversaries gather data to exfiltrate (steal). The culmination of this mapping process allows you to combine the asset value with the vulnerability score, generating a combined A/V score that reflects the potential impact of a vulnerability on a specific asset. If you used ordinal measures like high, medium, and low for your asset values, you will likely need to convert those into numbers like 1, 2, 3 in order to calculate them against the vulnerability score to determine a combined A/V score.

The subsequent stage involves establishing a connection between the identified threats and vulnerabilities. The template facilitates this by allowing the creation of dropdown lists populated from both the threat and vulnerability registers. Combining their respective scores generates a new T/V score, representing the potential impact of a threat exploiting a specific vulnerability.

With a register of threats, their corresponding vulnerabilities, and the vulnerabilities' impact on assets, we're now ready to construct the risk register itself. This is where we'll list, identify, and name each specific risk. I've offered up several fields for risk tracking in the template, but you can include any additional information like risk owner or discovery date. The template provides a field for risk categorization, drawing from a separate tab that lists business-level risk categories we discussed earlier, such as direct financial impact, production uptime, and CX. These business risk categories serve to help translate IT risk into business risk, to understand and manage the potential impact on the business itself, and to act as a reminder of the original purpose of registering risks.

Potential business-level risk categories

- Production uptime/business continuity
- Support innovation for growth
- Customer experience (CX)
- Operational efficiency
- Cyber risk reduction
- Legal, regulatory, and compliance
- Supply chain
- Resources and talent
- Environmental, Social, and Governance (ESG)
- Health and safety
- Trust and brand

After the business risk field comes the quantification data. This is where you put the actual dollar figure for impact and a percentage for probability that we discussed in the previous sections. These fields are where the rubber meets the road. It's where you translate those abstract concepts of impact and probability into concrete figures. This quantification section isn't just about numbers: it's about giving your risk register teeth. It transforms it from a theoretical document into a powerful decision-making tool. Assigning quantifiable values creates a common language that allows you to compare risks apples-to-apples, prioritize them based on their potential impact, and justify resource allocation for mitigation efforts.

The next section in the template is for the Risk Treatment Plan (RTP) discussed earlier. The treatment plan outlines how the risk will be handled, offering options like accept, avoid, transfer, mitigate – apply omitted control, mitigate – compensating control, and mitigate – create new control. Often people will include fields to track their risk exception process. I don't include those because that is all just a part of the RTP. It's important to note that any "exceptions" are essentially risk acceptances, and these should be made by the risk owner, not the line of business owner. The remaining columns include a free text field for a description of the Risk Treatment and Risk Mitigation Plans and fields to track control details, timeframes, etc. The treatment section ends with posttreatment impact and probability; this is where we put the quantified dollar amounts to show how much our treatment is buying down the risk. This is where the financial implications of your chosen controls become evident, showcasing the reduction in potential impact and/or likelihood of the risk. Remember, controls can influence both impact and likelihood, so be sure to capture both in your posttreatment quantification. In essence, this section encapsulates the tangible outcomes of your treatment strategy, demonstrating the effectiveness of your efforts.

The final sections of the template are where you'll likely begin your work, adding assets, threats, and vulnerabilities to calculate a combined A/T/V score that helps identify and qualify risks. You can tackle risks in a couple ways: you can create a single risk encompassing all threats to a vulnerability or create separate risks for each threat–vulnerability pair. You can do the same thing with the business impacts; for example, you could create one risk called data exfiltration that contains all the threats, vulnerabilities, or harms related to data exfiltration or you could create separate risks for each harm, that is, financial loss – data exfiltration, compliance – data exfiltration, and trust and brand – data exfiltration.

Once complete, you'll have a risk register populated with actual risks, mapped to both business risks and the associated assets, threats, and vulnerabilities. Combine this with the risk quantification methods we discussed earlier, and you'll have a powerful toolkit to identify, assess, and communicate your risks in the language of business.

6.13.1 Risk and Control Self-Assessments

Another great way to identify risks for your register is by conducting risk and control self-assessments (RCSAs). RCSAs are a proactive approach to risk management, allowing individuals to actively identify, assess, and manage risks in their specific areas across the organization. This represents a departure from the traditional top-down, audit-driven approach, fostering a more collaborative and insightful process that instills a sense of ownership and accountability throughout the company.

Conducting a risk and control self-assessment (*RCSA*) involves a series of important steps. First, you define the scope of the assessment, whether it's focused on a specific department, business function, or individual project. Then, identify the key objectives for that area, as this helps provide context for understanding how potential risks could affect those goals.

Next, brainstorm and document all possible risks that might prevent those objectives from being met. This includes considering internal and external factors, as well as emerging threats and vulnerabilities. Once the risks are identified, evaluate each one by assessing its likelihood and potential impact using a consistent rating system. This helps you prioritize risks and allocate resources effectively.

After that, document the existing controls in place to manage each risk and evaluate their effectiveness, again using a rating system. This step highlights any gaps or weaknesses in your current risk management strategy. Once those weaknesses are identified, action plans are developed to address them, with clear accountability and deadlines assigned for implementation. Finally, it's essential to regularly monitor both the effectiveness of these controls and the overall risk profile, making adjustments as new information or risks emerge.

RCSAs are versatile and can be applied across various departments. For example, a software development team might use an RCSA to identify risks related to code vulnerabilities, data breaches, or project delays. A finance department could assess risks related to fraud, financial reporting errors, and regulatory compliance, while an HR department might evaluate risks involving employee safety, data privacy, and discrimination.

Incorporating RCSAs into your risk management approach has several benefits. They increase risk awareness across the organization, empower employees to take ownership of risk management, and tap into the collective knowledge of the company to identify a broader range of risks. This leads to a stronger control environment, where issues are identified and addressed proactively.

For more help, you can download a simple RCSA template at CISO30.com or check out a detailed guide from the Institute of Risk Management at theirm.org.

6.14 INTEGRATING CYBER RISK WITH ENTERPRISE RISK

The integration of IT risk into enterprise risk management (ERM) can be a great help in aligning cybersecurity with the broader business strategy. When you start treating cyber risks as part of the broader risk landscape, it becomes easier to make informed, data-driven decisions about how to allocate resources and handle risk. Every organization has a risk threshold, whether it's clearly defined or not, and it's essential not to exceed that limit. Given that technology now touches every aspect of the business, addressing IT risks within silos and only in their individual departments is no longer viable.

That siloed approach to risk management can lead to inconsistencies in how risks are identified, assessed, and managed, creating blind spots and inefficiencies. Integrating IT risk into the larger picture can ensure a unified, consistent approach to managing all types of risk.

Although IT risk is just one piece of the overall risk puzzle, it has unique characteristics that need special attention. IT risk focuses on the risks tied to information technology, like cyberattacks, data breaches, or system outages. In contrast, enterprise risk covers all potential threats to the organization, including financial, operational, and strategic risks. IT risks can have a significant impact, disrupting operations, causing financial losses, and harming the brand. But enterprise risk management takes a wider view, considering how all risks together might affect the organization's ability to reach its goals.

The responsibility for managing these risks differs, too. IT risk is typically handled by the CISO and the IT security team, while enterprise risk management involves collaboration across the entire organization. Led by a CRO, ERM includes input from the board of directors, executive leadership, and various business units.

If your organization already has a CRO managing enterprise risk, it's often more efficient to align your cybersecurity strategy with the existing ERM program rather than building a new system from scratch. This integration ensures that cyber risks aren't treated as a separate issue but as part of the broader risk management strategy.

Assessing your organization's cybersecurity posture goes beyond just looking at your tech defenses. While IT plays a central role in cybersecurity, cyber risk needs to be embedded into the broader ERM framework. This requires strong backing from leadership, including the C-suite and the board, to foster a culture of cybersecurity throughout the entire organization.

To merge cyber risk into ERM effectively, it's necessary to develop a common language for discussing these risks. This ensures clear communication and mutual understanding from the boardroom to the IT department. Cyber risk assessments should lead to real, actionable steps. Share insights across the organization to pinpoint high-risk areas, develop controls, and monitor how well those controls are working.

Collaboration is essential. Encourage open communication and teamwork between leadership, risk management, and IT teams. Use consistent risk assessments, rating systems, and interconnected processes, and ensure that any changes in cybersecurity are quickly communicated across departments. IT teams should also understand the organization's risk appetite, as set by the board, so they can accurately assess which risks are acceptable and which are not.

By pulling together risk management data, you can create a comprehensive view of the organization's risk profile. This integrated approach allows cyber risk information to influence broader areas like vendor management and helps you evaluate risks at both the departmental and enterprise-wide levels.

Example

Scenario: A financial institution is planning a major digital transformation initiative, involving the migration of critical customer data and transaction systems to a cloud-based platform. While this initiative promises improved efficiency and scalability, it also introduces new cyber risks, such as data breaches, unauthorized access, and service disruptions.

Integration: The CISO collaborates with the ERM team to conduct a risk assessment, considering not only the technical vulnerabilities of the cloud platform but also the potential business impact of a cyber incident. This includes evaluating the potential financial losses due to service outages, regulatory fines for data breaches, and reputational damage affecting customer trust and market share.

The assessment results are then presented to the executive leadership and the board, highlighting the interconnectedness of cyber risk with other enterprise risks, such as operational resilience, financial stability, and regulatory compliance.

Key Integration Points

- **Risk Quantification:** The CISO and ERM team work together to quantify the potential financial impact of various cyber risk scenarios, enabling a more objective evaluation of the risks and their potential consequences.

- **Risk Mitigation Strategies:** The organization develops a multilayered risk mitigation strategy that includes technical controls (e.g., access control, intrusion detection, or encryption), operational procedures (e.g., incident response plans and employee training), and financial safeguards (e.g., cyber insurance).
- **Continuous Monitoring and Reporting:** The CISO and ERM team establish a process for ongoing monitoring and reporting of cyber risks, ensuring that the organization remains vigilant and can adapt its risk mitigation strategies.
- **Strategic Alignment:** The integration of cyber risk into ERM ensures that cybersecurity considerations are factored into strategic decision-making, enabling the organization to pursue its digital transformation goals while maintaining an acceptable level of risk.

6.14.1 COSO Alignment

The Committee of Sponsoring Organizations of the Treadway Commission (COSO) Enterprise Risk Management (ERM) framework is the cornerstone of risk management. COSO provides a structured approach to identifying, assessing, and managing risks, helping organizations strategically align their risk management efforts with overall business goals and risk appetite. What sets COSO apart is its ability to integrate risk management into every facet of an organization's operations and decision-making, fostering a culture where risk awareness is part of everyday activities, ultimately aimed at creating, preserving, and realizing value.

The COSO ERM framework has become the de facto standard for enterprise risk management, and it's highly likely that your organization's CRO and ERM team are already utilizing it or, at the very least, are familiar with its principles. For CISOs looking to align their cyber risk management strategies with the broader organizational goals, integrating the COSO framework into the cybersecurity program can be highly beneficial. Fortunately, NIST has provided specific guidance to facilitate this process. The NIST's October 2020 publication, "Integrating Cybersecurity and Enterprise Risk Management (ERM)", offers practical steps and best practices to blend cybersecurity risk management into the larger ERM framework.

By aligning COSO's approach with NIST's cybersecurity guidance, organizations can ensure that cybersecurity is not confined to the IT department but is treated as a strategic issue affecting the entire business. This alignment moves cybersecurity from being a reactive or isolated function to one that proactively identifies and addresses cyber risks in a way that supports overall business goals.

The integration of cyber risk into COSO ERM helps organizations move beyond a reactive, siloed approach to risk management. It enables them to proactively identify, assess, and respond to cyber threats in a manner that aligns with their strategic objectives and fosters a culture of risk awareness throughout the enterprise. This enables better-informed decision-making and positions cybersecurity as a strategic imperative that protects the organization's long-term objectives, ensuring that cybersecurity is recognized as a key priority across the entire enterprise.

6.14.2 Integrated Risk Management

Integrated risk management (IRM) is a comprehensive approach that breaks down traditional barriers between different risk functions, such as IT, operational, financial, and compliance risks, within an organization. It acknowledges that risks are interconnected and aims to create a culture where everyone in the organization understands how these risks impact business goals and works together to manage them effectively.

What makes IRM different from traditional risk management is its focus on these interdependencies, rather than looking at risks in isolation. For instance, a cyberattack doesn't just lead to financial losses: it can also disrupt operations, cause compliance failures, and damage the company's brand value. With IRM, organizations can see the full picture and develop strategies that address the broad range of consequences.

To implement IRM, companies should start by creating a cross-functional risk management team. This team should include representatives from IT, operations, finance, and compliance, ensuring collaboration and a well-rounded view of risks. Next, it's essential to establish a unified risk framework that uses a common language and consistent methodology to assess and report risks across the organization. Lastly, leveraging technology can help centralize risk data and automate processes, improving decision-making and efficiency.

One challenge to integrating IT risk into enterprise risk management (ERM) is the siloed mentality we discussed earlier. IT and risk management teams may operate separately, with different goals and communication styles, making collaboration difficult. Another obstacle is that nontechnical stakeholders might not fully understand IT risks or their potential impact, leading to an underestimation of the issue and insufficient resources being allocated to address it.

Limited resources can also complicate the integration of IT risk into ERM. With a finite amount of resources to dedicate to risk management, organizations must carefully prioritize and allocate them across both IT risk and broader ERM efforts.

Specific barriers arise when organizations classify risks based on individual departments, creating a fragmented view of overall risk. IT is often expected to manage risks that extend beyond their control, which can lead to frustration and inefficiencies. Additionally, risk identification in many organizations is reactive, focused on compliance or specific projects, rather than being part of a proactive, ongoing process.

To overcome these challenges, IT leaders should focus on several key actions. First, they must educate and engage business leaders, helping them understand the full spectrum of risks and empowering them to respond appropriately. Second, they should advocate for data-driven decision-making in risk management, moving away from gut instincts and anecdotal evidence. Third, organizations need to improve their overall risk maturity by fostering a culture of proactive risk management. Finally, it's important to build or enhance the organization's risk management capabilities, ensuring the right skills, training, and resources are in place to support an integrated approach.

By embedding IT risk into ERM, CISOs can achieve several important goals that align cybersecurity with broader business strategies. For one, they can demonstrate the business value of cybersecurity by quantifying the financial impact of cyber risks and the ROI of security measures. This enables CISOs to communicate in terms that resonate with business leaders, gaining trust and support from executive leadership. It also strengthens the case for increased budgets and resources by highlighting the potential financial losses from cyber incidents.

Ultimately, this integration positions the CISO as a strategic partner who not only manages cyber risks but also helps safeguard the overall success and resilience of the organization's operations.

6.15 DoCRA

In Chapter 5, we covered the concepts of duty of care and the standard of reasonableness. To quickly recap, duty of care is the legal responsibility to take reasonable steps to prevent harm to others. The standard of reasonableness measures whether the actions taken were appropriate given the circumstances.

Let me explain with a simple example: Imagine a busy café on a rainy day. The owner mops the floor to prevent slips but forgets to put up a "Wet Floor" sign. A customer slips and gets hurt. While the owner showed a duty of care by mopping, not warning customers about the wet floor could be seen as unreasonable, making them liable for the accident.

In cybersecurity, demonstrating both duty of care and reasonableness is essential for building a legally defensible security program. This is where the *Duty of Care Risk Analysis (DoCRA)* standard comes in. DoCRA offers a structured way to evaluate risks and safeguards, ensuring your actions align with legal and regulatory expectations.

What makes DoCRA stand out is that it addresses the interests of everyone potentially impacted by the risks, not just the organization. It helps balance the burden of implementing safeguards with the organization's mission, a key factor that regulators and courts consider when evaluating security practices.

DoCRA's reach goes beyond its own framework. It contributed a version of its methodology to the Center for Internet Security (CIS), which now uses it in CIS Risk Assessment Method (CIS RAM) 2.1. While DoCRA is embedded in CIS RAM, it can be applied to any framework, including NIST and ISO.

DoCRA has earned considerable recognition in legal circles. It's widely accepted by federal judges and regulators as a standard for demonstrating reasonable security practices. In fact, attorneys general often use DoCRA to help define what "reasonable security" means. By using DoCRA, you not only improve your risk management but also strengthen the legal defensibility of your security program.

6.15.1 DoCRA Principles and Practices

The three principles of DoCRA form the foundation of its approach to risk analysis. First, DoCRA is *mission-based*. This means that risk assessments should always be tied to the organization's core mission, goals, and responsibilities. The idea is to ensure that risk management isn't happening in isolation but is directly aligned with what the organization is trying to achieve. This approach ensures that decisions are not only technically correct but also support the broader business strategy.

The second principle is *equitability*. This focuses on considering the impact of risks on all stakeholders, ensuring that no one is overlooked. It's about fairness and making sure that every party affected by a risk is given due consideration. This inclusive approach helps foster trust among employees, customers, partners, and any others who may be impacted by the organization's decisions.

Finally, DoCRA emphasizes *transparency*. Risk management should be an open process, where methods and findings are clear and accessible to those who need them. Transparency helps build trust and accountability, both within the organization and externally, and ensures that decision-making is informed and well-grounded. It strengthens the organization's overall approach to risk by making it easier for everyone to see and understand the rationale behind key decisions.

The Three Principles of DoCRA

1. **Mission-Based:** Risk analysis should consider the organization's mission, objectives, and obligations.
2. **Equitable:** All parties potentially affected by risks should be considered equitably.
3. **Transparent:** Risk analysis should be conducted openly and transparently.

The Ten Practices of DoCRA

1. **Identify and Engage Stakeholders:** Identify and involve all relevant parties in the risk analysis process.
2. **Define the Mission:** Clearly define the organization's mission, objectives, and obligations.
3. **Identify Assets at Risk:** Identify and categorize assets that could be harmed.
4. **Identify Potential Harms:** Identify and categorize potential harms that could arise.
5. **Identify Potential Risk Actors:** Identify and categorize potential risk actors who could cause harm.
6. **Estimate Likelihood and Magnitude:** Estimate the likelihood and magnitude of potential harms.
7. **Identify Safeguards:** Identify existing and potential safeguards to reduce risks.
8. **Assess Safeguard Effectiveness:** Assess the effectiveness of safeguards in reducing risks.
9. **Assess Safeguard Burden:** Assess the burden of safeguards on the organization and stakeholders.
10. **Analyze Alternatives:** Analyze alternative safeguards and risk treatment options.

Risk Register 3.0 can be readily adapted to incorporate DoCRA principles and practices. Consider adding fields to capture stakeholder input, mission alignment, and safeguard burden assessments. This will enrich your risk analysis and provide a stronger foundation for demonstrating duty of care and reasonableness. With the use of DoCRA, you can elevate your risk management practices beyond mere compliance. You proactively address potential harms, demonstrate due diligence, and build a security program that stands up to legal scrutiny.

6.16 RISK FOR GLOBAL ENTERPRISES

Global CISOs of large, multinational enterprises may read this chapter and think that it's prohibitively difficult to apply this type of risk-centric focus, especially quantification, to their organization. With billions of dollars at stake, the scale of operations, the geographical spread, and the diverse regulatory requirements can make it appear too complex. However, these challenges, while significant, don't make it impossible to apply the risk-centric approach discussed throughout this chapter.

Large global organizations differ from smaller entities in several key ways when it comes to risk management. First, they often operate with multiple business units (BUs) across various regions, each with its own set of challenges, regulations, and risk appetites. Furthermore, the sheer volume of assets, vulnerabilities, and potential threats can make it difficult to gain a comprehensive view of the organization's risk posture. This scale can also mean that even significant cyber incidents may have a smaller impact on the overall financial performance, making it harder to prioritize cybersecurity investments.

6.16.1 Challenges in Implementing Risk Quantification

Regulatory Complexity: One of the primary challenges for large organizations is the regulatory environment. Different jurisdictions impose different rules, and what may be a critical compliance risk in one region might not be in another. For instance, the requirements for data protection in the European Union under General Data Protection Regulation (GDPR) differ significantly from those in the United States or Asia. This complicates efforts to quantify risk across the entire organization, as each jurisdiction introduces its own set of variables.

Scale of Operations: The vast size of a multinational corporation also presents logistical issues. Managing assets, vulnerabilities, and incidents across various BUs or regions can make it nearly impossible to create a unified, comprehensive risk register. Even with GRC tools, CMDBs, and vulnerability management software in place, tracking and registering every asset or vulnerability is a monumental task.

Materiality of Cyber Risk: For large enterprises with annual revenues in the billions, a data breach or ransomware event, even if it results in tens of millions of dollars in losses, might represent only a small percentage of their overall financial health. This makes it challenging to justify significant cybersecurity investments to mitigate risks that may not seem material from a board's perspective, even though these incidents could lead to serious operational, legal, or reputational consequences.

6.16.2 Adapting Risk Quantification Strategies for Global Enterprises

To overcome these challenges, CISOs in global enterprises can adapt their risk management approaches to better suit the scale and complexity of their organizations. Below are some strategies to tailor the ideas in this chapter for larger enterprises:

Break Risk Quantification into Smaller Units: One of the most effective ways to manage risk quantification for large enterprises is by breaking the organization down into more manageable parts. Each business unit (BU) should have its own risk register, asset inventory, and risk quantification process. By treating each BU as a smaller organization, CISOs can tackle risk quantification at a granular level, assessing inherent and residual risks, and crafting individual risk stories for each unit.

This BU-level approach simplifies the task of managing assets, vulnerabilities, and risks. For example, if a complete asset inventory is missing for one BU, that gap can be isolated, allowing other BUs to continue their risk management efforts without being impacted by that deficiency. This also allows CISOs to address specific compliance risks relevant to each region or jurisdiction, making it easier to comply with local regulations.

Leverage Enterprise Risk Management Practices: Larger organizations are more likely to have dedicated CROs and established ERM frameworks. CISOs should work closely with these teams to integrate cybersecurity into the broader enterprise risk management strategy. This integration often provides a head start in building an IRM strategy, as the CRO's office can provide valuable insights, resources, and tools to quantify and manage risk at an enterprise-wide level.

Blending Quantitative and Qualitative Factors: As mentioned earlier in this chapter, while it's crucial to quantify risk in financial terms, global CISOs must blend quantitative and qualitative factors when telling the risk story. For large enterprises, the financial impact of an incident might not cross materiality thresholds, but the operational, reputational, or regulatory risks might still warrant attention. Understanding the nonfinancial implications of cyber risk is essential for justifying security investments and for communicating effectively with the board. We'll discuss this more in Chapter 17 when we explore how to predetermine materiality.

Risk Storytelling for Large Enterprises: By approaching risk on a per-BU basis, CISOs can create localized risk stories that resonate with specific BU leaders. Once these individual stories are crafted, they can be bolted together to form an enterprise-wide risk narrative. This method helps create a cohesive risk story that reflects the broader enterprise's risk profile while acknowledging the unique challenges and risks faced by each BU.

Risk Acceptance for Large Organizations: It's also important to remember that risk acceptance is a valid strategy for large multinational enterprises. These companies often have the financial capacity to accept or self-insure cyber risks, particularly for lower-probability, high-impact events. The fact that a larger organization may choose not to invest in mitigating controls at the level a CISO prefers does not mean the risk is going untreated. CISOs can still tell the risk story and prepare the organization to finance these eventualities through risk acceptance and self-insurance strategies. As covered in Chapters 7 and 8, there are a range of self-insurance and financing options available to large enterprises, and CISOs should ensure that these are explored as part of their risk management strategy.

6.17 CONCLUSION

In this chapter, we've taken a deep dive into the complex world of risk, particularly as it relates to the ever-evolving cybersecurity landscape. We started by laying a solid foundation, defining what risk truly means, exploring its various forms, and drawing a clear line between risks and the threats that trigger them. We stressed the importance of precision, especially when it comes to distinguishing between likelihood and probability in risk assessments. We also highlighted the gap that often exists between the technical world of IT risk and the broader business perspective of enterprise risk, emphasizing the need for CISOs to bridge this gap by communicating in the language that resonates with the boardroom.

We then explored key concepts like risk appetite, tolerance, and limits, shedding light on the challenges of traditional qualitative risk analysis. We made a strong case for a more quantitative approach, one that uses data and statistical models to bring objectivity and clarity to risk assessments. We also acknowledged that numbers aren't everything; qualitative insights still play a vital role in crafting a comprehensive risk management strategy.

We introduced you to LECs, a powerful visualization tool that allows you to see and understand the financial side of risk. We also tackled the elephant in the room, tail risks and those dreaded "black swan" events. These are the unpredictable, high-impact events that can keep any CISO up at night. We stressed the importance of proactive preparedness and adaptability in the face of such uncertainties.

We also introduced the concept of material risk, the risks that truly matter to the business, and how it's central to a CISO's mission. And finally, we explored the critical need to integrate IT risk into the bigger picture of ERM. We looked at the common roadblocks that can hinder this integration and how overcoming them can benefit both the IT department and the business as a whole.

Lastly, we saw that while large enterprises present unique challenges in implementing risk quantification, these challenges can be addressed through careful segmentation, collaboration with ERM teams, and blending both quantitative and qualitative risk assessments. The strategies outlined in this chapter remain applicable to organizations of any size, including global corporations, and by taking a tailored approach, CISOs can effectively manage and communicate risk at every level of the enterprise.

The strategies and concepts we've covered in this chapter are the building blocks for any CISO looking to evolve from a tech-focused role to a strategic business partner. Once you've grasped the intricacies of risk management, you can move beyond just fixing technical problems and start speaking the language of the boardroom. You'll be able to quantify and communicate cyber risks in a way that resonates with executives, enabling them to make better decisions about how to allocate resources effectively. And by championing the integration of IT risk into ERM, you'll show that you understand the big picture, ensuring that security investments align with and support the company's overall strategy.

6.17.1 Key Takeaways

1. **The Foundational Concept of Risk in Cybersecurity**
   - Risk is the potential for loss or harm due to exposure to a threat and is quantified by the formula

     $$\text{Risk} = \text{Probability} \times \text{Impact}.$$

   - The distinction between *likelihood* (qualitative) and *probability* (quantitative) is crucial for effective risk assessment.
   - Understanding the nuances of related terms like threat, vulnerability, and exploit ensures accurate risk management and prioritization.
2. **Differentiating IT Risk from Enterprise Risk**
   - *IT risk* focuses on failures or breaches in information technology, whereas *enterprise risk* encompasses broader organizational concerns, including financial, operational, and reputational risks.
   - Boards require risks to be presented in business terms, focusing on their potential impacts on strategic objectives.
3. **Limitations of Qualitative Risk Analysis**
   - Qualitative risk assessments often lack precision and consistency, relying on subjective judgments and descriptive scales.
   - Tools like risk matrices provide quick visualizations but are limited by ordinal data, making them less effective for precise decision-making.
4. **Transitioning to Quantitative Risk Analysis**
   - Quantitative methods like FAIR and Monte Carlo simulations provide measurable and actionable risk insights.
   - These methods enhance decision-making by framing risks in financial terms and reducing subjectivity.
5. **Risk Quantification and Treatment Options**
   - Organizations can apply strategies such as *avoidance*, *transfer*, *mitigation*, or *acceptance* based on their risk appetite and resource availability.
   - Tools like LECs help visualize financial impacts, aligning treatment plans with risk tolerance and appetite.
6. **Black Swan and Tail Risks**
   - *Tail risks* represent rare but devastating events, often requiring scenario planning and robust resilience strategies.
   - *Black swan events* emphasize the need for adaptable and agile cybersecurity programs, as these events are inherently unpredictable but carry severe consequences.
7. **Integrated Risk Management**
   - IRM breaks down silos between IT, operational, financial, and compliance risks, fostering a unified approach.

- Tools like COSO ERM and integration with enterprise frameworks ensure cybersecurity is aligned with overall organizational goals.

8. **Enhancing the Risk Register**
   - A modern risk register incorporates separate registers for assets, threats, and vulnerabilities, ensuring clarity and actionable insights.
   - By mapping assets to vulnerabilities and threats, organizations can calculate risk scores and prioritize mitigation strategies effectively.

6.17.2 Reflection and Exploration Questions

Understanding the Concepts

1. Why is it so important to understand the difference between threat, vulnerability, and risk?
2. Why is it important to differentiate between likelihood and probability in risk assessment?

Connecting the Dots

3. How does integrating IT risk with enterprise risk improve organizational resilience and decision-making?
4. How do the limitations of qualitative risk analysis connect to the benefits of quantitative approaches?

Real-World Application

5. What examples from your organization or industry illustrate successful risk quantification and treatment?
6. How can tools like LECs and FAIR improve your risk communication with stakeholders?

Personal and Professional Growth

7. How can you apply the principles of risk quantification in your role to enhance decision-making?
8. What steps can you take to improve your organization's risk register and overall risk management processes?

Further Exploration

9. What additional tools or frameworks could complement your current risk management strategy?
10. How might emerging technologies like AI or blockchain influence future approaches to risk quantification and management?
11. How can you begin aligning risk management with strategic business goals?

Part 3

Risk Treatment

Part 3A

Transfer, Avoid, and Accept Risk

7 Cyber Liability Insurance

7.1 INTRODUCTION

Cyber liability insurance has emerged as a central pillar in the strategy of risk transfer, which is a major part of the broader Cyber Risk Buy-Down Investment Mix discussed in this chapter, building on the risk treatment options laid out in the last chapter. Risk transfer allows organizations to shift certain risks to external parties complementing strategies of risk mitigation and acceptance. Commercial cyber insurance is one of the primary mechanisms we have for leveraging risk transfer. This chapter will explore the investment thesis for commercial cyber insurance, presenting it not as a simple safeguard but as a strategic asset that can be leveraged for significant business value. Beyond just covering potential financial losses, cyber insurance offers access to specialized resources and expertise, helping organizations respond effectively to incidents.

Defining a Cyber Risk Investment Mix and transferring risk are two key steps in the risk strategy framework introduced in the last chapter. A well-defined Cyber Risk Investment Mix ensures that organizations allocate their resources effectively, balancing proactive security measures with financial strategies that address residual risk. Risk transfer, as a core component of this mix, provides a safety net that strengthens overall resilience while enabling companies to focus on strategic growth rather than being entirely consumed by risk avoidance. Without a structured approach to these elements, organizations may either over-invest in security controls without financial fallback or under-invest, leaving themselves exposed to catastrophic losses.

However, realizing the full value of cyber insurance requires more than just purchasing a policy. Organizations must adopt a methodical approach to procurement, ensuring the policy fits their specific needs and aligns with their risk management objectives. This chapter introduces a structured framework for evaluating, procuring, and operationalizing cyber insurance to maximize its effectiveness.

Through this exploration, readers will gain insights into balancing the Cyber Risk Investment Mix, leveraging cyber insurance strategically, and establishing a robust foundation for organizational resilience in an increasingly volatile environment.

7.2 CYBER RISK BUY-DOWN INVESTMENT MIX

To develop a resilient cybersecurity posture, organizations must embrace a comprehensive approach to risk treatment. This involves balancing three primary strategies: risk acceptance, risk mitigation, and risk transfer. Together, we call this the *Cyber Risk Investment Mix*. Right-sizing and optimizing this mix are a critical conversation for CISO 3.0s as they engage with executive leadership and the board.

Risk transfer is a large part of how we manage the financial and operational impacts of cyber threats. This strategy, often executed through mechanisms like cyber liability insurance or contractual agreements, shifts some of the organization's

DOI: 10.1201/9781003510789-11

risk to external entities. Unlike mitigation, which reduces the likelihood or impact of a threat, risk transfer ensures that when incidents occur, their financial burden is shared or entirely covered by third parties.

The key to leveraging risk transfer lies in understanding its place within the Cyber Risk Investment Mix:

1. **Risk Acceptance:** Determining which risks align with the organization's risk appetite and can be tolerated without additional action.
2. **Risk Mitigation:** Implementing controls to reduce the likelihood or impact of risks within acceptable thresholds.
3. **Risk Transfer:** Shifting specific risks beyond the organization's capacity or appetite to third parties like insurers.

Striking the right balance among these strategies requires a detailed understanding of the organization's risk landscape. First, organizations must identify their inherent risk by assessing the baseline risk present in the absence of any controls. Next, they need to evaluate their residual risk to understand what risk remains after applying existing controls. With this information, they can determine the risk exceedance, which is the gap between residual risk and the organization's defined risk appetite, as explored in Chapter 6. Based on this exceedance, organizations can pinpoint how much risk should be transferred to maintain a balanced exposure and align with their broader business objectives.

Risk transfer is an indispensable tool for addressing risk exceedance and protecting the organization against high-impact scenarios. While commercial insurance is a common form of risk transfer, organizations should also explore additional mechanisms. Contractual provisions allow liability to be shifted to vendors or third parties through service agreements, such as indemnity clauses in technology vendor contracts or cloud service agreements. Outsourcing specific functions, such as security monitoring or data recovery services, transfers operational risks to managed service providers who assume responsibility for those functions.

Another example includes using third-party payment processors for financial transactions, which shifts the risk of fraud and regulatory compliance to the payment processor. Leveraging partnerships with cybersecurity firms for managed detection and response services can also serve as a risk transfer mechanism, where these firms are contractually obligated to provide remediation in the event of an incident. Finally, adopting indemnification clauses in contracts with strategic partners can further transfer liabilities related to breaches originating from their systems.

It's up to the CISO 3.0 to drive discussions about the right sizing and balancing the Cyber Risk Investment Mix. These discussions should address whether the organization is effectively balancing acceptance, mitigation, and transfer strategies. Chief information security officers (CISOs) must evaluate if the current risk transfer mechanisms align with the overall business goals and risk appetite. Furthermore, they need to ensure that the organization is prepared to reassess its mix as risks evolve and external factors change.

By reframing risk transfer as a strategic investment rather than a reactive measure, CISOs can ensure it is fully integrated into the organization's cybersecurity and business objectives. This proactive approach not only enhances resilience but also builds trust with stakeholders who seek clarity and confidence in the organization's risk posture.

7.3 COMMERCIAL CYBER INSURANCE

Nowadays, organizations are constantly under siege from a relentless onslaught of cyber threats. Ransomware attacks, data breaches, and system meltdowns, the risks are endless, and the financial consequences can be severe. That's where *cyber liability insurance* comes in, offering businesses one of the most effective tools for transferring cyber risk.

As cybersecurity becomes a higher priority for organizations, the role of CISOs in shaping cyber insurance strategies is growing in importance. With the global cyber insurance market valued at $14 billion in 2023 and expected to surpass $29 billion by 2027, the demand for specialized standalone cyber insurance policies is on the rise. CISOs are being asked to evaluate cyber risks, quantify potential impacts, and collaborate with CFOs and insurance brokers to determine the right level of coverage for their organizations.

This shift is driven in part by the increasing frequency and severity of cyberattacks. Data breaches and other incidents have made organizations more vulnerable, pushing insurers to improve policy clarity. This, in turn, helps risk managers better understand their options and manage costs more effectively. In many industries, standalone cyber insurance has become essential, with companies requiring their partners to carry this type of coverage, highlighting again the close connection between cyber risk and business risk.

CISOs are in a unique position to contribute meaningfully to these discussions. While the decision to buy cyber insurance has traditionally fallen to chief financial officers (CFOs) or chief risk officers (CROs), CISOs bring valuable insights about the organization's specific vulnerabilities and the security controls in place to mitigate them. They can also help ensure that the terms of the insurance policy align with the organization's needs, particularly when it comes to incident response resources.

CISOs' involvement in the insurance process goes beyond selecting policies. They must also ensure the organization has the right security measures in place to meet the requirements of the insurer and to facilitate a smooth response in the event of a cyber incident. By engaging early in insurance discussions, CISOs can help avoid potential delays during a crisis, ensuring the organization has access to the right resources when they are needed most.

For the modern CISO, cyber insurance isn't just a safety net: It's a key part of their overall strategy. It's about understanding the risks your organization faces and making smart choices about which ones to transfer to an insurer. It's about finding that sweet spot where you have enough coverage to protect your company's finances without breaking the bank on premiums. This involves not only selecting the appropriate coverage but also preparing for the financial and operational challenges that come with responding to cyber incidents. In doing so, CISOs help ensure their organizations remain resilient in the face of an ever-changing threat landscape.

But let's be real, cyber insurance isn't as simple as buying a security product. It's a complex world of coverage options, application processes, and fine print. To truly leverage its power, you need to dive deep, understand the nuances, and align your insurance strategy with your overall risk management approach. It's about being proactive, not reactive, and making sure you're prepared for the worst, even if you hope it never happens.

7.4 THE CHANGING INSURANCE MARKET

While the overall number of ransomware incidents has seen a slight decline in the recent past, the average ransom cost has greatly increased, underscoring the escalating financial threat posed by cyberattacks. Organizations now face a complex set of risks, where fewer but more impactful attacks can lead to crippling financial losses.

Despite the growing recognition of cyber insurance as the "gold standard", many companies still rely on more generalized coverage. According to recent research, only a small percentage of organizations have comprehensive cyber insurance policies, and many decision-makers aren't fully aware of the details of their coverage. Additional findings further indicate that while the majority of enterprises have some form of cyber insurance, only about a quarter of those have a standalone policy specifically tailored to cyber risks.

The cyber insurance market itself is continuing to change. While insurance premiums are gradually decreasing, insurers are increasingly requiring more co-insurance and higher retentions (deductibles). This trend highlights the shared responsibility for risk management between insurers and organizations. The cyber insurance market is like a rollercoaster these days. Premiums are going down, retentions are going up, coverage is getting tighter, and underwriters are asking more and more questions.

Anecdotally, I've found that for many organizations without cyber insurance policies, the barrier isn't always cost – it's often a lack of understanding. Many businesses simply don't grasp the value proposition of cyber insurance or how it fits into their overall risk management strategy. More than that, they often struggle with navigating the procurement process. Knowing what coverage they need, how to present their risk profile to insurers, and how to negotiate terms can feel daunting. For CISOs, this knowledge gap represents a chance to lead the conversation, clarify the value of cyber insurance, and guide their organizations through the complexities of obtaining coverage that truly meets their needs.

7.4.1 The Value Proposition of Cyber Insurance

I do a lot of speaking engagements on cyber insurance, and one of the most frequent questions I get is, "Should I still be investing in cyber insurance, and is it actually valuable". The short answer is yes. Not only is it a major part of a balanced risk treatment strategy, but it's also mandatory for a lot of organizations. Not that any regulatory compliance requires insurance, but most organizations have contractual obligations to carry cyber insurance. Even if that wasn't the case, insurance continues to be a solid investment in terms of risk buy-down. Let's imagine you have a policy that is $2 million coverage for a $100,000 annual premium; what other controls could you buy for $100,000 that would buy down your risk by $2 million? There are not a lot of tools that are that impactful at that price point. Not to mention, the cost of a cyber insurance policy is often substantially lower than the potential financial losses from a cyber incident. In addition, cyber insurance can provide access to specialized expertise and resources for incident response and recovery, which can be invaluable for managing the fallout after an attack. This combination of financial protection and access to expert assistance makes cyber insurance a compelling value proposition for organizations seeking to manage their cyber risks effectively.

You and your organization need to determine in advance your primary motivation for purchasing cyber insurance: Is it primarily as a financial backstop, or do you also want a partner to provide leadership and support for incident response? This decision will influence the type and extent of coverage required and inform your overall risk management strategy. Later in the chapter, we will investigate the pros and cons of leveraging insurance for incident response leadership, helping you align with your organization's specific needs and risk appetite.

7.5 A FRAMEWORK FOR PROCURING CYBER INSURANCE

To address the complexities of procuring cyber insurance, I have developed a four-phase framework I call CLIF (cyber liability insurance framework). This framework guides organizations through the entire process, ensuring they understand and are able to realize the value of the policy, choose the right coverage, and are prepared to make a claim if necessary.

1. **Discovery:** This phase involves engaging key stakeholders, such as the CFO, CRO, CISO, general counsel, and procurement team, to assess the organization's current cyber risk profile, coverage needs, and financial goals. A thorough self-assessment is needed to identify vulnerabilities, quantify risks, and determine the appropriate level of coverage. This can be done through various tools and services, including cybersecurity ratings companies that provide objective assessments of an organization's security posture. Organizations should also consider their existing coverage tower, which includes other insurance policies like general liability, property, and professional liability, to avoid purchasing duplicate coverage. They should also explore alternative risk transfer mechanisms like self-insurance or captive insurance, where they create their own insurance entity.
2. **Application and Prequalification:** This phase involves completing a detailed application questionnaire and potentially undergoing a risk assessment by the insurer. It's essential to provide accurate and complete information to avoid claim denials or fraud allegations. Organizations should be prepared to answer questions about their security practices, incident response capabilities, and risk management procedures. Engaging with security experts or legal counsel can be helpful in navigating this process.
3. **Coverage Comparison and Purchase:** Once you receive quotes from different insurers, this phase involves comparing coverage options, negotiating terms, and selecting the policy that best suits your organization's needs. It's important to understand policy language, including exclusions and sublimits, to ensure that the policy meets the organization's needs.

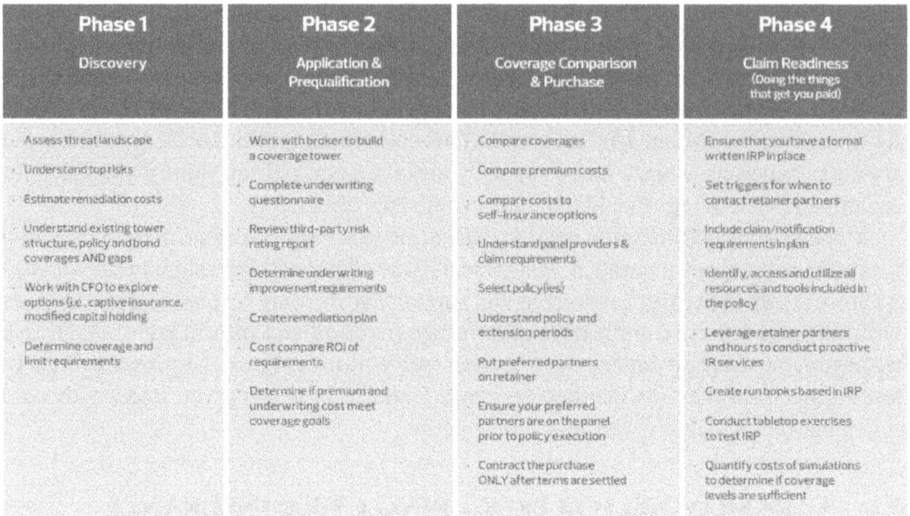

FIGURE 7.1 CLIF Phases.

4. **Claim Readiness:** This final phase focuses on aligning your incident response plan with your insurance policy. This includes understanding the claims process, building your incident response team, and ensuring that you are prepared to take the necessary steps to get paid in the event of a cyber incident. This includes understanding the insurer's panel of preapproved vendors and negotiating to include preferred partners if needed.

7.6 PHASE 1: DISCOVERY AND SELF-ASSESSMENT

The initial phase of the CLIF, discovery, focuses on understanding your organization's unique cyber risks and insurance requirements. This phase is vital for ensuring your cyber insurance strategy aligns with your business goals (Figure 7.1).

To begin, engage all relevant stakeholders within your organization. These may include the CFO, CRO, CISO, general counsel, and procurement team. The CFO, in particular, is instrumental in assessing financial implications and determining insurance coverage needs. Collaborating early on can ensure all perspectives are considered, and the final decision aligns with the organization's overall goals.

Risk management is an ongoing process, and conducting a self-assessment is essential to gain a clear understanding of your current cybersecurity posture. This process helps identify vulnerabilities, potential risks, and gaps in existing controls. You can establish a baseline for determining your cyber insurance needs by assessing your risk exposure, risk appetite, and financial capabilities.

Key questions to address during the self-assessment phase include the following:

- What are our top cyber risks, threats, and control gaps?
- Can we quantify the potential financial impact of these risks?
- Is our organization prepared to meet the underwriting requirements of potential insurers?

Consider utilizing third-party cybersecurity ratings companies to gain an objective, data-driven assessment of your organization's security posture. These companies provide valuable insights into how external entities, including insurance providers, perceive your security controls. If you can proactively address your identified weaknesses, you can improve your overall security posture and potentially enhance your insurability.

> **Pro Tip:** Insurance providers also use these third-party ratings to assess where you need improvement on your controls, so doing this puts you ahead of the game.

7.6.1 Determine and Understand Your Coverage Needs

Cyber risk may be covered under various types of insurance, such as property, general, and professional liability; business interruption; or crime policies, as well as other standard forms of coverage. It's important to know what non-cyber coverage you have, including general liability, business income, and commercial property insurance. By knowing your existing coverage, you can avoid unnecessary spending on duplicate policies. Collaborating with your CFO is key in evaluating your coverage options and determining the appropriate target amounts for each policy to ensure comprehensive protection without overspending (Figure 7.2).

Once you understand what additional coverage you need, it's time to engage your broker to build your insurance coverage tower. Your coverage tower is made up of multiple coverages with multiple policies, often from different insurance companies, that create a solid and seamless coverage structure or "tower" to better ensure that your organization is protected in the event of an incident. Your broker can help you spread out the coverage you need according to the options available to you while also helping you hit your coverage targets in alignment with your business's financial goals (Figure 7.3).

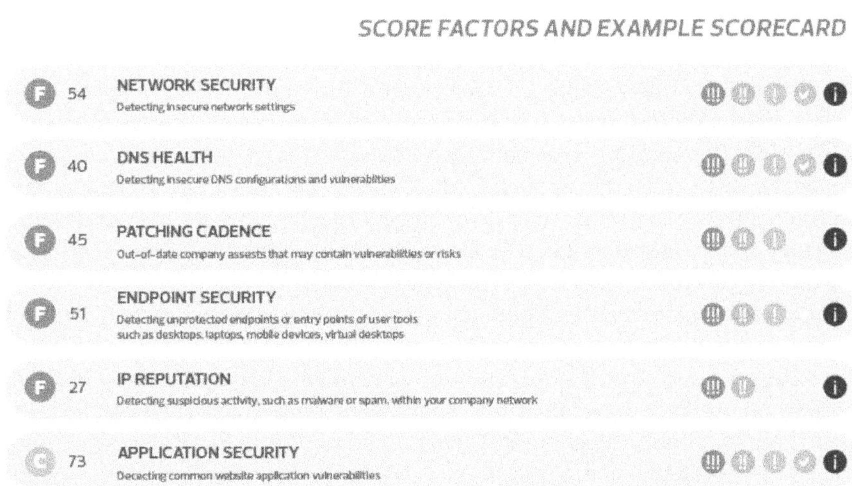

FIGURE 7.2 Scorecard.

FIGURE 7.3 Tower.

7.7 PHASE 2: APPLICATION AND PREQUALIFICATION

After discovery and self-assessment (and remediation), you're ready to apply for your coverage quote. Your broker will provide you with a coverage questionnaire/application. Consider this application an additional assessment. If some of the answers to the questions in this application are "no", you might be tempted to go to another insurance company to avoid addressing the provider's requirement. It's to your advantage to leverage this application and take the opportunity to level up your controls and make the improvements necessary to get the coverage you need.

Make sure that you answer all the questions in the application as fully and truthfully as possible. Incorrect or untruthful answers can be grounds for claims denial or, worst-case, insurance fraud! If there's a question you don't have the answer to, figure out the right answer before submitting your application. That may mean additional remediation. Don't fall subject to insurance fraud by filling in an answer purely for the sake of completing the questionnaire. It's becoming increasingly popular for underwriters to use absolutist language in these application questionnaires. Terms like "always", "never", and "all" make it increasingly difficult to truthfully respond in the affirmative to these queries and require organizations to amend their responses with details about their exception policies and compensating controls. You are also generally not required to submit your responses on the provided form. More and more organizations are copying application questions into a Word document where they can append yes and no questions with additional qualification information. It would be highly unlikely to see a broker, provider, or underwriter reject that type of response, and that additional disclosure informs the underwriter's risk calculus and helps shield you from claims denials.

Suppose you're unsure how to draft appropriate and factual responses to application questions. In that case, it may be wise to engage third-party assistance from security solutions specialists like CDW, whose experts understand policy language and can help craft appropriate responses.

Once you've submitted your application, the underwriter may respond with any additional controls they want you to have in place.

7.7.1 Cybersecurity Controls

The key to insurability, mitigation, and resilience is preparation. Start preparing for the underwriting process early. Without responses in the top control categories, coverage offered and insurability may be in question. Evaluating your cybersecurity maturity against these top controls will help you identify where improvements are needed. You cannot be too prepared. With that in mind, you should expect that the underwriting process will be rigorous and require you to answer progressively more detailed questions from underwriters.

It is common for the underwriter to "request" that you make controls improvements. Those may include, but are not limited to, the controls found in Figure 7.4.

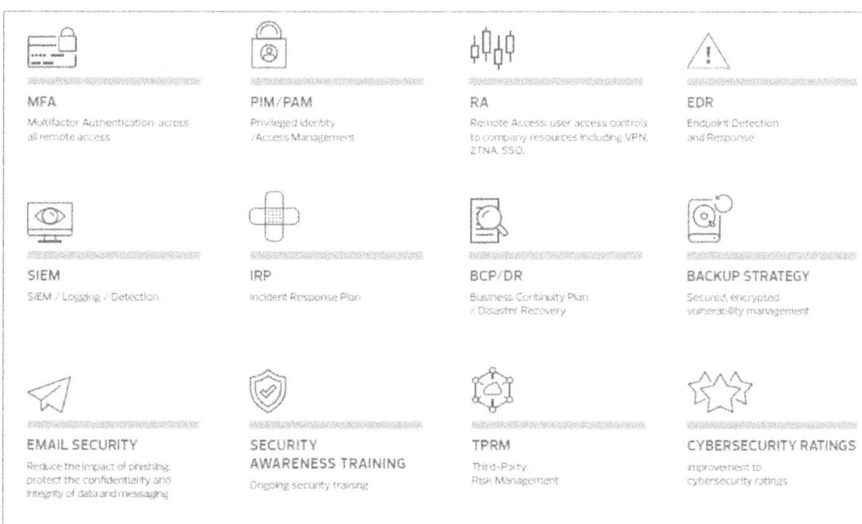

FIGURE 7.4 Controls List.

While these are standard security program leading practices, you don't necessarily have to implement all these. You do have the option to negotiate a subset of improvements or to provide an alternative attestation. Working with security experts to improve your security posture will go a long way to negotiating these requirements. A third-party vetted plan of action is often enough to signal to providers that you are taking the necessary steps to reduce your exposure.

7.8 PHASE 3: COVERAGE COMPARISON AND PURCHASE

Cyber coverage includes both first-party and third-party hazards. First-party cyber liability insurance covers the costs associated with responding to data breaches or cyber incidents that directly affect your organization's systems. Third-party cyber liability insurance provides protection in cases where you are sued due to a breach on a client's or partner's systems that impacts your business. Just to clarify, this coverage is paid to you in either situation, whether it happened directly to you or happened to a partner and impacted you (Figure 7.5).

Example

First-party coverage typically includes the following:

- Cyber incident response
- Extortion
- Business interruption (e.g., system failure, denial-of-service attack, and customer attrition loss)
- Digital data loss/recovery
- Cybercrime (e.g., computer fraud, funds transfer fraud, and invoice fraud)
- Social engineering fraud
- Reputational loss (e.g., brand restoration, adverse media publication, and review bombing).

	A. FIRST PARTY COVERAGE	SUBLIMIT OF INSURANCE	RETENTION	
☒	1. Breach Response Costs	$3,000,000	$10,000	
☒	2. Cyber Extortion Costs	$3,000,000	$10,000	
☒	3. Data Loss Costs	$3,000,000	$10,000	
☒	4. Business Interruption Loss	$3,000,000	$0	Waiting Period 8 hours
☒	5. Contingent Business Interruption Loss	$3,000,000	$0	Waiting Period 8 hours
	B. THIRD PARTY COVERAGE	SUBLIMIT OF INSURANCE	RETENTION	
☒	1. Breach Liability	$3,000,000	$10,000	
☒	2. Regulatory Penalties	$3,000,000	$10,000	
☒	3. PCI Assessments	$3,000,000	$10,000	
☒	a. PCI Recertification Costs	$50,000	$10,000	
☒	4. Media Liability	$3,000,000	$10,000	
☐	5. Technology Errors and Omissions Liability	N/A	N/A	

ITEM 8 RETROACTIVE DATE: N/A (full prior acts)

FIGURE 7.5 Policy.

Third-party coverage typically includes the following:

- Breach liability
- Payment card loss
- Regulatory proceedings
- Media liability
- Vendor/technology errors and omissions
- Outsourced provider business interruption.

Insurers will often break out and quantify separate coverage limits, retentions (deductible), and waiting periods for each of these items.

7.8.1 Understanding Policy Language

It's important to understand policy language because a $2-million cyber liability policy doesn't necessarily cover every loss scenario to the full $2 million. Which events have lower limits, higher retentions, or are completely excluded are all defined in policy language that is often hard to read and understand. It's not uncommon for organizations to be surprised when they attempt to file a claim and discover that they aren't covered as well as expected and aren't getting what they thought they were paying for. Again, engaging a security firm that understands policy language can help, but below are some terms you should be familiar with and look out for.

Acts of War/Terrorism: Almost all insurance policies exclude losses or damages caused by acts of war or terrorism. This has changed recently in the wake of the Mondelez and Merek cases. According to the *New York Times*, both organizations tried to file claims under their all-coverage policies for losses stemming from the 2017 NotPetya attack. Since NotPetya originated as a Russian attack on Ukraine, their carriers denied the claims under the Acts of War clause. While those companies did recently win settlements, those cases haven't 100% settled what constitutes an act of war; those cases fell in favor of the victim companies since the policies didn't explicitly identify cyberattacks. That exception wouldn't apply to cyber liability policies, and providers are now adding the cyber language to most Acts of War exclusions. It's important to understand what language is included in the war exclusions across all your policies to ensure that you are covered in the event of direct or indirect nation-state action.

Sublimit: A sublimit is a cap within your insurance policy that limits coverage for a specific type of loss. It's part of the overall coverage, not an addition to it. For example, a $3 million policy might have a $250,000 sublimit specifically for cybercrime, limiting how much you can claim for that type of incident.

Retention: This is essentially the deductible you're responsible for. It can be a set amount per claim or across all claims within a specific coverage area.

Waiting Period: The amount of time that must pass during an outage before you can make a claim. For instance, if your policy has an 8- to 10-hour waiting period, a 1-day outage may not be covered.

Extended Reporting Period (ERP): A designated window of time after a claims-made policy expires, during which you can still file a claim as if the policy were still in effect.

Mergers and Acquisitions (M&A) Coverage or Representation: This coverage is designed to protect against claims related to corporate M&A. Some policies automatically cover newly acquired entities, some allow you to add them for a small fee, while others exclude them entirely.

Retroactive Date: Many claims-made policies include a retroactive date, which eliminates coverage for claims related to actions that occurred before that date, even if the claim is filed within the policy period.

Full Prior Acts: A type of claims-made policy that does not have a retroactive date, meaning it covers claims related to acts that occurred at any time before the policy began.

Duty to Defend: This refers to an insurer's obligation to defend you in a claim covered under the policy. As long as there's potential coverage under the policy, the insurer is required to step in and cover defense costs.

When a lawsuit includes claims that are covered and others that are not, there can be disputes about how much the insurer has to pay for defense costs. Some policies are more favorable to policyholders, stating the insurer will cover all costs if any part of the lawsuit is potentially covered. Others say the insurer will only pay for the costs it believes are covered, unless otherwise determined through negotiation, arbitration, or court decision.

These disputes are less common under a "duty to defend" policy, where the insurer assumes the responsibility for defending any third-party claims. In these cases, if any part of the claim is potentially covered, the policy typically covers all defense costs. Under a "duty to reimburse" policy, where the insurer reimburses you for defense costs rather than providing a defense, disputes over cost allocation are more likely.

Duty to Pay/Reimburse: The insurer is responsible for paying defense costs but not for directly defending the policyholder.

Partial Subrogation Waiver: This waiver limits the insurer's right to take legal action against a third party responsible for the loss, particularly if you signed an agreement limiting their liability before the breach occurred.

If your insurer pays out a claim, they may attempt to recover the amount from a third party responsible for the breach, known as subrogation. However, many contracts with data management vendors limit their liability, which can lead to issues with your insurer. To avoid conflicts, it's wise to negotiate a partial "waiver of subrogation" provision in your cyber policy. This ensures that your insurer won't claim you impaired their rights if you had a preexisting contract limiting the liability of the vendor. Some insurers will agree to this, but others won't. If yours doesn't, it may be worth shopping around for a policy that does.

Extension Period: If you choose to cancel your policy, you can sometimes pay a multiple of your premium to be able to claim events that are discovered later but occurred during your coverage period. This can be important if your new policy doesn't cover full prior acts.

Right to Audit: It's increasingly common to see providers include requirements for regular audits and penetration testing. This can usually be performed by a third party of your choosing as long as that company also isn't providing you with a managed security service. Some providers will include provisions that allow their firm or selected third-party auditors to assess your program or environment to validate the controls and risk exposure you represented in your application. This provision can be periodic, annual, or triggered by an event.

7.8.2 Understanding Exclusions

Broadly written exclusions can undermine many of your key reasons for obtaining cyber insurance in the first place. Exclusions of special note here may include breach of contract, physical destruction of operational technology (OT), and officer liability. Make sure that you negotiate physical destruction of OT and coverage for your officers in your coverage contract or that you have these covered by a separate policy. The verdict in the 2022 ex-Uber CISO case reinforces the need for officer coverage. Officer coverage needs to be both named and positional. Positional coverage covers C-level officers while they are in their roles within the company, and named coverage identifies executives by name for coverage after they leave the organization. Obviously, officer liability coverage helps shield against civil but not criminal cases like *United States v. Joseph Sullivan.*

Some policies exclude coverage for any liability related to a breach of contract. This can be problematic for organizations that collect and store sensitive information, whether it's from customers, patients, or business partners, under agreements requiring confidentiality. Many businesses purchase cyber insurance specifically to protect themselves in the event of a privacy breach that leads to claims under these agreements.

Unfortunately, it's often tricky to get clear answers from brokers or providers about vaguely worded exclusions. That's where your legal team comes in: They can review and mark up (or "redline") these exclusions to help you better understand what's actually covered and what you're agreeing to. It's worth the effort to ensure you're not left exposed by confusing or ambiguous language in the policy.

Ensuring that your program is claim-ready and aligned with your policy is covered in phase 4; however, it is important to discover and understand the claims language within your terms in advance of signing on the dotted line.

The following terms are generally included in the policy definitions, but it's increasingly common to find additional claims info on things like panel partners, hotlines, filing forms, etc., outside the policy document, either in separate files or on the provider's website. All of those ideas are further discussed in phase 4.

Providing Notice: It is recommended that you work with your legal team to determine when you want or need to notify the insurer (especially if you don't end up making a claim) to determine what constitutes a "reasonable time period" for providing notice since nondisclosure can result in claim denial. You should also review the language around notification requirements for potential claims to determine if your policy contains a clause stipulating your "duty to disclose" events for which you do not intend to file a claim.

Period of Restoration: Loss and costs resulting from a breach are generally covered only during the period of restoration, which is typically 120 days, meaning costs that run past a quarter are not covered. This can often be negotiated out to 180 days or more.

Once you parse and understand the terms of your policy, you will likely discover that some terms don't meet the requirements you defined in phase 1. When this happens, don't forget everything is negotiable.

7.8.3 Negotiate

On that note, you can negotiate and redline nearly every part of a policy, but you must do it prior to purchase. This is a contract. In the same way, you would have your legal team redline a contract, you can redline an insurance policy. That doesn't mean that underwriters and providers will agree to these terms, but your broker can often help guide you toward those terms to which individual providers are most amenable.

As discussed in this chapter, common negotiation points include the following:

- Coverage types
- Underwriting requirements (controls)
- Sublimits
- Prior acts
- Subrogation waiver
- Length of restoration period
- Exclusion language
- Panel providers
- Waiting periods
- Duty to defend
- Premium cost
- Retention cost

I'm not a fan of requests for proposals (RFPs), and I feel that most organizations should avoid them whenever possible. We'll discuss why later in Chapter 10. Cyber insurance, however, is the exception. Lately I've seen some larger firms leverage an RFP-style process to request bids for coverage. This creates a competitive situation that can change the power dynamic and help reduce some of the underwriting back-and-forth. So many organizations just "take what they can get", which is also why I see so many organizations dissatisfied with their cyber insurance. At the end of the day, you should be treating cyber liability acquisition like the procurement of any other product. You need to negotiate and make sure that you are customizing your coverage to fit your needs. Even if it raises the premium a little, it is often worth it to adjust things like your waiting period or sublimits and ensure that you are able to get what you're paying for.

7.9 PHASE 4: CLAIM READINESS

I've seen too many organizations scramble to figure out their insurance coverage after a breach has already occurred. It's like trying to build a lifeboat in the middle of a storm. That's why claim readiness is so important. The terms of your policy will include details about the process of filing claims. Ensuring that your organization is claim-ready and that your response activities are aligned to the claims process is key to actualizing the value of your insurance investment in the event of an incident. By aligning your incident response plan with your insurance policy, you can ensure a smooth claims process, minimizing the financial impact of a cyberattack. Aligning your formal incident response plan (IRP) with your policy is also vital to getting your claim validated. Make sure your IRP includes the things that get you paid in the event of an incident. For example, if your policy includes a breach hotline, then include guidance in your IRP and playbooks around if and when to call that hotline.

Pro Tip: Make sure you're not leaving free resources on the table. Many cyber insurance providers contract to offer their policyholders access to online portals full of risk reduction, incident response (IR) planning, vulnerability management, and other readiness tools. Some providers offer annual premium rebates for cyber improvement projects. These resources can generally be found in the policy terms, but sometimes, you must go looking for them on the provider's websites.

7.9.1 Understanding Panel Providers

An insurance panel is a group of service providers that an insurance company has approved to offer services to its customers. If you're working with providers on the panel, the insurance company will cover the costs for those services. This process is also known as credentialing, and each insurance company creates and manages its own panel of approved providers.

The brokers and providers have started to position their panel partnerships as a value-added benefit and suggest that there are cost and overhead benefits in not having to contract and retain response partners. However, I believe it is still best to build your own panel team because your insurance provider's partners might not always be functioning in your best interest. The insurance providers value consistency and predictability when it comes to payout cost. Consequently, when it comes to getting on panel, partners need to be able to deliver repeatable services at a similar cost for every engagement. This panel provider is working for the insurance company, and you are getting a cookie cutter, one-size-fits-most response. That might be fine in some cases for some people, but I'd rather have a partner that works for me where my satisfaction is the primary driver.

This idea can be clarified by applying this logic to other types of insurance. If heaven forbid, your house burned down, you probably wouldn't call your insurance carrier and ask them to come to the table with the fire marshal, ask them to help you find a builder, and get their input about your blueprints and building plans. So, waiting until an event happens to have a provider randomly "assign" resources via hotline or even point you toward a list of providers you haven't previously contracted with can be an extra burden at an already complicated time. Not onboarding your panel partners in advance can be a huge hindrance since just executing contracts, Master Service Agreements (MSAs), and Non-Disclosure Agreements (NDAs) can take days or weeks, especially since these types of contracts are usually executed as a tri-party agreement with you, the partner provider, and your external legal firm. You generally can't wait that amount of time to begin response activities. The result would be that you are forced either to agree rapidly or to take on work at risk.

Furthermore, ensuring that your forensics partner can access your data and leverage your tool stack is the key to a quick response. Breaches where organizations needed to procure and/or roll out thousands of endpoint agents before the IR partner can begin their investigation are not unheard of. Those types of surprises can be expensive and time-consuming but are entirely avoidable if you preplan with your forensics partner. Beyond that, major cyber incidents require a lot of internal and external stakeholders to work together as a team; like any team, they need practice to perform well. Knowing your partners in advance and including them in your playbooks and tabletop exercises can help ensure that your IR is smooth and that your program is claim-ready. Think of your IR team as a sports team. Can you imagine what would happen if you tried to play a high-pressure championship game with a team that had not only never practiced but had also never even met? That is very much what many of the breach responses I've been a part of feel like.

Again, policyholders are generally not obligated to contract for services with the insurer's panel providers. Even though insurers would prefer that you use their panel providers, you can almost always arrange for your preferred partners to be added to the panel for your policy. You do generally need to make that request before executing your policy, and they will need to adhere to the policy provider's requirements for rates and coverage.

7.9.2 Put Retainers in Place

If you opt to select your own partners in advance, one of the best ways to ensure that your partner team is appropriately on-boarded is to put those partners on retainer.

Your response team should include partners for forensics, incident response, third-party legal, a breach coach, and public relations, but you don't necessarily have to select one of each. You can select your own legal and forensics partners and let your provider help you find a breach coach or comms specialist if the need arises.

The benefit to putting your own retainers in place is, of course, that you get to choose who you prefer to work with, not someone the insurer prefers to work with. Having retainers in place also means you're doing your due diligence to work with a team that is familiar with you and your business. Being helped by a team of people who are actually on your team is well worth the additional cost.

Incident response retainers should include the following:

- A service-level agreement (SLA)
- A block of hours (active/proactive IR)
- The ability for unused hours to be repurposed
- An onboarding plan to ensure that legal paperwork is in place in advance
- A break-in-case-of-emergency playbook to engage the IR partner

7.9.3 Coordinate with Legal

While you should engage your general legal counsel back in step #1 (discovery), it is important to ensure that your legal team remains aligned with your formal incident response plan. In-house counsel is great for proactive IR activities like procuring policies and setting up, but in your response program, you will likely want to engage third-party external counsel for any reactive IR activities. Leveraging your third-party counsel to work as an intermediary to your IR and forensic partner can provide a layer of attorney–client privilege that could be vital if your incident were to lead to litigation. Your in-house legal team most likely won't have cyber breach experience; you will want to find a firm that can provide you with breach coaching.

Even though cases like the E.D. Virginia Capital One ruling limit the extent of attorney–client privilege around forensic reports, it is still best practice to have all your IR contracts, statements of work, retainers, etc., executed as tri-party agreements where your external counsel contracts with the partner provider on your behalf. It is important to note that your attorney initiates attorney–client privilege and includes only legal advice or communications, not facts such as the timeline of breach events.

7.9.4 Ransom Brokers and Extortion Payments

One of the key reasons organizations take on cyber insurance is to help cover ransom payments in the event of extortion attacks. Studies indicate that ransomware accounted for as much as 75% of all cyber claims in 2021. It's common for insurers to make extortion payments, but even though it is an option, paying extortion should still be considered last among the list of options and should be taken on as a business decision. Decisions regarding extortion payments should be carefully evaluated on an individual basis by executive leadership, legal counsel, the breach response team, IR firm, and broker, ensuring that they understand the payment risks.

The risks of extortion payments include the following:

- Non-recovery from double encryption or restoration failure
- Additional demands across multiple extortion vectors
- Legal sanctions violations
- Repeat targeting
- Bolstering the ransomware economy

Also, extortion payments should never be made in an attempt to cover up an incident.

If you decide to make ransom payments, avoid detailing the specifics of those payments in your written IRP. When an organization is writing down their formal

IRP and aligning it to their policy, they may be tempted to write down plans for paying ransoms, obtaining/storing cryptocurrency to facilitate extortion payments, or engaging/retaining brokers. Those things should NOT be included in your formal IRP. You should not be planning in advance to be breached and to pay ransoms. It would be hard to explain in court why your plan to protect your customer/employee/business data included plans for paying a ransom.

If you find that you do need to make a ransom payment, your insurance provider or breach coach should be able to help you engage a ransom broker. Ransom brokers can help you negotiate payments, including costs, terms, and deadlines. The broker will also facilitate the payment and the decryption key exchange to help ensure that everything works as expected.

It's becoming common for threat actors to threaten data or encryption key deletion if victims involve brokers and/or law enforcement. That should not deter you from engaging a broker for assistance. If you do engage a broker, be sure to choose a reputable firm recommended by your provider or law firm. Recently, many ransom operators have started pushing victims toward their "preferred partner" brokers because they have pre-negotiated discounts and payment details. According to an individual associated with the Darkside ransomware group, certain data recovery firms they collaborate with expedite the decryption process and may offer substantial discounts, sometimes ranging from 20 to 40%. This individual also indicated that the entire recovery procedure, from initial contact to data restoration, can be completed within a single day.

Also, be sure to effectively "hide" your insurance policy and IRP from bad actors. They are often prime targets for ransom groups because they provide a lot of valuable information, including exactly how much they can ask for.

7.10 CONCLUSION

The cyber insurance world is changing, and not in the way we might have hoped. Insurers are tightening their belts, demanding higher deductibles, and scrutinizing every detail of your security posture before they'll even consider coverage. It's a tough market out there, and the days of simply buying a policy and calling it a day are over.

Don't get me wrong, cyber insurance is still a major piece of the puzzle. It's a safety net that can protect your organization from financial ruin in the face of a cyberattack. But it's not a magic shield. You can't just rely on insurance to bail you out if your security is weak. It's about building a strong defense *and* having a solid backup plan.

It's important to understand that cyber insurance should be tailored to each organization, as every company faces its own set of unique risks and requirements. The key is to craft a policy that's tailored to your specific situation. Whether you're looking for a financial safety net or a partner to help you navigate the chaos of an incident, make sure you're getting the coverage that's right for you.

The CISO 3.0 gets this. They understand that cyber insurance is a strategic decision, not just a checkbox on a compliance list. They take a risk-based approach, carefully assessing their organization's vulnerabilities and quantifying the potential impact of a breach. They work closely with insurers, communicating their needs clearly and negotiating for the best possible coverage.

But it doesn't stop there. The CISO 3.0 also knows that the devil is in the details. They meticulously review policy language, making sure they understand the exclusions and limitations. They align their incident response plans with their insurance coverage, ensuring a smooth and efficient claims process if the worst happens.

In the end, cyber insurance is a valuable tool, but it's just one part of a comprehensive cybersecurity strategy. It's about building a strong defense, being prepared for the unexpected, and having a plan to get back on your feet if you get knocked down. Your best bet is to combine proactive risk management with a well-crafted insurance policy to create a blended risk treatment strategy.

7.10.1 Key Takeaways

1. **The Critical Role of Cyber Insurance**
   - Risk transfer is a fundamental part of the Cyber risk buy-down investment mix.
   - Cyber liability insurance is an essential tool for transferring financial risk associated with cyber threats, such as ransomware and data breaches.
   - With the cyber insurance market projected to exceed $29 billion by 2027, the strategic role of CISOs in shaping insurance policies and aligning them with organizational goals is increasingly significant.
2. **The Evolving Insurance Market**
   - Premiums are decreasing, but higher retentions (deductibles) and co-insurance requirements are becoming more common, emphasizing shared responsibility in risk management.
   - Many organizations lack comprehensive policies due to knowledge gaps about the value and intricacies of cyber insurance, presenting an opportunity for CISOs to lead informed discussions.
3. **Framework for Procuring Cyber Insurance (CLIF)**
   - **Discovery:** Involves assessing the organization's cyber risk profile and determining coverage needs through stakeholder engagement and self-assessment.
   - **Application and Prequalification:** Entails completing detailed insurer questionnaires and preparing for rigorous underwriting processes.
   - **Coverage Comparison and Purchase:** Focuses on comparing policy options, negotiating terms, and thoroughly understanding exclusions and sublimits.
   - **Claim Readiness:** Aligns incident response plans with insurance policies, ensuring efficient claims processes and access to critical resources during incidents.
4. **Importance of Policy Language**
   - Understanding exclusions, sublimits, and other policy terms ensures coverage aligns with organizational risk management needs.
   - Negotiating terms before purchase, such as restoration periods and subrogation waivers, prevents surprises during claims.

5. **The Strategic Involvement of CISOs**
   - CISOs bridge the gap between cybersecurity and financial risk management by ensuring policies reflect the organization's specific vulnerabilities and controls.
   - Their role extends to preparing the organization for underwriting requirements and integrating cyber insurance into broader risk management strategies.
6. **Panel Providers and IR**
   - Insurers often recommend panel providers, but organizations may negotiate to include their preferred partners for tailored IR.
   - Building relationships with response partners before incidents occur ensures smoother response efforts and better alignment with organizational needs.
7. **Cyber Insurance as a Complementary Tool**
   - While cyber insurance provides a financial safety net, it cannot replace robust cybersecurity practices and proactive risk management.
   - Combining insurance and other risk transfer strategies with strong security measures ensures comprehensive resilience against cyber threats.

7.10.2 Reflection and Exploration Questions

Understanding the Concepts

1. Why is risk transfer an important part of the cyber investment mix?
2. How does the CLIF framework help organizations navigate the complexities of cyber insurance?

Connecting the Dots

3. How does the evolving cyber insurance market connect to broader discussions of risk transfer and treatment?
4. In what ways do the strategic roles of CISOs in cyber insurance align with their responsibilities discussed in earlier chapters?

Real-World Application

5. Can you identify examples of organizations where cyber insurance successfully mitigated financial losses from a breach?
6. How might understanding policy language and exclusions prevent disputes or claim denials in real-world scenarios?

Personal and Professional Growth

7. How can you apply the CLIF framework to assess and enhance your organization's readiness for cyber insurance procurement?
8. What steps can you take to improve your understanding of insurance policy terms and negotiation strategies?

Further Exploration

9. What additional skills or resources would enhance your ability to integrate cyber insurance into your organization's risk management strategy?
10. How might you begin aligning cyber insurance policies with organizational goals?

8 Self-Insurance and Risk Financing

8.1 INTRODUCTION

In Chapter 6, we laid out the four main ways to handle risk: accept it, avoid it, transfer it, or mitigate it. We've already talked about transferring risk through cyber insurance, which is basically paying someone else to take on some of the financial burden if things go wrong. Now, let's get our hands dirty with the other two options: accepting and avoiding risk. These strategies put you in the driver's seat, taking a more active role in managing your company's cyber risk exposure.

For the forward-thinking chief information security officer (CISO), understanding how to finance risk is key. It's about knowing your company's risk appetite, how much risk you're willing to live with, and its financial strength. It's about making smart choices about which risks to keep and which to offload.

We'll dig into self-insurance, where you set aside money to cover potential cyber losses. We'll look at going all-in on self-insurance and also using a dedicated reserve fund. We'll also explore other ways to finance risk, like captive insurance companies and bonds.

We'll also guide you through the nitty-gritty of cost–benefit analysis (CBA). We'll look at both the direct and indirect costs of a cyber incident and how this analysis can help you decide where to put your resources and how to best reduce your risk. We'll stress the importance of the CISO and chief financial officer (CFO) working hand-in-hand to figure out the company's risk appetite, financial capabilities, and the best way to finance cyber risk.

We'll also tackle risk avoidance and how to reduce your exposure to cyber threats proactively. We'll look at real-world examples and discuss the trade-offs involved. Because while avoiding risk is ideal, it's not always realistic in today's interconnected digital world. Sometimes, you have to take calculated risks to keep the business moving forward.

By the end of this chapter, you'll have a toolbox full of strategies for managing cyber risk. You'll be able to make more data-driven decisions that not only protect your company but also align with its unique needs and risk tolerance.

8.2 RISK ACCEPTANCE

Risk acceptance in the cyber context involves acknowledging a certain level of cyber risk without implementing specific mitigation or transfer measures. This decision is guided by the company's cyber risk appetite, which reflects the amount of cyber risk it is willing to tolerate to achieve its business objectives.

It's important to understand that risk acceptance is the default position, and if you do nothing, you are effectively accepting the risk. Inaction doesn't make the

DOI: 10.1201/9781003510789-12

risk disappear, it simply means you're accepting it without a plan to cover potential losses. Often, when someone says, "I can't afford cyber insurance", I remind them that cyber insurance is a transfer strategy. If you don't transfer the risk to insurance or mitigate it with controls, you are accepting the risk. At that point, the question becomes: How will you cover the losses when something goes wrong? Without a clear plan for addressing potential financial impacts, those cost land on the balance sheet and organizations are left exposed to significant disruptions of their cash flow, potentially threatening their overall financial stability.

It's important to recognize that accepting cyber risk implies a likelihood of incurring losses. While materiality is often measured as a percentage of revenue, it's the impact on free cash flow that is most important during a cyber incident. For instance, if a company has a $10 million cyber risk tolerance, it accepts that losses up to this amount are possible. Planning for how to finance these losses is essential; otherwise, they will almost certainly impact the balance sheet.

CISOs should collaborate with CFOs to determine how best to protect free cash flow. Strategies include commercial cyber insurance, contingent financing, self-insurance, and alternative risk transfer mechanisms. When you proactively address the financial implications of risk acceptance, you can ensure that your organization has the resources to recover from a cyber incident without jeopardizing its long-term financial health. Throughout the remainder of this chapter, we'll explore some of the options available to finance accepted risk. Also note that allocating resources for accepted risk is step number six in the risk strategy framework introduced in Chapter 6.

Note that risk acceptance isn't inherently negative. In fact, as we discussed in Chapter 6, your risk appetite is a form of risk acceptance. Risk acceptance is a very important part of the cyber risk investment mix, but it must be fully understood and strategically planned for. Risk acceptance should be an informed choice, aligned with the organization's risk appetite and business goals, rather than a passive result of inaction.

8.2.1 Business Line CBA for Cyber Risk

A business line cost-benefit analysis (CBA) is a tool for CISOs and CFOs to make decisions based on data to address cyber risk financing and resource allocation. This analysis involves evaluating the potential costs and benefits of various risk mitigation strategies for each business line within the organization. Understanding the financial implications of cyber risk at a granular level permits organizations to make truly strategic decisions that align with their overall business objectives and risk appetite.

A CBA enables targeted resource allocation. Since different business lines have varying levels of cyber risk exposure and potential financial impact, a CBA allows resources to be allocated precisely where they're most needed. This ensures that high-risk areas receive adequate protection while avoiding overspending on low-risk areas.

A CBA also facilitates informed risk acceptance by helping quantify the potential financial losses tied to each cyber risk. That empowers the CISO and CFO to choose which risks to accept, considering the company's risk tolerance and financial capabilities.

A CBA assists in justifying return on investment (ROI). Cybersecurity investments can be substantial, and a CBA furnishes the CFO with the data needed to justify these expenditures. The CISO can also gain budget and buy-in from showcasing the potential cost savings of preventing cyber incidents or the revenue generated from improved security and customer trust.

A CBA can also promote data-driven decision-making. It offers a structured and data-driven approach to cyber risk management, moving away from decisions based on gut feelings or anecdotal evidence. This helps ensure cybersecurity investments align with the organization's financial goals and risk appetite.

The CISO and CFO work together to identify the most critical business lines based on their revenue generation, customer impact, and potential for disruption in the event of a cyber incident. Subsequently, the CISO conducts a thorough risk assessment for each business line, identifying potential cyber threats, vulnerabilities, and the likelihood and impact of potential incidents.

The CFO then collaborates with the CISO to quantify the potential financial losses associated with each identified cyber risk, encompassing both direct costs (like incident response and data recovery) and indirect costs (such as lost business and brand damage).

Together, the CISO and CFO evaluate various risk mitigation strategies, including cybersecurity investments, insurance coverage, and risk transfer mechanisms. They analyze the costs and benefits of each strategy, considering its effectiveness in reducing the financial impact of a cyber incident.

The CFO then conducts a detailed CBA, comparing the costs of each risk mitigation strategy against the potential financial losses it aims to prevent. This analysis considers both quantitative factors (like monetary costs and probability of occurrence) and qualitative factors (like customer trust).

Based on the results of the CBA, the CISO and CFO can select which risk mitigation strategies to implement for each business line. They then allocate resources accordingly, ensuring that the organization is adequately prepared to manage cyber risks.

8.2.1.1 Example

Consider one of my clients, a manufacturing company with two main business lines: production and sales. The CISO and CFO conducted an assessment and found that the production line was most vulnerable to cyberattacks due to its reliance on interconnected industrial control systems. Then, they conduct a CBA to help determine the potential financial impact of a cyber incident on the production line. The production impact was estimated to be $10 million, while the impact on the sales line was estimated to be an additional $2 million.

Based on this analysis, the CISO and CFO were able to decide how much of that risk to accept. They ultimately decide to invest in additional cybersecurity measures for the production line, such as network segmentation, intrusion detection systems, and employee training. They also opted for a higher level of cyber insurance coverage with lower waiting periods for the production line to mitigate the financial risk further. In the end, after the transfer and mitigations, they still accepted more than $2 million in risk to the production and sales lines.

By conducting a business line CBA, the CISO and CFO were able to make data-driven decisions that protected the organization's financial well-being while also ensuring that critical business operations remained resilient.

8.3 CAPITAL REALLOCATION FOR CYBERSECURITY

Capital reallocation in cybersecurity is a dynamic process where the *CISO* and *CFO* work together to identify areas where current security investments might not be delivering the best value. It's not only about fixing problems, but they also take a forward-looking approach, assessing emerging cyber threats to see if additional resources are needed to address them. This requires a deep dive into the organization's current risk landscape, considering both the evolving threat environment and the effectiveness of existing security measures.

Once they've pinpointed areas for improvement, the CISO and CFO explore alternative security investments that could deliver a better ROI or align more closely with the company's risk tolerance. This could mean reallocating funds from outdated systems or technologies to more modern, effective solutions that offer stronger protection. It might also involve adjusting how risks are managed, such as shifting from mitigation strategies to risk transfer (like using insurance) or vice versa.

To make informed decisions, the CFO works closely with the CISO to quantify the potential financial impact of both current and emerging cyber risks. This involves conducting risk assessments, analyzing historical data, and looking at industry benchmarks to translate these risks into clear financial terms that fit within the company's broader business strategy.

Armed with this data, the CISO and CFO can make well-informed, data-driven decisions about how to reallocate resources best. They might increase investments in some areas, scale back in others, or explore new ways to transfer risk, such as cyber insurance. The goal is always to optimize security spending, ensuring that the company is well-protected against cyber threats while making the most efficient use of its resources.

From the CFO's perspective, capital reallocation in cybersecurity is more than just responding to threats: It's about strategically aligning security investments with the company's financial goals and risk appetite. The CFO plays a crucial role in balancing the need for strong security with the imperative to meet financial objectives. This involves making sure that every dollar spent on security is justified, delivers value, and contributes to the overall financial health of the company while also considering the potential costs of security incidents.

The CFO also evaluates the cost-effectiveness of security solutions, comparing the costs and benefits of different approaches and assessing how they impact the company's financial performance. Their goal is to get the most value from every dollar invested in cybersecurity.

By working together, the CFO and CISO prioritize investments that address the most critical risks and provide the greatest protection for the company's assets and reputation. The CFO might also explore risk transfer options like cyber insurance to help mitigate the financial fallout from a potential attack. This involves carefully weighing the cost of premiums against the potential losses and making sure the coverage aligns with the company's risk tolerance. Alternatively, they might set aside funds to create an internal "rainy day" fund specifically for cybersecurity incidents, reducing reliance on external insurance.

Capital reallocation is not a one-time event. Instead, it's an ongoing process. The CISO and CFO should continuously monitor the effectiveness of their security investments and adjust as needed to address new threats or changing business priorities. This ongoing evaluation ensures that the organization's security efforts stay aligned with its financial goals and risk tolerance.

8.3.1 Examples of Capital Reallocation

Shifting from on-premises infrastructure to cloud-based security solutions: This can lead to cost savings and improved scalability, enabling the organization to adapt to changing needs more effectively.

Investing in advanced threat detection and response tools: These tools can help identify and mitigate threats more quickly, reducing the potential financial impact of a cyberattack.

Reallocating funds from reactive incident response to proactive risk mitigation: This can help prevent incidents from occurring in the first place, saving the organization considerable costs in the long run.

Increasing investments in security awareness training: Educating employees about cyber risks can reduce the likelihood of successful attacks, protecting the organization's assets and value.

A CISO 3.0 should be involved in capital reallocation by providing the CFO with valuable insights into cyber threats and the potential financial impact of security risks. This partnership extends to collaborating with the CFO to develop a risk-based approach to resource allocation, ensuring investments are prioritized based on their potential impact. The CISO 3.0 must also clearly articulate the business case for security investments, demonstrating their ROI in terms that resonate with financial stakeholders. By continuously monitoring the effectiveness of security measures and recommending adjustments to resource allocation as needed, the CISO 3.0 ensures that the organization's financial resources are deployed strategically. This proactive approach safeguards the organization against cyber threats while also supporting its overall business growth and objectives.

8.4 SELF-INSURANCE FOR CYBER RISK

Self-insurance for cyber risk involves setting aside funds to cover potential losses from cyberattacks or data breaches. This approach isn't a universal fix and demands thoughtful planning and customization to fit the specific needs of each situation. Organizations can choose from various types of self-insurance, including pure self-insurance, where the organization assumes full financial responsibility for all cyber losses up to a predetermined limit; funded reserves, where a dedicated fund is established for cyber risk; or utilizing a captive insurance company, a subsidiary owned by the parent organization, to provide self-insurance for cyber risk.

Large organizations with significant financial resources and diversified risk portfolios are often well-positioned to self-insure for cyber risk. They have the capacity to absorb losses and the expertise to manage their risk exposure. Organizations with a predictable loss history, meaning a track record of few or low-impact cyber incidents, may find self-insurance a cost-effective option. Organizations with mature risk management programs and dedicated cybersecurity teams are better equipped to handle the complexities of self-insurance.

The appeal of self-insurance lies in potential cost savings compared to traditional cyber insurance premiums, especially for organizations with a low risk profile. Self-insured organizations have more control over their risk management and claims handling processes, allowing for greater flexibility in customizing coverage to meet specific needs.

That said, self-insurance is not without its drawbacks. It exposes the organization to potentially sizable financial losses in the event of a major cyber incident. Managing a self-insurance program requires expertise in risk assessment, claims handling, and financial modeling, which may not be readily available in every organization. Regulatory compliance is also a concern, as organizations must ensure that their self-insurance arrangements adhere to applicable regulations.

Before deciding on self-insurance, organizations must carefully consider their risk appetite, financial capacity, and risk management maturity. If the organization has the appetite to accept the financial risk of a major cyber incident, the financial capacity to absorb potential losses without jeopardizing its stability, and the risk management maturity to effectively manage a self-insurance program, then self-insurance might be a viable option. You should consult with risk management experts and conduct a thorough CBA before making a final decision.

8.4.1 Pure Self-Insurance

Pure self-insurance is a risk financing strategy in which an organization chooses to retain the financial risk of cyber incidents instead of transferring it to an insurance company. In essence, the organization acts as its own insurer, setting aside funds to cover potential losses. In pure self-insurance, the organization assumes full financial responsibility for all cyber losses. Pure self-insurance can be an attractive option for organizations with a low-risk profile, strong financials, and a willingness to bear the full financial burden of a cyber incident.

It's important to note that pure self-insurance carries a higher degree of financial risk, as the organization is solely responsible for covering all losses. Consequently, it's important to have a robust risk assessment process in place to accurately estimate potential losses and ensure that the organization has sufficient funds to cover them.

Self-insurance can be strategically combined with commercial insurance to offer a balanced risk management approach. This hybrid model allows organizations to retain control over smaller, more predictable losses, while transferring the risk of larger, potentially catastrophic events to an external insurer. For example, a company might decide to self-insure for losses up to a predefined threshold, often referred to as a self-insured retention (SIR) amount, and then rely on commercial insurance to cover any costs that exceed that threshold. This approach enables the organization to minimize insurance premiums while still safeguarding against significant financial impacts from rare but severe incidents.

By structuring the coverage in this way, businesses can optimize their risk financing strategy, leveraging the cost-efficiency of self-insurance while ensuring that they are protected against unforeseen, large-scale losses. The collaboration between the CISO, CFO, and legal teams becomes essential in determining the right balance between self-insured retention levels and the scope of commercial insurance coverage, aligning with the company's overall risk appetite and financial goals.

The process of self-insuring should begin with a risk assessment to identify and quantify the organization's potential cyber risks. This involves evaluating the likelihood and impact of various cyber incidents, such as data breaches, ransomware attacks, and business interruption.

Based on its risk appetite and financial capacity, the organization then determines if they want to establish a self-insured retention, representing the maximum loss it's prepared to bear in a cyber incident. Subsequently, funds are allocated to a dedicated reserve account to cover potential cyber losses up to the SIR. This fund should be sufficient to cover anticipated losses and readily accessible in case of an incident.

Finally, in the event of a cyber incident, the organization takes charge of managing the response and recovery efforts, as well as covering the financial losses up to the SIR.

Unlike traditional insurance, pure self-insurance doesn't involve an external insurance policy or premium payments. Instead, the organization maintains an internal reserve fund dedicated to covering potential cyber losses. This approach gives the organization full control over the claims process, from investigation and remediation to payout.

8.4.1.1 Example

One of my large financial services clients, with a robust balance sheet and a mature cybersecurity program, decided to adopt a partial self-insurance strategy for cyber risks. They recognized that their strong financial position and proactive risk management practices allowed them to retain a portion of the risk comfortably.

After conducting a thorough risk assessment and evaluating their financial capacity, they established an SIR of $10 million. This meant that they would cover the first $10 million of any cyber loss while purchasing a traditional insurance policy to cover losses exceeding that amount. This approach allowed them to benefit from the potential cost savings of self-insurance for smaller incidents while still having the protection of an insurance policy for catastrophic events that could exceed their risk tolerance.

While pure self-insurance can be a cost-effective strategy for some organizations, it's important to be aware of its potential drawbacks. The biggest drawback is the higher financial risk, as the organization is solely responsible for covering all losses up to the SIR. An accurate risk assessment is, therefore, critical to ensure that the SIR is set appropriately, and the reserve fund is adequate. The organization would also need to have internal expertise to manage cyber incident response and claims effectively.

Self-insurance can be a practical option for organizations with a solid financial foundation and a lower risk profile. However, it takes careful planning, thorough risk assessments, and effective management of cyber incidents to make it work. If these factors aren't in place, a hybrid approach or traditional cyber insurance might be a better fit.

8.4.2 Captives for Cyber Coverage

A *captive insurance company* is a subsidiary that's fully owned and controlled by its parent organization. Its main role is to provide insurance coverage for the parent company and, in some cases, other affiliated businesses. In the context of cyber insurance, a captive can be set up specifically to cover cyber risks or added to an existing captive that handles other types of insurance.

Using a captive offers flexibility, allowing companies to take on more risk themselves while still having coverage. It can also make the company's primary coverage more appealing to commercial insurers. Captives have been shown to work effectively for retention layers ranging from $250,000 to $200 million, according to recent industry data.

This approach gives companies more control over their insurance programs and can offer tax benefits. Plus, it can be customized to fit a company's specific cyber risk profile. Captives are especially helpful for organizations dealing with hard-to-insure or emerging cyber risks.

Captives are a common solution for health insurance. If your organization already has a captive established for health coverage, it may be feasible to extend its capabilities to include cyber risk. This can streamline operations and potentially reduce costs associated with managing separate insurance programs. It's also important to carefully assess the feasibility and potential implications of adding cyber risk to an existing captive structure.

The process of setting up a captive insurance company starts with the parent company establishing the captive, typically in a location that offers favorable regulatory and tax conditions. Once the captive is created, the parent company contributes funds to build its reserves. From there, the captive underwrites insurance policies for the parent, which can include cyber insurance. To further protect itself from large losses, the captive may also purchase reinsurance. If a cyber incident occurs, the captive manages the entire claims process, from investigation to payout.

Setting up a captive requires careful planning and involves several steps. It begins with a feasibility study to determine whether establishing a captive is a smart financial decision. Next, the company selects a jurisdiction with favorable conditions for captives and secures the necessary regulatory approvals. After that, the captive is funded to meet regulatory requirements and cover potential losses. The final step involves designing insurance policies that meet the parent company's specific needs and setting up a management team and board of directors to oversee the captive's operations.

A captive operates as a separate legal entity from the parent company, with its own management and board. One of the key benefits is the ability to tailor insurance coverage to the parent company's unique risks. The captive also has direct access to the reinsurance market, which provides extra financial protection. Additionally, in certain jurisdictions, captives may offer tax benefits.

Companies choose to set up captives for a variety of reasons. It can lead to cost savings by cutting out the middleman (commercial insurers) and allowing the parent company to keep underwriting profits and investment income. Captives also provide more flexibility, offering customized coverage for risks that might be hard to insure or expensive in the commercial market. With a captive, the parent company has greater control over its insurance program, including claims handling and risk management. Plus, access to the reinsurance market offers further financial protection. Lastly, captives may offer tax advantages depending on the chosen jurisdiction.

8.4.2.1 Example

One of my tech company clients was already utilizing a captive insurance structure for their employee health benefits. As their cyber risk exposure grew, we explored the feasibility of extending their existing captive to include cyber coverage.

After careful analysis, we determined that incorporating cyber risk into their captive was a viable option. This allowed them to leverage their existing infrastructure and expertise, streamlining operations and potentially reducing costs associated with managing separate insurance programs.

By adding cyber risk to their captive, they gained more control over their cyber insurance program, tailoring coverage to their specific needs and risk profile. This proved particularly beneficial as they faced emerging cyber threats unique to the tech industry that were not adequately addressed by traditional insurance policies.

To ensure full coverage and manage potential catastrophic losses, the client worked closely with their insurance broker to structure a hybrid solution. They retained a portion of the cyber risk within their captive while also partnering with a commercial insurer to provide excess coverage for losses exceeding their self-insured retention. This collaborative approach allowed them to optimize their risk financing strategy, balancing the benefits of self-insurance with the financial protection of traditional insurance.

8.4.3 Funded Reserves

Funded reserves, in cyber risk management, refer to the establishment of a dedicated financial pool specifically earmarked to cover potential losses stemming from cyber incidents. Funded reserves act as a pre-allocated financial buffer to address a broader range of cyber risks. Funded reserves act as a financial safety net, helping organizations absorb the financial shock of a cyber incident and avoid disruptions to their operations. The size of this reserve is meticulously determined through a combination of the organization's risk assessment and financial modeling, reflecting its specific risk appetite and financial capacity.

Funded reserves often work hand-in-hand with SIR self-insurance, where the funded reserve holds the money to cover all the losses up to a certain limit. You can also set up a funded reserve in conjunction with commercial insurance or in lieu of any insurance at all. Funded reserves by themselves can offer more flexibility with partial self-insurance, as they can be used to cover a broad range of cyber risks, including both first-party losses (e.g., data recovery and business interruption) and third-party losses (e.g., legal fees and regulatory fines). It's important to ensure that the funded reserve is adequately sized to cover potential losses based on the organization's risk profile. It's important to regularly review and adjust the reserve to account for shifts in the threat landscape, changes to the organization's risk profile, and its financial position. You'll also want to consider investment options for the reserve funds that provide a reasonable return while keeping enough liquidity for quick access in case of a cyber incident. Establishing and maintaining a funded reserve encourages organizations to perform regular risk assessments and financial modeling, leading to a more proactive and strategic approach to managing cyber risks.

8.4.3.1 Example

I once worked with a large retailer who recognized that they were exposed to cyber risks such as data breaches and ransomware attacks and decided to establish a funded reserve of $3 million. This reserve was held separately from its operating capital and specifically earmarked to cover potential cyber losses. In the event of a data breach, the company could then draw from this reserve to cover the costs of incident response, customer notification, credit monitoring services, and any potential legal fees, ensuring minimal disruption to its core business operations and financial stability. To accelerate the funding of this reserve, they strategically leveraged any excess annual budget from the IT and security departments, recognizing that the potential cost of a cyber incident far outweighed the temporary impact of these budget adjustments.

I recently heard about a software development company, CISO, that managed to get security development services calculated as part of the cost of goods sold. So, each time a new product entered development, they would receive an incremental budget increase to cover the security development for the new product. The CISO decided to use those incremental increases to fund a reserve and eventually had a fairly sizable rainy day fund without having to draw from the primary security budget.

8.4.4 Cyber Risk Pools

The escalating cost of cyber insurance and the challenges in obtaining adequate coverage have prompted many local governments to explore alternative solutions. One such alternative gaining traction is the formation of cyber risk pools. These consortiums of local governments offer a form of self-insurance within a group, typically administered by a third-party manager. The pooling of resources allows for more affordable coverage compared to traditional private market insurance, making it a viable option for budget-constrained municipalities.

Beyond cost savings, cyber risk pools offer additional benefits. They create a collaborative environment where similar entities can engage in meaningful discussions about cybersecurity risks and collectively develop best practices. This mutual support system fosters a sense of shared responsibility and encourages continuous improvement in cybersecurity measures.

The combination of participation in risk pools and adherence to stringent cybersecurity risk mitigation strategies can lead to a reduction in overall costs. Moreover, it can enhance the credit quality of public sector entities by demonstrating a proactive and responsible approach to managing cyber risk.

8.4.4.1 Example: League of Minnesota Cities Insurance Trust

The League of Minnesota Cities Insurance Trust (LMCIT) exemplifies the benefits of a cyber risk pool. It offers standalone first-party cyber insurance to its members, covering data security breaches, response costs, expenses incurred from attacks, data restoration, and hardware replacement. The LMCIT further incentivizes good security practices by offering higher aggregate limits to members who meet specific requirements, such as employee training, computer use policies, and regular data backups. The LMCIT also addresses other cyber risks through separate coverage plans, such as municipal liability insurance for claims from external parties and property insurance for wire transfer fraud. The trust actively supports its members with loss control initiatives, including incident response planning, fraud prevention training, and cyberattack prevention strategies. It also provides expert consultation on cyber policy creation, staff training, and technology procurement. The LMCIT, established by the Minnesota State Legislature but operating as an independent nonprofit for 50 years, serves over 800 member municipalities and is governed by a board of directors comprising local officials.

In some cases, risk pools may incentivize members to strengthen their security controls by offering lower deductibles to those with robust cybersecurity measures. This creates a positive feedback loop, encouraging continuous investment in cybersecurity and fostering a culture of proactive risk management.

However, it's important to acknowledge that cyber risk pools are not a one-size-fits-all solution. They may not be available in all jurisdictions, and some municipalities might face challenges meeting the minimum security control requirements for membership. Nevertheless, for those who can participate, cyber risk pools represent a promising avenue for addressing the growing challenges of cyber insurance affordability and accessibility.

While the majority of current risk pools cater to local governments, the concept of collective self-insurance is not limited to the public sector. We're also witnessing the emergence of similar models within the private sector, particularly among cooperatives (co-ops). Even though the LMCIT was established by the Minnesota State Legislature, it demonstrates that any group or consortium, public or private, can theoretically create a risk pool and collaborate to self-insure. This highlights the adaptability and potential of the risk pool model to address the rising costs and complexities of cyber insurance across various sectors.

8.4.5 Bonds as a Cyber Risk Mitigation Tool

Bonds offer another way to finance risk, but they aren't the right fit for every organization. At their core, bonds are a form of debt where an investor lends money to a company or government, and in return, the borrower agrees to pay back the principal on a set date, along with regular interest payments. Essentially, it's a contract where the borrower promises to repay the loan with interest over time.

Corporate bonds are essentially loans that companies obtain from investors rather than from banks. For modern organizations, issuing corporate bonds could be an innovative strategy for financing cyber risk. By issuing bonds specifically designated for cyber risks, companies could create a dedicated pool of funds that is reserved exclusively for responding to cyber incidents. This approach ensures that financial resources would be available when a breach or cyber event occurred, eliminating the need to hastily pull funds from general operating budgets. Having predictable and earmarked resources for incident response could be a great way to accept more risk.

Corporate bonds also offer *predictable costs*, with fixed interest rates and maturity dates, making them a stable financing tool for planning cybersecurity expenditures. For *CISO 3.0* professionals, whose roles are expanding to include business strategy and financial management, this predictability allows for more strategic budgeting, which is particularly important in managing both short-term threats and long-term cybersecurity resilience.

Issuing bonds could also help diversify a company's funding sources, reducing reliance on traditional methods like insurance or internal reserves. Depending on the organization's risk profile and market conditions, bonds can even prove to be a more cost-effective method for covering cyber risk compared to the high premiums associated with cybersecurity insurance. In certain situations, corporate bonds might offer a way to manage the *limitations of insurance coverage*, especially in the face of increasingly complex cyber events that insurance policies may not fully cover.

The process of issuing corporate bonds for cybersecurity purposes involves several steps. First, the company evaluates its financial needs and creditworthiness to structure the bond offering. An investment bank typically underwrites the bond issue, helping determine key terms such as the interest rate and maturity date while assisting in marketing the bonds to potential investors. Following this, credit rating agencies like Moody's or S&P assess the company's creditworthiness, influencing the bond's interest rate. Once issued, the bonds are sold to investors and can be traded in secondary markets.

Corporate bonds can also be tailored as *self-insurance mechanisms*. For example, *contingent capital bonds* can be structured so that payment is triggered when a specific cyber event occurs, such as a data breach that exceeds a particular financial threshold. Meanwhile, *catastrophe bonds (Cat Bonds)*, which we will discuss more later, are typically used for extreme events like natural disasters and can be adapted for large-scale cyber risks, providing financial coverage for catastrophic breaches.

Utilizing corporate bonds for cyber risk management could not only provide a dedicated financial tool for security initiatives but also signal to investors that the organization is taking a proactive and strategic approach to cybersecurity. This could bolster *investor confidence*, as the successful issuance of such bonds demonstrates that the company views cybersecurity as a priority and is actively managing its exposure to risks.

In considering this approach, companies should remain aware of the challenges, such as ensuring *investor appetite* for cyber risk bonds and maintaining accurate *risk modeling* to determine bond terms. Additionally, companies need to navigate *legal and regulatory frameworks* governing bond issuance to ensure compliance and transparency.

8.4.6 Surety Bonds

When an organization is said to be "bonded", it typically means they have obtained a surety bond. A surety bond is a three-party contract involving the principal (the organization seeking the bond), the obligee (the entity requiring the bond), and the surety (the insurance company issuing the bond). The bond essentially guarantees that the principal will fulfill its obligations to the obligee. If the principal fails to do so, the surety will compensate the obligee for any financial losses incurred up to the bond amount.

Organizations might need to be bonded for a variety of reasons, such as the following:

- Licensing and Permits: Many industries require businesses to be bonded to obtain licenses or permits to operate. This ensures that they have the financial backing to compensate clients or the public if they fail to meet their obligations.
- Contractual Agreements: In some cases, businesses might need to be bonded as part of a contractual agreement. This provides assurance to the other party that the business will fulfill its contractual obligations.
- Employee Dishonesty: Organizations handling large sums of money or sensitive information might obtain fidelity bonds to protect against employee theft or fraud.

While surety bonds are not specifically designed to cover cyber risks, they can indirectly contribute to an organization's overall cyber risk management strategy. Being bonded can demonstrate to clients, partners, and stakeholders that the organization is financially responsible and committed to meeting its obligations, enhancing trustworthiness, which can be crucial after a cyber incident. If a cyber incident results in financial losses for clients or partners, a surety bond could provide some level of compensation, helping maintain business relationships and mitigate brand damage. Some surety bonds might even require the principal to implement certain security measures or internal controls to qualify for the bond. This can indirectly encourage organizations to adopt better cybersecurity practices, thereby reducing their overall cyber risk exposure.

Most surety bonds do not explicitly cover losses directly caused by cyberattacks, primarily focusing on financial losses due to the principal's failure to fulfill contractual or regulatory obligations. The impact of surety bonds on cyber risk is often indirect and may not provide adequate financial protection in the event of a major cyber incident. Therefore, organizations should not rely on surety bonds as a replacement for dedicated cyber insurance coverage, which is specifically designed to address the unique risks associated with cyberattacks.

8.4.6.1 Example

I worked with a large financial institution a few years ago. We were reviewing their insurance policy to determine what was covered, what was excluded, and what had lower sublimits. Our goal was to ensure they had adequate coverage for the risks they were most concerned about. When we identified a gap in their fraud coverage, the CISO and CFO mentioned they had a surety bond, believing it would address this shortfall. However, after consulting with the legal team, it became clear that their surety bond likely wouldn't cover cyber fraud and social engineering perpetrated by external threat actors. As a result, they needed to renegotiate their insurance policy for a higher sublimit for those specific types of harm.

8.4.7 Cyber Catastrophe Bonds

Companies can also issue specialized bonds, often referred to as "cyber catastrophe bonds" or "cyber cat bonds" for short, which are specifically designed to cover potential financial losses from cyberattacks or data breaches. These are known as insurance-linked securities (ILS) and are issued by insurance companies and underwritten by reinsurance companies. Insurers and reinsurers usually create cat bonds through a special purpose vehicle, which is a company specifically set up to issue these bonds. They are one of the ways cyber insurance companies protect themselves from major losses to widespread events.

The process typically works as follows: The company issues bonds to investors, outlining the specific cyber risks covered, the terms of the bond (maturity date and interest rate), and the trigger events that would lead to bondholders absorbing the financial losses. Investors then purchase the bonds, essentially providing the company with upfront capital to cover potential cyber losses. The company makes regular interest payments to the bondholders. Suppose a predefined cyber incident occurs and meets the specified trigger conditions. In that case, the bondholders lose a portion or all of their principal investment, which is then used to cover the company's losses. If no trigger event occurs before the bond's maturity date, the company repays the principal amount to the bondholders.

Cyber bonds offer several advantages. They provide diversification of risk financing, offering an alternative to traditional insurance and self-insurance, allowing companies to tap into the capital markets for risk financing. They can also help companies with strong credit ratings save money, as these businesses may be able to issue bonds at lower interest rates compared to the cost of traditional insurance premiums. They also facilitate the transfer of risk, shifting the financial risk of a cyber incident to bondholders. Finally, they offer flexibility and can be customized to cover specific cyber risks and trigger events.

However, cyber bonds also have some disadvantages. They can be complex to structure and issue, requiring specialized expertise. They are also subject to market risk, as the value of cyber bonds can fluctuate depending on market conditions and investor sentiment. Lastly, they may have limited availability and may not be readily available or affordable for all companies, particularly those with lower credit ratings.

Since cyber cat bonds are insurance-linked securities, you would need an insurance or reinsurance company to issue them. It could be possible for a company with a captive to issue cyber cat bonds, but I haven't seen anybody do that yet. It is, however, common for corporations to sponsor corporate cat bonds via captives or reinsurers that cover natural disasters and the like. So, there is no reason an organization couldn't sponsor a cyber cat bond.

8.5 RISK AVOIDANCE IN CYBERSECURITY

Now that we've explored risk acceptance strategies, let's take a look at risk avoidance. Avoidance in cyber risk means eliminating a particular cyber risk altogether. This involves proactive measures taken by the CISO to prevent potential cyberattacks or data breaches.

Here are specific examples of how a CISO might help an organization avoid cyber risk:

Discontinuing Legacy Systems: Outdated systems are often more vulnerable to cyberattacks due to lack of security patches and support. A CISO can advocate for phasing out or replacing legacy systems with modern, secure alternatives.

Limiting Exposure to High-Risk Regions: If operating in regions with heightened cyber threats or lax cybersecurity regulations, a CISO might recommend limiting the organization's presence or activities in those areas to minimize exposure.

Strict Data Access Controls: By implementing strict access controls and data segmentation, a CISO can limit the potential impact of a breach by ensuring that sensitive data is only accessible to authorized personnel.

Vendor Risk Management: Third-party vendors can introduce vulnerabilities into an organization's network. The CISO can establish rigorous vendor risk management processes to assess and mitigate the risks associated with third-party relationships.

Employee Training and Awareness: A well-informed workforce is the first line of defense against cyberattacks. The CISO can create and roll out cybersecurity training initiatives to equip employees with knowledge on recognizing threats and following security best practices.

While risk avoidance is a desirable goal, it's not always the most feasible or practical option. Eliminating certain activities or avoiding specific markets may limit business opportunities or hinder innovation. For instance, discontinuing e-commerce operations could protect against cyberattacks targeting online transactions but would also eliminate a sizable revenue stream.

Completely avoiding cyber risk is often impossible in today's interconnected digital world. Organizations need to engage with the digital world to remain competitive, and this inherently involves a certain level of risk exposure. Therefore, risk avoidance should be viewed as a tool in the cyber risk management toolkit rather than as a standalone solution.

The CISO should be the driving force behind balancing the need for risk avoidance with the organization's business objectives. Careful evaluation of the potential impact of different risks and the feasibility of avoidance measures can guide the organization toward a risk management strategy that aligns with its overall goals and risk appetite.

8.6 DETERMINING THE RIGHT MIX OF RISK TREATMENT METHODS

Creating a risk treatment plan (RTP) is the fourth step of the risk strategy framework and is the key to determining risk buy. In order to create an RTP as described in Chapter 6, we have to determine which treatment options will best address the risk. Remember that our options include accept, avoid, transfer, mitigate by applying an omitted control, mitigate with a compensating control, mitigate by creating a new control, or some combination of these. Choosing the right mix of risk treatment methods can be tricky and demands a nuanced understanding of your organization's unique risk profile, financial constraints, and strategic objectives. It's a careful balance between the potential impact of the risk and the cost and effectiveness of available mitigation or transfer options. As my friend Dave Cardwell likes to say, "Don't buy a $100 fence for a $5 dog". While that is an important point, any dog owner knows that protecting your dog isn't a purely financial decision, and your emotional attachment to your dog may drastically outweigh the cost of the fence. The same is true for cyber risks: You have to consider both the quantitative and qualitative aspects of the risk. This makes for a complex decision that requires a complete understanding of the risk and its impact on the business.

Begin by understanding your organization's risk appetite. What level of risk is the organization willing to accept? This foundational understanding will help determine whether risk acceptance is a viable option and, if so, to what extent. Engaging in open discussions with key stakeholders, including the board and executive leadership, is essential to define the organization's risk appetite clearly.

Next, it is imperative to assess your financial capacity. How much can your organization afford to lose in the event of a cyber incident? This involves analyzing the potential financial impact of various cyber risks, considering both direct costs and indirect costs. This analysis helps evaluate the feasibility of self-insurance and determine appropriate coverage limits for insurance policies or bonds.

A thorough understanding of your organization's risk profile is also necessary. Identify the greatest cyber risks facing your organization. This involves conducting risk assessments, identifying vulnerabilities, and evaluating the likelihood and impact of potential cyberattacks.

Compliance with regulatory requirements is a nonnegotiable aspect of risk management. Are there any industry-specific or regional regulations that mandate specific risk management practices? Ensuring your chosen strategies comply with all applicable laws and regulations is imperative.

After all that, detailed CBA will allow you to compare the different risk treatment options. This involves weighing the financial costs of each strategy against the potential financial losses it aims to prevent. Consider both quantitative factors and qualitative factors. The goal is to identify the most cost-effective and efficient approach for your organization, striking the right balance between risk mitigation and financial prudence.

Selecting the optimal combination of risk treatment methods is crucial for managing cybersecurity risks effectively. The decision requires thoughtful collaboration between key stakeholders, including the CISO, CFO, chief risk officer (CRO), and general counsel, to balance the organization's risk appetite with its operational and financial goals.

A productive conversation with the business about risk treatment should focus on balancing risk mitigation, risk transfer, and risk acceptance. This involves calculating the cost of potential losses against the investment required to mitigate or transfer them. Regarding risk mitigation, the focus should be on determining the necessary controls to reduce risk to an acceptable level. For risk transfer, the conversation should center on identifying risks that can be shifted to third parties through insurance or outsourcing. Finally, regarding risk acceptance, the discussion should define which risks the organization is willing to bear and options for financing those accepted losses.

Examples

- **High-Risk, High-Impact Scenario:** For a critical system handling sensitive customer data, the optimal risk treatment approach might involve a combination of risk mitigation (through robust security controls and employee training) and risk transfer (through a cyber insurance policy with adequate coverage limits).
- **Low-Risk, Low-Impact Scenario:** For a noncritical system with limited exposure to sensitive data, risk acceptance or a lower level of insurance coverage might be more appropriate.
- **Emerging Risks:** In cases where emerging threats pose uncertain financial impact, a combination of proactive risk mitigation (such as investing in new security technologies) and risk transfer (through specialized insurance coverage or captive insurance) might be the most effective approach.

Remember, cyber risk management is an ongoing process. As your organization grows and the threat landscape changes, regularly reassess your risk treatment strategy and make adjustments as needed.

8.7 CONCLUSION

In this chapter, we got into the nuts and bolts of how to use self-insurance and other creative strategies to pay for cyber events. We kicked things off by talking about the difference between accepting risk (knowing it's there and choosing to live with it) and avoiding it altogether (doing whatever it takes to eliminate it). We stressed how important it is to understand your company's risk appetite and financial situation before making these calls.

We then dug deeper into self-insurance, where a company basically becomes its own insurance company. We looked at different ways to do this, from taking on all the risk yourself to setting up a special fund to cover potential losses. We also talked about the pros and cons of self-insurance and how it can save you money, but it also means you're on the hook if something major happens.

Next, we explored some alternative ways to finance risk, like captive insurance companies and bonds. These can be great options for companies looking to spread their risk or tackle those hard-to-insure cyber threats.

We also spent some time on CBA, looking at both the direct and indirect costs of a cyber incident. This kind of analysis is needed to figure out where to invest your security budget and how to protect your company best. And we can't forget about the importance of the CISO and CFO working together on this: It's a partnership that can make or break your cyber risk financing strategy.

Finally, we talked about risk avoidance. While it sounds great in theory, it's not always practical in the real world. Sometimes, you have to take calculated risks to achieve your business goals. It's all about finding the right balance.

So, what's the takeaway? Cyber risk financing isn't a universal solution. Every company is unique, and the best approach depends on your specific risks, budget, and goals. Self-insurance and other alternative strategies can be powerful tools, but they're not without their challenges. It takes careful planning and execution to make them work.

The bottom line is this: The CISO and CFO need to be in sync when it comes to managing cyber risk. By understanding the different financing options and making data-driven decisions, you can protect your company's financial health while also supporting its growth and innovation. It's about being a strategic leader and ensuring that every cybersecurity investment delivers real value.

8.7.1 Key Takeaways

1. **Risk Financing and Treatment Options**
    - Effective cyber risk management requires a nuanced understanding of risk financing, balancing *acceptance*, *avoidance*, *transfer*, and *mitigation*.
    - *Risk acceptance* involves preparing for potential losses and aligning them with organizational risk appetite, financial capacity, and strategic goals.
    - *Risk avoidance* focuses on eliminating potential cyber threats entirely, although it is often limited by practical and business constraints.
2. **Self -Insurance Options**
    - **Self-Insurance:** Involves setting aside funds to cover potential cyber losses. It can be a cost-effective strategy for organizations with strong financials and mature risk management programs but requires careful planning and risk assessment.
    - **Pure Self-Insurance:** The organization assumes full financial responsibility for all cyber losses up to a predetermined limit (SIR). It offers greater control but also carries higher financial risk.
    - **Captive Insurance:** A subsidiary owned by the parent organization to provide self-insurance or alternative risk financing options. It offers flexibility, customization, and potential tax benefits.
    - **Capital Reallocation:** A dynamic process where the CISO and CFO continuously adjust security investments to address evolving threats and optimize resource allocation. It ensures that security spending aligns with the organization's financial goals and risk appetite.

- **Funded Reserves:** A dedicated financial pool specifically allocated to cover potential cyber losses. It acts as a financial buffer to absorb the impact of cyber incidents.

3. **Alternative Risk Financing Strategies**
   - **Risk pools:** Groups of entities share resources to collectively manage cyber risks, promoting cost savings, collaboration, and improved risk management.
   - **Bonds:** Cyber catastrophe bonds and other instruments enable organizations to transfer specific risks to investors while diversifying financing strategies.
4. **Cost–Benefit Analysis**
   - CBAs enable organizations to quantify the financial implications of cyber risks and evaluate the ROI for mitigation strategies.
   - Collaboration between CISOs and CFOs is essential for prioritizing resource allocation and ensuring that cybersecurity efforts align with financial goals.
5. **Dynamic Capital Reallocation**
   - Continuous adjustments in cybersecurity investments ensure optimal resource use while addressing evolving threats.
   - CISOs and CFOs collaborate to align security budgets with business objectives, balancing the trade-offs between short-term costs and long-term resilience.
6. **The Role of CISOs in Risk Financing**
   - Modern CISOs must bridge the gap between technical expertise and financial strategy, advocating for informed decision-making.
   - Their role extends to designing, implementing, and continuously improving risk treatment plans that protect organizational assets and align with strategic objectives.

8.7.2 Reflection and Exploration Questions

Understanding the Concepts

1. What are the main differences between risk acceptance and avoidance?
2. How do funded reserves and captives contribute to effective risk financing?

Connecting the Dots

3. How do the principles of CBA in this chapter connect to previous discussions on quantitative risk assessment?
4. In what ways does capital reallocation support broader organizational resilience?

Real-World Application

5. Can you identify examples where self-insurance strategies effectively mitigated cyber risks in your industry?

6. How might a CISO collaborate with a CFO to tailor risk financing options to an organization's unique needs?

Personal and Professional Growth

7. How can you apply the principles of dynamic capital reallocation to improve cybersecurity investments in your organization?
8. What steps can you take to better understand and utilize bonds or risk pools for cyber risk management?

Further Exploration

9. What additional skills or resources could enhance your ability to design and implement a comprehensive risk financing strategy?
10. How might changes in the regulatory landscape influence the adoption of self-insurance or alternative financing strategies in the future?

Part 3B

Risk Mitigation

9 Developing a 3.0 Program Strategy

9.1 INTRODUCTION

We've covered a lot of ground so far, exploring the many facets of the CISO 3.0 role. We've talked about the importance of understanding the business side of things, managing risk like a pro, and communicating effectively with everyone from the IT team to the boardroom. Now, it's time to put all those pieces together and create a cybersecurity strategy that truly makes a difference.

This is where the real work begins. It's about taking your deep knowledge of cyber threats and business risks and turning it into a concrete plan. We're not talking about a dusty binder full of compliance checklists; we're talking about a dynamic, data-driven strategy that aligns with your company's goals and delivers real value.

In this chapter, we'll show you how to do just that. We'll look back into the world of risk quantification, where you'll learn to measure and prioritize cyber risks. We'll also guide you through assessing your current security program so you can identify its strengths and weaknesses and pinpoint areas for improvement.

But it's not only about identifying problems but also about solving them. We'll help you build a roadmap that prioritizes the most impactful security initiatives, ensuring that every dollar you spend delivers maximum risk reduction. And finally, we'll equip you with the communication skills you need to win over key stakeholders and secure the budget and buy-in your program deserves.

By the end of this chapter, you'll have the tools and knowledge to transform your cybersecurity program from a reactive, compliance-driven operation into a proactive, strategic powerhouse. You'll being to translate your security initiatives into the language of the boardroom, demonstrating the value of your work, and positioning yourself as a trusted advisor who's essential to the company's success. That's the essence of a CISO 3.0, a cybersecurity leader who understands not only the technology but also the business and who can bridge the gap between the two to create a truly secure and resilient organization.

9.2 THE ROLE OF RISK MITIGATION IN THE CYBER RISK INVESTMENT MIX

Risk mitigation represents the third pillar of the cyber risk investment mix, complementing the activities of risk acceptance and risk transfer. Once organizations determine how much risk remains after these activities, they can focus on mitigation strategies to bring their residual risk in line with their defined risk tolerance and appetite.

DOI: 10.1201/9781003510789-14

Understanding the target level of residual risk is a critical first step. By quantifying inherent risk (the baseline level without controls) and subtracting the risk reduced through acceptance and transfer, organizations can identify their mitigation target. This clarity enables the creation of a strategic roadmap to develop and implement the necessary controls.

For example, an organization with an inherent risk of $100 million may reduce $50 million through acceptance and transfer activities, leaving a residual risk of $50 million. If their risk appetite is $20 million, the risk exceedance is $30 million. This quantification enables organizations to clearly define their mitigation target, guiding the development of a strategy to close the gap.

With a defined mitigation target, organizations can create a strategic roadmap to develop the necessary controls. This roadmap prioritizes initiatives based on their potential to reduce risk efficiently and effectively. By leveraging risk quantification tools and methodologies, CISOs can assign financial values to risks and model the impact of proposed controls. This ensures that mitigation efforts focus on areas with the highest impact and align with the organization's strategic objectives.

For example, an organization may identify gaps in multifactor authentication (MFA) or endpoint detection and response (EDR). Risk quantification can determine the financial impact of these gaps, revealing that implementing MFA could reduce unauthorized access risks by $10 million annually while providing a high ROI. Similarly, upgrading EDR capabilities might address a broader range of threats but at a higher cost, requiring careful prioritization. By evaluating these initiatives through a quantified lens, organizations can prioritize actions that deliver the most significant risk reduction relative to their investment.

A well-crafted mitigation strategy leverages risk quantification data to align initiatives with business priorities. For example, focusing on high-impact areas, addressing the most probable threats, and aligning controls with strategic objectives ensure that every dollar spent delivers maximum value. Quantified roadmaps allow CISOs to prioritize initiatives, justify investments to stakeholders, and communicate the ROI of cybersecurity efforts in business terms. Risk quantification forms the foundation for data-driven decision-making by providing concrete financial evidence of the value of mitigation activities. This allows organizations to justify investments by demonstrating their potential to reduce risk in measurable terms, aligning with the expectations of executives and board members. Additionally, it helps prioritize actions by focusing resources on initiatives that offer the highest ROI and align with broader business goals.

Effective communication is another critical benefit of a quantified strategy. It translates technical cybersecurity efforts into business terms, showing their direct impact on financial performance and organizational resilience. For instance, if a security gap in privileged access management is estimated to expose the organization to $5 million annually in potential losses, investing $1 million in a solution that reduces the risk by $4 million demonstrates a clear financial advantage. This type of analysis provides clarity and confidence, empowering CISOs to advocate for the resources needed to address critical vulnerabilities while aligning their initiatives with the organization's strategic priorities.

Risk mitigation is not solely about implementing technical controls: It also involves aligning cybersecurity efforts with business strategy. This means understanding the organization's risk appetite, industry-specific threats, and operational priorities. By integrating risk mitigation activities into the broader business context, CISOs can ensure that security investments support key initiatives, protect critical assets, and enable growth.

For example, a company launching a new product might prioritize securing intellectual property and ensuring the integrity of its supply chain. Similarly, a financial institution might focus on safeguarding customer data and meeting regulatory compliance requirements. Aligning mitigation strategies with these objectives ensures that cybersecurity becomes a business enabler rather than a reactive expense.

By embedding these principles into a strategic mitigation roadmap, organizations can ensure that their cybersecurity programs are not only effective but also aligned with overarching business objectives, enabling informed decision-making and fostering a culture of proactive risk management.

In this chapter, we'll explore methodologies for using risk quantification to create strategic roadmaps that guide the implementation of risk-reducing controls. These strategies not only help prioritize actions but also foster meaningful conversations with executive leadership about cybersecurity as a business enabler, ensuring alignment with organizational goals.

9.3 CONTINUOUS IMPROVEMENT

Continuous improvement lies at the core of what it means to be a CISO 3.0. The traditional model of point-in-time assessments, where everything appears in order during an audit but deteriorates shortly after, is outdated and dangerous. To truly safeguard an organization, the CISO 3.0 must champion a culture of continuous monitoring and refinement, not only of threats but also of controls, posture, and the entire security program. This approach ensures the organization remains agile, resilient, and aligned with business objectives.

A cybersecurity program built on continuous improvement doesn't just react to incidents: It anticipates and evolves. By consistently analyzing and enhancing processes, technologies, and policies, CISOs can mitigate risks more effectively, foster a proactive security culture, and ensure alignment with both industry standards and organizational goals.

9.3.1 Key Pillars of Continuous Improvement in Cybersecurity

Continuous improvement is the foundation of a resilient and adaptive cybersecurity program. It ensures organizations stay ahead of threats, align security initiatives with business goals, and foster a culture of proactive risk management. Below are the key pillars of continuous improvement.

9.3.1.1 Monitoring Performance Metrics

Continuous improvement begins with the regular review and monitoring of cybersecurity performance metrics, ensuring that the program remains effective against evolving threats. Organizations should refine key performance indicators (KPIs) to provide actionable insights into operational impact and effectiveness. For instance, an e-commerce company shifted from tracking generic incident counts to metrics like "time to contain" and "time to recover". This change identified areas of underperformance, such as increased phishing attacks, and led to targeted training and enhanced email filtering, reducing phishing success rates by 30%. Proactive monitoring, combined with data-driven decision-making, creates a feedback loop that drives continuous improvement, helping organizations allocate resources effectively and adapt confidently to a changing risk landscape.

9.3.1.2 Leveraging Stakeholder Feedback

A successful cybersecurity program integrates input from employees, customers, and partners. Feedback loops refine policies, training, and practices, ensuring they remain relevant and effective. For example, a company struggling with low compliance rates in security awareness training conducted a survey and learned that employees found the sessions disengaging. By introducing interactive, scenario-based sessions, the organization increased participation and improved phishing test results. Incorporating stakeholder perspectives ensures that programs are user-friendly, adaptable, and effective across all organizational levels.

9.3.1.3 Driving Innovation through Advanced Tools

Innovation is essential for maintaining cutting-edge cybersecurity capabilities. Exploring and adopting advanced tools enhances detection, prevention, and response mechanisms to address evolving challenges. For instance, a machine-learning-based anomaly detection system allowed an organization to improve insider threat detection while reducing false positives, saving time and resources. Regularly evaluating emerging technologies ensures that cybersecurity programs remain adaptive and aligned with the latest advancements.

9.3.1.4 Real-Time Threat Intelligence Integration

The ever-changing threat landscape necessitates ongoing monitoring of emerging vulnerabilities, attack vectors, and adversarial tactics. Real-time threat intelligence allows security teams to anticipate and mitigate risks proactively. For example, a financial institution using multiple threat intelligence feeds updated its intrusion detection rules based on identified attack patterns, significantly reducing successful intrusion attempts. Continuous threat monitoring ensures defenses evolve alongside the threat landscape, minimizing exposure to potential risks. We'll explore cutting-edge technologies, including continuous threat monitoring, in Chapter 12.

9.3.1.5 Developing Professional Expertise

A cybersecurity program is only as strong as its team. Investing in continuous professional development ensures cybersecurity professionals remain informed about the latest threats, tools, and best practices. For instance, an organization facing advanced persistent threat (APT) campaigns trained its incident response team in threat hunting, enabling them to detect and neutralize APT activity before significant harm occurred. Ongoing training builds a more adaptable team capable of addressing complex challenges. We'll delve into capability development in the next chapter.

9.3.1.6 Fostering Cross-Department Collaboration

Effective cybersecurity programs thrive on collaboration across business units. Partnerships with HR, legal, operations, and other departments embed security into every organizational facet. For example, a product development team worked with the security team to integrate cybersecurity into the design process of new features, reducing vulnerabilities and improving product integrity. Cross-department collaboration fosters a unified security approach, ensuring all stakeholders contribute to a robust cybersecurity posture. This topic is discussed further in Chapter 11.

By addressing these pillars, organizations can build a continuous improvement culture that strengthens their cybersecurity programs and ensures sustained resilience in an evolving threat environment.

9.3.2 Benefits of Continuous Improvement

Continuous improvement in cybersecurity delivers a multitude of advantages that go beyond simply managing threats. It enhances the overall resilience of an organization and ensures its security program evolves in line with both technological advancements and business priorities.

First, continuous improvement strengthens business alignment. By regularly assessing and refining the cybersecurity program, CISOs can ensure that security initiatives directly support organizational goals. For example, aligning risk management strategies with key business objectives – such as securing intellectual property during a product launch – demonstrates that cybersecurity is an enabler of growth rather than a cost center.

Second, this approach fosters proactive risk management. Real-time monitoring and iterative updates enable teams to anticipate and address risks before they escalate into major incidents. For instance, an organization that continuously refines its threat detection processes can identify and mitigate ransomware attempts in their early stages, preventing operational disruptions and financial loss.

Third, continuous improvement bolsters stakeholder confidence. Transparent reporting and consistent enhancements reassure executives, employees, and customers that the organization is committed to maintaining robust defenses. By demonstrating measurable progress in security metrics, CISOs can secure buy-in for future initiatives while reinforcing trust across the organization.

Finally, it optimizes resource allocation. Continuous evaluation ensures that investments are directed toward high-impact areas, maximizing the ROI for cybersecurity expenditures. For example, a data-driven approach might reveal that upgrading endpoint detection tools yields greater risk reduction than pursuing less critical initiatives, ensuring resources are spent where they deliver the most value.

By embedding continuous improvement into the cybersecurity program strategy, the CISO 3.0 not only enhances the organization's defenses but also strengthens its alignment with business goals. This proactive, iterative approach ensures the program remains effective, agile, and poised to tackle both current and future challenges. In the following sections, we'll explore assessments and how to incorporate risk quantification into those assessments to build a more robust, data-driven foundation for your security strategy.

9.4 PROGRAM ASSESSMENTS

One of the best ways to start the continuous improvement journey is with a program assessment. We're going to begin this section with an explanation of assessments, and I know this will be a review for many of you. However, I find it's common for CISOs to confuse audits and assessments, and many don't understand the purpose of an assessment or how to leverage the results to truly understand their gaps, develop strategies and roadmaps to address those gaps, or communicate the exposures uncovered. Beyond that, it's really rare to see CISOs drive value from assessments by earning budget and buy-in. So, we are going to start with a definition of what an assessment is and isn't.

A program assessment is a thorough review of an organization's cybersecurity efforts, designed to pinpoint strengths, weaknesses, and areas that need improvement. It's designed to highlight strengths, weaknesses, and areas needing improvement. A good assessment should be part of an ongoing process that helps organizations stay ahead of threats and keep their security measures effective, not just a point in time measurement.

9.4.1 Assessment versus Audit

While both assessments and audits aim to evaluate an organization's security posture, they differ in their purpose, scope, and approach.

Assessment: According to the National Institute of Standards and Technology (NIST), an assessment is "the process of identifying risks to organizational operations (including mission, functions, image, reputation), organizational assets, individuals, other organizations, and the Nation, resulting from the operation of an information system".

Audit: NIST defines an audit as "an independent review and examination of records and activities to assess the adequacy of system controls, to ensure compliance with established policies and operational procedures".

In essence, assessments are more focused on identifying and understanding risks, while audits are primarily concerned with compliance and adherence to established policies and procedures. Assessments are typically conducted internally or by trusted third-party advisors, whereas audits are often performed by independent external auditors.

NIST Definitions

Cybersecurity Framework: "A layered structure indicating what kind of programs can or should be built and how they would interrelate".

Cybersecurity Standard: "A rule, condition, or requirement".

9.4.2 Common Security Assessments: NIST CSF and CIS-18

Two of the most widely used cybersecurity assessment frameworks are as follows:

NIST Cybersecurity Framework (CSF): This is a voluntary set of standards, guidelines, and best practices designed to help organizations manage cybersecurity risks. It's built around five key functions: Identify, protect, detect, respond, and recover. One of the reasons the NIST CSF is so popular is because it's flexible and it can be customized to fit an organization's unique needs and risk profile. It's especially helpful for companies that are new to cybersecurity or those looking to establish a solid foundation for their security program.

CIS Controls (Formerly CIS Top 20): This is a prioritized set of 18 security controls designed to protect organizations against the most common cyberattacks. These controls are based on industry best practices and are regularly updated to address emerging threats. The CIS Controls offer a more prescriptive and actionable approach, providing clear guidance on specific security measures that organizations should implement. They are well-suited for organizations that require a more structured and prioritized approach to cybersecurity, especially those with limited resources or those operating in highly regulated industries.

The goal of a security assessment is to uncover cyber risks that may have gone unnoticed, providing a clear understanding of vulnerabilities, assessing the strength of existing defenses, and offering a path to a stronger, more resilient security posture.

Each assessment is unique, customized to fit the specific needs and priorities of the organization. It typically involves a thorough review of the people, processes, and technologies that make up the organization's cybersecurity landscape. This might include reviewing policies, speaking with staff to gauge security awareness, running vulnerability scans, and even conducting penetration tests that simulate real-world attacks.

Assessments can be carried out using qualitative or quantitative approaches or a combination of both. A qualitative approach focuses on the human side, using interviews, surveys, and observations to get a sense of the organization's security culture and practices. A quantitative approach uses data analysis and metrics to provide a more measurable, objective view of the risks and vulnerabilities.

How often assessments are conducted depends on the organization's risk profile and industry requirements. Higher-risk organizations, or those going through significant changes, should be assessed more frequently. At the very least, an annual assessment is recommended to ensure that the cybersecurity program stays relevant and well-aligned.

The assessment results in a final report that serves as a strategic guide for improving the organization's security posture. This report outlines the identified risks, any gaps in existing controls, and specific recommendations for improvement. It's not just a technical document: It's designed to clearly communicate complex security issues in a way that resonates with stakeholders across the organization.

A successful assessment starts with careful planning, defining its scope, goals, and methodology to align with the organization's needs. Key stakeholders are engaged early, ensuring their input is considered, and their support is gained for any recommended changes.

During the data collection phase, information about the organization's cybersecurity environment is gathered through interviews, surveys, and technical evaluations. This step aims to create a comprehensive picture of the organization's current security posture.

The data is then analyzed to pinpoint risks, vulnerabilities, and gaps in existing controls. The focus is not only on identifying weaknesses but also on understanding how they could impact the organization's objectives and risk tolerance.

Once the analysis is complete, a tailored report is prepared, summarizing the findings and offering actionable recommendations. The report is crafted to ensure that the insights are meaningful and relevant to each stakeholder, regardless of their technical background.

In the end, a remediation plan is put into action, addressing the identified vulnerabilities and enhancing the organization's cybersecurity defenses. This plan lays out specific steps, timelines, and responsible parties to ensure timely and effective improvements.

A security assessment should be more than just a risk discovery tool, used correctly it can be a transformative process that helps guide an organization toward a stronger more business-aligned future.

9.4.3 The Assessment Process

1. **Planning:** Define the scope, objectives, and methodology of the assessment. Identify key stakeholders and secure their buy-in.
2. **Data Collection:** Gather information about the organization's cybersecurity program through interviews, questionnaires, and technical assessments.
3. **Analysis:** Analyze the collected data to identify risks, vulnerabilities, and control gaps.
4. **Reporting:** Prepare a report detailing the findings, including recommendations for improvement.
5. **Remediation:** Develop and implement a plan to address the identified weaknesses and enhance the organization's cybersecurity posture.

Assessments are like a health checkup for your security program. They can reveal hidden weaknesses and provide a roadmap for improvement. But it's not enough to just get the diagnosis: You need to take action. I've seen assessments gather dust on shelves, their valuable insights ignored. The CISO 3.0 uses assessments as a catalyst for change, driving continuous improvement and strengthening the organization's security posture.

9.4.4 Standards versus Frameworks

While NIST CSF is a framework offering guidelines and best practices, standards like ISO 27000 and CMMC (Cybersecurity Maturity Model Certification) are more prescriptive, outlining specific requirements that organizations must meet to achieve certification. These standards go beyond risk identification and assessment, focusing on the implementation and maintenance of a robust information security management system (ISMS).

ISO 27000: A series of international standards that provide a systematic approach to managing sensitive company information, ensuring its confidentiality, integrity, and availability.

CMMC: A US Department of Defense (DoD) framework that assesses and certifies the cybersecurity maturity of defense contractors and their supply chain.

Organizations might pursue certification to a standard like ISO 27000 for several reasons. First, certification offers a tangible way to demonstrate their commitment to security to customers, partners, and regulators. This can be a powerful signal of the organization's dedication to protecting sensitive information. Second, gaining a competitive advantage is another motivator. In certain industries, certification might be a prerequisite for doing business or can set an organization apart in the marketplace. Third, the certification process itself often leads to a more mature and effective risk management program. The rigorous requirements of achieving certification can help organizations identify and address weaknesses in their security posture. Finally, meeting regulatory requirements is sometimes a driving factor. In some cases, certification might be mandated by law or regulation, making it a necessity for compliance.

By conducting regular security assessments and leveraging established frameworks like NIST CSF and CIS Controls or pursuing certification to standards like ISO 27000 or CMMC when appropriate, organizations can proactively manage cyber risks, enhance their security posture, and demonstrate their commitment to cybersecurity to stakeholders, including insurers and potential business partners.

9.5 LAYERING RISK QUANTIFICATION INTO STRATEGY

Here is where the fun starts, and we move beyond simply explaining traditional assessments to look at how to incorporate quantification and next-generation insights into the outputs. As I'm sure many of you already know, traditional security assessments, while valuable, often suffer from similar shortcomings. Frameworks like NIST CSF and CIS-18 provide a structured approach to identifying vulnerabilities and control gaps, but they fall short of quantifying the financial impact of these risks. This leaves CISOs struggling to determine if their investments are truly aligned with the enterprise risk management strategy, whether they are prioritizing efforts against the costliest risks, or if they have the right balance of risk mitigation, transfer, and acceptance.

These gap assessments do little to help CISOs justify their investment choices and communicate them effectively to executives and the board. To bridge this gap, it's essential to layer risk quantification on top of security assessments. This involves assigning financial values to identified risks, allowing for a more strategic and data-driven approach to cybersecurity program development.

9.5.1 How to Leverage Risk Quantification for Strategic Planning

We explored how to perform risk quantification in Chapter 6. We are now going to look at how to restructure and reuse the information we discovered from the loss exceedance curves to build a strategic roadmap and communicate the value and ROI of our program and initiatives.

We will begin with inherent risk, which, as you'll remember, is the risk of just doing business without any controls in place. Inherent risk can be easily found via one of the productized cyber risk quantification (CRQ) platforms, like X-Analytics®, as we discussed in Chapter 6. Let's use our example company, Vandelay Industries, and imagine they have an inherent risk of $144 million per year. We can also use those tools to get a percent rating of the certainty of that loss, and we can also scale from high to low probability, so we can see what minimal and maximal loss scenarios might look like. In this scenario, we say the minimum is $50 million and the maximum is $350 million.

What we can do next is perform a risk assessment like NIST CSF or CIS-18, which will give us an idea of the controls they have in place. We can plug those assessment results right into the tools, and it will show us how much the risk is reduced by the controls discovered in the assessment. This will give us our residual risk, the amount of risk left over after we've applied our current control set. In this scenario, we'll say that Vandelay's current controls are buying down their risk by $60 million per year.

If we then take the annual budget of the security program, which we'll call $4.5 million for both tools and people, we can do our first set of calculations to show the value of our current program. If we have a $4.5 million budget and that buys down our risk by $60 million, then we can simply divide $60 million/$4.5 million = $13.33. So, for every dollar invested in Vandelay's security program, they get back $13.33 in risk reduction. If we want to figure out what that is in terms of ROI, then we just need to calculate ROI. The basic formula for ROI is

- ROI = (Gain from Investment – Cost of Investment)/Cost of Investment
- or (60,000,000 – 4,500,000)/4,500,000,
- which is 1233.33%.

That is a pretty incredible ROI and tells an amazing story about the value of Vandelay's current security program. However, the $60 million risk reduction, when subtracted from the inherent risk, still leaves a residual risk of $84 million:

- Inherent Risk – Risk Reduction = Residual Risk
- or $144,000,000 – 60,000,000 = $84,000,000

$84 million is still a significant amount of risk per year, so Vandelay has to decide what they want to do with that risk. Will they just accept all of it? Perhaps, but we looked at how to determine risk appetite in Chapter 6, and we'll assume for this

FIGURE 9.1 Risk Quantification.

example they determined that their risk appetite is $25 million, but they have a risk tolerance of up to $50 million.

If they are willing to accept $50 million in risk, that would still leave a risk exceedance because the $84 million residual risk minus the $50 million risk tolerance leaves $34 million in risk that must be dealt with:

$$\textit{Residual Risk} - \textit{Risk Tolerance} = \textit{Risk Exceedance}$$
$$\textit{or } \$84{,}000{,}000 - \$50{,}000{,}000 = \$34{,}000{,}000$$

So, what should Vandelay do with that remaining $34 million in risk? They can now have a business-level conversation about how to invest in risk reduction. They could increase their risk transfer and purchase more insurance; they could increase their risk appetite and choose to accept more risk; they could change their business practices to avoid some risk; or most commonly, they will increase the investment in controls to mitigate the risk (Figure 9.1). Regardless of what they choose to do, they are now having a business-level discussion about why and how to invest in security.

Once the investment strategy is determined, we can focus on mitigations. To do this, we will return to the risk assessment and look to see what gaps were identified. In this scenario, we'll imagine that there were 12 control deficiencies that need to be addressed. Typically, we would have to choose which of those to address and then just make educated guesses about which order to prioritize the initiatives. However, we can again apply our risk quantification to first earn buy-in and then budget for these initiatives. We can go back to our tool, and in the same way we applied the assessment to determine how much risk our controls brought down, we can model how much risk would be bought down if we were to apply certain controls (Figure 9.2).

FIGURE 9.2 Risk Buy-Down.

FIGURE 9.3 Risk Buy-Down Plan.

In this scenario, we have 12 initiatives, and we determined they would cost $3.7 million over 3 years but reduce our risk by $36 million. We can use the same calculations we used to determine the ROI of our program to determine the ROI of our proposed initiatives. In this case, the 12 initiatives would buy down $9.73 for every dollar invested, which is an 872.97% ROI. We can also use the same residual risk calculation to determine that our $84 million risk, when bought down by $36 million, would leave us with a new residual risk of $48 million, which is within the defined $50 million risk tolerance. It's not below the $25 million risk appetite, so there is still work that can be done, but if those 12 initiatives were undertaken, Vandelay would be within its risk tolerance (Figure 9.3).

This changes the dynamic when incidents occur. Instead of a cyberattack being an automatic resume-generating event for the CISO, we can now say, "Yes, we had an event; it cost us $10 million, which was well within our risk appetite, and as described in Chapters 7 and 8, we covered the bulk of that loss with cyber insurance and financed the rest with our rainy-day fund". That is a much different discussion than most CISO 2.0s are able to have following a breach.

Now that we've earned buy-in for our investment strategy, we still have to prioritize the remediations and create a roadmap.

9.6 CREATING A QUANTIFIED ROADMAP

With buy-in secured for your investment strategy, the next step is to prioritize remediations and craft a well-structured roadmap. This roadmap will serve as your strategic guide, outlining the sequence and timeline for implementing security initiatives to achieve your risk reduction goals.

9.6.1 Security Program Strategy versus Security Program Roadmap

Before we get into the specifics, let's clarify the distinction between a security program strategy and a security program roadmap:

Security Program Strategy: This outlines the overarching goals and objectives of your cybersecurity program. It defines the desired security posture for the organization and sets the direction for your efforts. The strategy should be aligned with the organization's risk appetite and business goals.

Security Program Roadmap: This is a more tactical document that details the specific initiatives and projects required to achieve the strategic goals. It outlines the timeline, resource allocation, and dependencies for each initiative.

9.6.2 Leveraging Risk Quantification for Prioritization

Risk quantification is essential for prioritizing initiatives on your cybersecurity roadmap. When you assign financial values to the risks you've identified and the potential improvements to your controls, you can focus on the areas that will be most impactful for reducing risk. This approach helps ensure that your investments align with your broader risk management strategy while also maximizing the ROI for your cybersecurity efforts.

9.6.2.1 Example

Let's revisit our Vandelay Industries example. Suppose their risk assessment identified 12 control deficiencies. Leveraging risk quantification tools, Vandelay can model the potential risk reduction associated with addressing each deficiency. This allows them to prioritize initiatives based on their impact on residual risk, ensuring they tackle the most pressing issues first.

For instance, implementing MFA might reduce the risk of unauthorized access by $10 million, while improving patch management practices might reduce the risk of vulnerabilities being exploited by $5 million. Based on this quantification, Vandelay might prioritize MFA implementation as it offers a greater risk reduction potential.

However, risk reduction isn't the only factor to consider. Suppose Vandelay's gap assessment also revealed the need for privileged identity and access management (PIM/PAM) and data access governance (DAG). Which should they implement first?

Again, we can use our CRQ tool to model the risk reduction for each control set. In this scenario, let's say PIM/PAM reduces risk by $9.3 million per year versus $8.8 million for DAG. While PIM/PAM offers slightly higher risk reduction, other factors might influence the decision.

Perhaps cost is a major concern. Data governance might cost $682,000 over 3 years, while PIM/PAM comes in at $827,000. Or maybe the focus is on maximizing ROI. Even though it's cheaper, data governance could still yield a higher ROI.

Alternatively, Vandelay might want to target specific threats. The CRQ tool can model risk reduction for particular harm categories. If ransomware is a major concern, and data governance only impacts ransomware risk by $3.1 million while PIM/PAM reduces it by $7.2 million, the latter becomes a clear priority.

Finally, let's consider business alignment. We can break down the 3-year total cost of ownership (TCO) for each initiative into capital expenditures (CAPEX) and operating expenses (OPEX). If the CFO has a near-term earnings per share (EPS) target, which is influenced by earnings before interest and taxes (EBIT), prioritizing the more CAPEX-heavy initiative can help achieve this goal. CAPEX is depreciated over time and doesn't directly impact EBIT, while OPEX is deducted immediately (Figure 9.4).

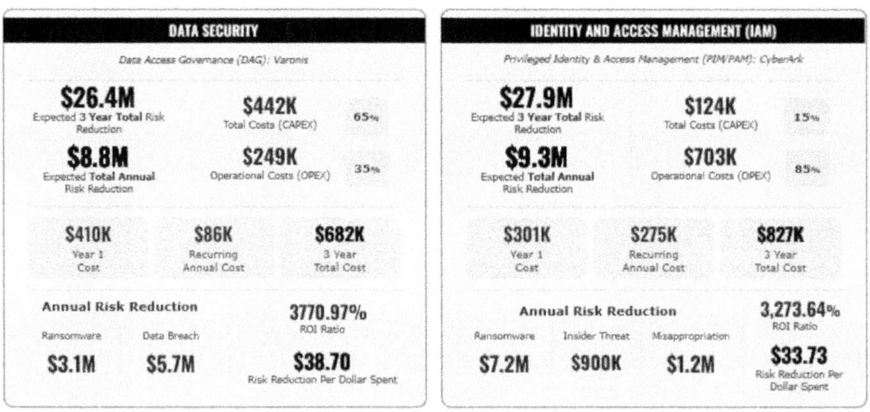

FIGURE 9.4 Controls Investment Prioritization.

By strategically choosing the initiative with a higher CAPEX component, we can align security initiatives with the business's financial drivers, showcasing the CISO 3.0's ability to contribute to the company's overall success.

9.6.3 Building the Roadmap

Once you've prioritized your initiatives, the next step is to build a clear roadmap. Start by setting realistic timelines for each project, considering any dependencies, resource availability, and potential challenges that might come up. It's essential to communicate these timelines clearly so everyone is on the same page and expectations are managed across teams.

Next, allocate the necessary budget and assign the right people to each initiative. This means carefully evaluating the financial impact of each project and ensuring you have the skilled personnel to make it happen. Be sure to define key milestones and deliverables for each initiative so you can track progress and hold teams accountable. Make sure these milestones follow the SMART criteria (specific, measurable, achievable, relevant, and time-bound) to provide a solid framework for evaluating success.

Finally, make sure the roadmap is communicated clearly to key stakeholders, including executives and the board. Being transparent about your plan helps build understanding and support for your cybersecurity efforts. Regular updates on progress and any necessary adjustments will show your commitment to staying proactive and reinforce trust with stakeholders.

9.6.3.1 Example

Let's build Vandelay Industries' roadmap incorporating the insights from our risk quantification and prioritization exercise.

Year 1

- Total cost: $2.52 million
- Total risk reduction: $27.8 million
- Overall ROI: 1003%
- Overall risk reduction per dollar spent: $11.03

1. **Implement MFA across All Critical Systems**
    - Why: High risk reduction aligns with initial prioritization
    - Risk reduction: $10 million per year
    - Cost: $500,000 over 1 year (primarily OPEX)
    - ROI: 1900%
    - Risk reduction per dollar spent: $20
    - Timeline: Q1–Q2
    - Resources: Internal IT team, potential external consultants for implementation and training
    - Milestones:
        - Q1: Complete vendor selection and solution design
        - Q2: Pilot implementation and user training
        - Q3: Full rollout and integration with existing systems
    - NIST CSF: Identify (ID.AM), protect (PR.AC)

2. **Enhance Security Awareness Training for Employees**
   - Why: Proactive measure to reduce overall risk exposure, addresses human factor vulnerabilities
   - Risk reduction: $1 million per year
   - Cost: $200,000 over 1 year (primarily OPEX)
   - ROI: 400%
   - Risk reduction per dollar spent: $5
   - Timeline: Ongoing, with quarterly refreshers
   - Resources: Internal training team, potential external vendor for content development and delivery
   - Milestones:
     - Q1: Develop and launch initial training program
     - Q2: Conduct phishing simulations and assess employee readiness
     - Q3 and Q4: Quarterly refresher training and ongoing awareness campaigns
   - NIST CSF: Protect (PR.AT)
3. **Implement PIM/PAM**
   - Why: Highest risk reduction among identified gaps, addresses ransomware risk
   - Risk reduction: $9.3 million per year
   - Cost: $827,000 over 3 years (mix of CAPEX and OPEX)
   - ROI: 1027% (annualized)
   - Risk reduction per dollar spent: $11.25 (annualized)
   - Timeline: Q3–Q4
   - Resources: Dedicated project team, potential external consultants for implementation and integration
   - Milestones:
     - Q3: Complete vendor selection and solution design
     - Q4: Pilot implementation and testing
     - Year 2 Q1: Full rollout and integration with critical systems
   - NIST CSF: Identify (ID.AM), protect (PR.AC)
4. **Upgrade EDR Capabilities**
   - Why: Enhances threat detection and response capabilities, improves the ability to identify and contain attacks
   - Risk reduction: $3 million per year
   - Cost: $400,000 over 2 years (primarily OPEX)
   - ROI: 650% (annualized)
   - Risk reduction per dollar spent: $7.50 (annualized)
   - Timeline: Q2–Q3
   - Resources: Internal security team, potential external vendor for solution deployment and support
   - Milestones:
     - Q2: Evaluate and select EDR solution
     - Q3: Deploy and configure EDR across endpoints
     - Q4: Integrate EDR with SIEM and incident response processes
   - NIST CSF: Protect (PR.PT), detect (DE.AE), respond (RS.AN)

5. **Strengthen Network Security through Segmentation and Micro-Segmentation**
    - Why: Limits lateral movement and reduces the attack surface in case of a breach, improving overall network security
    - Risk reduction: $4.5 million per year
    - Cost: $600,000 over 2 years (mix of CAPEX and OPEX)
    - ROI: 650% (annualized)
    - Risk reduction per dollar spent: $7.50 (annualized)
    - Timeline: Q4 – year 2 Q1
    - Resources: Network security team, potential external consultants for architecture design and implementation
    - Milestones:
        - Q4: Design network segmentation strategy
        - Year 2 Q1: Implement segmentation and micro-segmentation
        - Year 2 Q2: Monitor and optimize network security
    - NIST CSF: Protect (PR.AC, PR.PT)

Year 2

- Total Cost: $2 million
- Total risk reduction: $19.8 million
- Overall ROI: 890%
- Overall risk reduction per dollar spent: $9.90

6. **Implement DAG**
    - Why: Cost-effective solution with good ROI, addresses specific risk areas related to data protection and compliance
    - Risk reduction: $8.8 million per year
    - Cost: $682,000 over 3 years (mix of CAPEX and OPEX)
    - ROI: 1193% (annualized)
    - Risk reduction per dollar spent: $12.90 (annualized)
    - Timeline: Q1–Q2
    - Resources: Data governance team, potential external consultants for tool selection and implementation
    - Milestones:
        - Q1: Define data classification and access policies
        - Q2: Implement data access governance tools and processes
        - Q3: Monitor and enforce data access policies
    - NIST CSF: Identify (ID.GV), protect (PR.AC, PR.DS)
7. **Conduct a Third-Party Risk Assessment**
    - Why: Proactive measure to identify and mitigate supply chain risks, reduces potential vulnerabilities introduced by vendors
    - Risk reduction: $1.2 million per year
    - Cost: $150,000 over 1 year (primarily OPEX)
    - ROI: 700%
    - Risk reduction per dollar spent: $8.00
    - Timeline: Q2–Q3
    - Resources: Risk management team, potential external vendor for assessment, and vendor management support

- Milestones:
  - Q2: Identify and prioritize critical third-party vendors
  - Q3: Conduct risk assessments and develop mitigation plans
  - Q4: Implement ongoing vendor risk management program
- NIST CSF: Identify (ID.RA, ID.SC), protect (PR.IP)

8. **Enhance Incident Response Capabilities**
   - Why: Essential for preparedness and minimizing the impact of potential incidents, improves the ability to respond effectively to cyberattacks
   - Risk reduction: $500,000 per year
   - Cost: $300,000 over 2 years (primarily OPEX)
   - ROI: 233% (annualized)
   - Risk reduction per dollar spent: $3.33 (annualized)
   - Timeline: Ongoing
   - Resources: Incident response team, potential external consultants for training and tabletop exercises
   - Milestones:
     - Q1: Update incident response plan and playbooks
     - Q2 and Q4: Conduct tabletop exercises and simulations
     - Q3: Evaluate and enhance incident response tools and technologies
   - NIST CSF: Identify (ID.RP), protect (PR.RP), detect (DE.DP), respond (RS.RP), recover (RC.RP)
9. **Deploy Advanced Threat Intelligence and Analysis Tools**
   - Why: Improves proactive threat detection and response, enables faster identification and mitigation of emerging threats
   - Risk reduction: $1.5 million per year
   - Cost: $350,000 over 2 years (primarily OPEX)
   - ROI: 900% (annualized)
   - Risk reduction per dollar spent: $10 (annualized)
   - Timeline: Q3–Q4
   - Resources: Security operations team, potential external vendor for tool selection and integration
   - Milestones:
     - Q3: Evaluate and select threat intelligence platform
     - Q4: Integrate threat intelligence into security operations and incident response
   - NIST CSF: Identify (ID.RA), detect (DE.AE, DE.DP), respond (RS.AN)
10. **Establish a Robust Vulnerability Management Program**
    - Why: Identifies and remediates vulnerabilities before they can be exploited, reduces the attack surface, and strengthens overall security
    - Risk reduction: $1 million per year
    - Cost: $500,000 over 2 years (mix of CAPEX and OPEX)
    - ROI: 300% (annualized)
    - Risk reduction per dollar spent: $4 (annualized)
    - Timeline: Q1–Q2
    - Resources: Security team, potential external vendor for vulnerability scanning and remediation support
    - Milestones:

    - Q1: Implement vulnerability scanning and assessment tools
    - Q2: Develop and implement vulnerability remediation processes
    - Ongoing: Continuous vulnerability scanning and patching
  - NIST CSF: Identify (ID.RA, ID.VM), protect (PR.MA, PR.PT)

Year 3

- Total cost: $650,000+
- Total risk reduction: $3.2 million+
- Overall ROI: 392%+
- Overall risk reduction per dollar spent: $4.92+

11. **Implement Data Loss Prevention (DLP) Solutions**
    - Why: Addresses data protection concerns and potential regulatory compliance, safeguards sensitive data from unauthorized access or exfiltration
    - Risk reduction: $2.2 million per year
    - Cost: $450,000 over 3 years (mix of CAPEX and OPEX)
    - ROI: 1366% (annualized)
    - Risk reduction per dollar spent: $14.66 (annualized)
    - Timeline: Q2–Q3
    - Resources: Data security team, potential external consultants for solution design and implementation
    - Milestones:
      - Q2: Define data protection policies and requirements
      - Q3: Implement and configure DLP solutions
      - Q4: Monitor and fine-tune DLP policies
    - NIST CSF: Protect (PR.DS)
12. **Conduct Regular Penetration Testing and Vulnerability Assessments**
    - Why: Continuous monitoring and improvement of security posture identifies and addresses vulnerabilities before they can be exploited
    - Risk reduction: $1 million per year (ongoing benefit)
    - Cost: $100,000 per assessment (biannually)
    - ROI: 900% (annualized)
    - Risk reduction per dollar spent: $10 (annualized)
    - Timeline: Ongoing, at least biannually
    - Resources: Internal or external penetration testing team
    - Milestones:
      - Q1 and Q3: Conduct penetration tests and vulnerability assessments
      - Q2 and Q4: Remediate identified vulnerabilities and report findings
    - NIST CSF: Identify (ID.RA, ID.VA), protect (PR.MA, PR.PT)

This revised roadmap reflects a strategic approach to risk mitigation, balancing immediate needs with long-term goals. It prioritizes initiatives with the highest risk reduction potential, cost-effectiveness, and alignment with business objectives. It demonstrates the CISO 3.0's capacity to adapt and advance the security program based on continuous monitoring and assessment of the threat landscape.

These same principles can be communicated in a traditional spreadsheet like the one below Figure 9.5.

Initiative	Year	Why	Risk Reduction (Annual)	Cost (Total)	ROI (Annualized)	Risk Reduction per Dollar Spent (Annualized)	Timeline	Resources	Milestones	NIST CSF
Implement MFA	1	High risk reduction, aligns wit	$12,000,000	$600,000	1900%	$20	Q1-Q2	Internal IT team,	Q1. Vendor se	Identify (ID.AM), Protect (PR.AC)
Enhance security awareness t	1	Proactive measure, addresse	$3,600,000	$240,000	1400%	$15	Ongoing	Internal training	Q1: Launch pr	Protect (PR.AT)
Implement PIM/PAM	1	Highest risk reduction, addre	$11,160,000	$992,400 (over	1024%	$3.72	Q3-Q4	Dedicated team,	Q3. Vendor se	Identify (ID.AM), Protect (PR.AC)
Upgrade EDR	1	Enhances threat detection &	$4,800,000	$480,000 (over	900%	$5.00	Q2-Q3	Internal security	Q2: Select ED	Protect (PR.PT), Detect (DE.AE), Respond (RS.AN
Strengthen network security	1	Reduces attack surface, limit	$4,500,000	$600,000 (over	650%	$3.75	Q4 - Y2 Q1	Network security	Q4: Design str	Protect (PR.AC, PR.PT)
Implement Data Access Gove	2	Cost-effective, addresses dat	$10,560,000	$813,600 (over	1200%	$4.30	Q1-Q2	Data governance	Q1: Define pol	Identify (ID.GV), Protect (PR.AC, PR.DS)
3rd-party risk assessment	2	Mitigate supply chain risks	$3,000,000	$180,000	1567%	$16.67	Q2-Q3	Risk manageme	Q2: Identify ve	Identify (ID.RA, ID.SC), Protect (PR.IP)
Enhance incident response	2	Preparedness & minimizing i	$4,800,000	$360,000 (over	1233%	$6.67	Ongoing	Incident respons	Q1: Update IR	Identify (ID.RP), Protect (PR.RP), Detect (DE.DP),
Deploy threat intelligence tool	2	Proactive threat detection & r	$1,500,000	$350,000 (over	328%	$2.14	Q3-Q4	Security operatic	Q3: Select pla	Identify (ID.RA), Detect (DE.AE, DE.DP), Respond
Vulnerability management pro	2	Identify & remediate vulnerat	$1,000,000	$500,000 (over	100%	$2	Q1-Q2	Security team, p	Q1: Implemen	Identify (ID.RA, ID.VM), Protect (PR.MA, PR.PT)
Implement DLP	3	Data protection & compliance	$2,200,000	$450,000 (over	388.89%	$1.83	Q2-Q3	Data security tea	Q2: Define pol	Protect (PR.DS)
Penetration testing & vulnerab	3	Continuous monitoring & imp	$1,000,000	$100,000 (bi-anr	900%	$10	Ongoing	Internal/external	Q1 & Q3: Ass	Identify (ID.RA, ID.VA), Protect (PR.MA, PR.PT)

FIGURE 9.5 Spreadsheet.

9.7 EARNING BUDGET AND BUY-IN WITH THE QUANTIFIED ROADMAP

The quantified roadmap transforms the traditional budget discussion from a plea for resources to a strategic conversation about risk and ROI. This shift in perspective helps you to gain buy-in from executive leadership and secure the necessary budget for your cybersecurity program.

9.7.1 Changing the Discussion

Historically, CISOs have struggled to justify their budget requests, often resorting to fear-mongering or relying on compliance requirements. I often hear from CISOs that they can't get enough money to hire enough top talent and/or build a complete program. When I suggest that they may be miscommunicating their risk, I often get pushback that executives don't care, understand, or feel the risks are an important priority. That is because they haven't earned executive buy-in and haven't done enough to build a culture of security. If I had a dollar for every time a CISO said to me, "I'd love to do that but there's no way I can get budget for that", I'd have enough money to fund all those poorly positioned initiatives.

Earning budget and buy-in is like sales. You're selling the value of your program and the initiatives within it. The quantified roadmap provides a compelling business case for cybersecurity investments, demonstrating the financial impact of cyber risks and the potential ROI of proposed initiatives. This data-driven approach resonates with executives who are accustomed to making decisions based on financial metrics and quantifiable outcomes.

The quantified roadmap provides the CFO and other executive leaders with the key insights they need to make well-informed decisions regarding cybersecurity investments. It allows them to strategically prioritize investments by comparing the risk reduction potential and ROI of different initiatives. This ensures that resources are directed toward the areas that offer the greatest value and protection to the organization.

The roadmap acts as a bridge between cybersecurity and business goals, demonstrating how security investments directly align with and support the organization's strategic objectives and risk tolerance. This alignment is key to securing executive buy-in and fostering a shared understanding of the importance of cybersecurity in achieving business success.

The quantified data provides concrete evidence to justify budget requests, replacing fear-based arguments with tangible financial benefits. This empowers CISOs to confidently advocate for the resources needed to implement critical security initiatives and build a resilient cybersecurity posture.

Lastly, the roadmap's clear milestones and metrics provide a mechanism for tracking progress and measuring the effectiveness of cybersecurity investments over time. This transparency allows executives to monitor the impact of their decisions, make data-driven adjustments as needed, and ensure that the security program remains aligned with business priorities.

Securing budget and buy-in used to feel like an uphill battle. I'd spend hours crafting presentations, only to be met with blank stares or skepticism. But then I discovered the power of the quantified roadmap. By speaking the language of the boardroom, dollars, and cents, I was finally able to get my message across and secure the resources we needed to protect the organization.

Take a moment to reflect on your own experiences in securing budget and buy-in for cybersecurity initiatives. Have you encountered resistance from executive leadership? How could you start to leverage a data-driven approach and a focus on ROI to improve your communication and advocacy efforts?

9.7.2 Communicating the Quantified Roadmap to Executive Leadership

When presenting a quantified roadmap to executive leadership, your message needs to be clear and impactful. The key is to translate the technical details of cybersecurity into something the boardroom understands like financial implications. Help them visualize the potential consequences of cyber risks, not only in terms of data breaches but also in ways that hit home: lost revenue, operational downtime, and opportunity costs.

Next, quantify the risk reduction potential of each proposed initiative, demonstrating how these actions directly contribute to mitigating the identified threats. This provides executives with a clear and measurable understanding of the value proposition, connecting cybersecurity investments to the protection of the organization's financial health.

You can bolster your case by calculating the ROI for each initiative and for the overall cybersecurity program. Shifting the narrative from costs to returns resonates powerfully with executives who are constantly seeking to maximize value and optimize resource allocation. Show them that investing in security is not only a necessary expense but also a strategic decision that can yield considerable financial benefits.

Emphasize the alignment between your roadmap and the organization's overarching strategic objectives. Highlight how cybersecurity investments directly support key business initiatives and enable growth. This positions security not as a roadblock but as a strategic enabler, fostering a sense of shared responsibility for risk management across the entire executive team.

To ensure transparency and accountability, clearly define the milestones and metrics you'll use to track progress and measure the program's success. This not only shows your commitment to making data-driven decisions but also establishes a solid framework for continuous evaluation and improvement. Keeping executives in the loop about the program's effectiveness allows them to make informed, agile adjustments to the overall strategy as needed.

This approach transforms the conversation from a technical monolog into a strategic dialog, positioning the CISO 3.0 as a trusted advisor and a key player in safeguarding the organization's future.

9.7.2.1 Examples

- *"Investing in MFA will reduce our risk of unauthorized access by $10 million annually, resulting in a 1900% ROI"*. This statement clearly quantifies the financial benefit of implementing MFA, making it a compelling investment for executives.

- *"Our proposed roadmap will reduce our overall cyber risk exposure by $36 million over three years, bringing us well within our risk tolerance".* This demonstrates the strategic value of the roadmap and its alignment with the organization's risk management goals.
- *"By prioritizing initiatives with the highest ROI, we can maximize the value of our cybersecurity investments and ensure we're focusing on the areas that matter most to the business".* This highlights the data-driven approach to decision-making and resource allocation.

By presenting the quantified roadmap in a clear and concise manner, focusing on the business impact and financial benefits, you can effectively communicate the value of your cybersecurity program and secure the buy-in and budget needed to protect your organization.

9.7.3 Combining Qualification and Quantification for a Powerful Narrative

While risk quantification paints a compelling picture in the language of dollars and cents, it's just one brushstroke on the canvas of your cybersecurity narrative. Remember, not every stakeholder thinks in purely financial terms. Qualitative benefits, the often-overlooked "soft" side of security, can add depth and resonance to your case, swaying hearts and minds where numbers alone might fall short.

Consider this: A well-implemented single sign-on solution doesn't just mitigate the risk of credential theft: It streamlines employee workflows, boosting productivity and morale. Automating routine security tasks liberates your team from the mundane, allowing them to focus on strategic initiatives that drive innovation. These qualitative wins, while less tangible than a risk reduction figure, can be just as impactful and often even more persuasive.

By weaving together quantification and qualification, you craft a narrative that's both compelling and persuasive. You appeal to the CFO's love for ROI while also resonating with the COO's focus on operational efficiency. It's a multifaceted approach that showcases the true value of your cybersecurity program, extending far beyond mere risk mitigation.

Keep in mind that every stakeholder has their own priorities. To get their support, tailor your message to address what matters most to them. If the CIO values agility, focus on how your security measures can drive faster innovation. If the HR director is concerned with employee experience, emphasize how your initiatives will boost productivity and streamline processes. Be sure to frame the benefits in a way that resonates with their specific needs so you can shift the perception of cybersecurity from a cost center to a strategic advantage.

In the end, the most persuasive argument isn't just about numbers or narratives: Combining quantitative and qualitative benefits creates a more persuasive business case for cybersecurity investments. This method helps highlight the value of cybersecurity and demonstrates how your program protects the bottom line. That's the kind of story that resonates, inspires action, and secures the investment you need to build a truly resilient enterprise.

9.7.3.1 Example

"Implementing SSO will not only reduce our risk of unauthorized access by $2 million annually but also improve employee productivity by an estimated 10%, leading to cost savings and increased efficiency". This statement exemplifies the dual benefits of single sign on (SSO), appealing to both financial and operational concerns, thereby strengthening the case for its implementation.

By strategically combining risk quantification with qualitative benefits, you can craft a compelling and persuasive argument for your cybersecurity program. That argument will help secure the necessary budget and buy-in to protect your organization and enable success.

9.8 CONCLUSION

In this chapter, we've tackled the art of crafting a solid cybersecurity strategy, one that's fit for a CISO 3.0. We've stressed the importance of using risk quantification to make smart decisions, not just relying on gut feelings or checklists. It's about taking a hard look at your current security program, figuring out where it's strong and where it needs some work.

The goal here is to build a roadmap that's more than just a to-do list. It's about prioritizing the actions that will have the biggest impact on reducing risk and protecting the company's bottom line. And let's not forget the importance of communication. You need to be able to clearly articulate the value of your cybersecurity investments to the people who hold the purse strings.

If you can leverage these tools, you'll be well on your way to transforming your security program into a next-generation strategic business partner. It's about moving beyond the old-school, compliance-focused mindset and adopting a data-driven, risk-based approach that truly aligns with the business. When you can do that, you become more than a CISO – you become a business leader who understands the value of security and can make a real difference in the company's success.

9.8.1 Key Takeaways

1. **Building a Strategic Cybersecurity Program**
   - A CISO 3.0 strategy goes beyond compliance to focus on aligning cybersecurity with business goals through risk quantification and data-driven decision-making.
   - Effective strategies prioritize investments with the highest financial and operational impacts, enabling organizations to minimize risks while supporting growth and innovation.
2. **Role of Program Assessments**
   - Regular assessments, such as those based on NIST CSF or CIS Controls, are essential for identifying control gaps, vulnerabilities, and improvement opportunities.

- Assessments differ from audits, in that they focus on identifying risks and opportunities for strengthening security, rather than solely ensuring compliance.

3. **Quantified Roadmaps for Decision-Making**
   - A quantified roadmap integrates financial data into traditional security assessments, enabling organizations to prioritize initiatives that yield the greatest risk reduction per dollar spent.
   - By aligning security investments with organizational goals and demonstrating their ROI, these roadmaps transform budget discussions into strategic dialogs.
4. **Strategic Roadmap Components**
   - A robust roadmap includes clear timelines, budget allocations, resource assignments, milestones, and SMART (specific, measurable, achievable, relevant, time-bound) objectives.
   - Prioritization criteria should include risk reduction potential, cost-effectiveness, ROI, and alignment with organizational strategies.
5. **Combining Quantitative and Qualitative Benefits**
   - While financial data provides compelling evidence, qualitative benefits – such as improved productivity, streamlined processes, and enhanced customer trust – strengthen the overall business case for cybersecurity investments.
   - Tailored messaging ensures that both technical and nontechnical stakeholders understand the value of security initiatives.
6. **Continuous Improvement**
   - Iterative refinement of cybersecurity strategies ensures they remain aligned with evolving threats, technological advancements, and organizational priorities.
   - Performance metrics, stakeholder feedback, and emerging tools are vital for driving continuous improvement.
7. **Effective Communication and Advocacy**
   - Communicating cybersecurity priorities in business terms helps secure executive buy-in and budget approvals.
   - Quantified roadmaps bridge the gap between cybersecurity efforts and business objectives, highlighting the strategic value of security investments.

9.8.2 Reflection and Exploration Questions

Understanding the Concepts

1. How can you leverage your risk quantification skills to redefine the approach to cybersecurity program development?
2. How do program assessments contribute to identifying and addressing security gaps?

Connecting the Dots

3. How does integrating quantitative data into security roadmaps connect with earlier discussions on risk quantification?

4. In what ways do qualitative benefits complement quantitative data in building a business case for cybersecurity?

Real-World Application

5. Can you identify examples of how quantified roadmaps have transformed budget discussions or justified security investments in your organization?
6. How might the SMART framework improve the planning and execution of security initiatives?

Personal and Professional Growth

7. How can you apply the principles of quantified roadmaps and prioritization to your own cybersecurity program?
8. What steps can you take to refine your ability to communicate cybersecurity initiatives in business terms?

Further Exploration

9. What additional tools or frameworks could enhance the integration of risk quantification into your security assessments and roadmaps?
10. How might emerging threats and technologies shape the future of cybersecurity program strategies?
11. What could you do to begin quantifying your security strategy to better align cybersecurity efforts with organizational goals and secure stakeholder buy-in?

10 Security Tactics and Capabilities

10.1 INTRODUCTION

Every year, it seems like there's a new "must-have" gadget promising to solve all your security woes. It's easy for organizations to fall into the trap of accumulating all the latest security tools. While this may seem proactive, it usually leads to bloated tool stacks, redundant capabilities, and underutilized solutions, a phenomenon known as "shelfware". One of the most common mistakes I see chief information security officers (CISOs) and their teams make is shopping for and buying tools without a clear plan for operationalizing them and integrating them into a cohesive program or capability.

I worked a few summers as an auto mechanic in my youth and learned a valuable lesson about tools. It was easy to get lured in by the shiny new specialty tools on the Snap-on truck, even if they were for jobs I rarely did or had never attempted. The temptation to buy that fancy timing belt tension tool, hoping it would magically enable me to tackle timing belt replacements, was strong. But without the experience or knowledge of the procedure, that tool would likely gather dust in my toolbox while the payments continued to drain my paycheck.

This same principle applies to cybersecurity. This tendency toward "widget-ism", prioritizing tools over skills and capabilities, is a primary challenge in modern cybersecurity. It results in unused shelfware and underutilized licenses that drain annual budgets. Another issue is what I call the "dashboard ratio", or the number of dashboards and tools each engineer and analyst must manage. When this ratio exceeds roughly 3 to 1, the security program's overall capability level, efficacy, and value begin to decline.

The savvy CISO 3.0 knows that true security isn't about the number of tools you have but about the capabilities you can deploy. It's about understanding your organization's unique risks and building a security program that can actually address them.

In this chapter, we're going to ditch the "tool first" mentality and focus on building real security capabilities. We'll show you how to assess your current strengths and weaknesses, identify the gaps in your defenses, and strategically acquire solutions that actually make a difference. We'll also take a hard look at the traditional RFP process and why it often leads to expensive shelfware. Then, we'll offer some alternative approaches that prioritize value and align with your business goals.

It's time to stop chasing shiny objects and start building a security program that's as effective as it is efficient.

DOI: 10.1201/9781003510789-15

10.2 BUYING CAPABILITIES, NOT TOOLS

The CISO 3.0 embodies a paradigm shift in cybersecurity procurement, moving away from the traditional "tool-centric" approach to a "capabilities-based" strategy. They understand that security isn't about amassing a collection of shiny new tools but about strategically building and acquiring the right capabilities to address the organization's unique risks and support its business objectives.

I've seen too many security teams fall into the trap of "shiny object syndrome", chasing the latest and greatest tools without a clear strategy. It's like buying a Ferrari when all you need is a reliable minivan to get the kids to school. The CISO 3.0 focuses on building capabilities, not just collecting tools. It's about understanding the organization's needs and finding the right solutions to address them.

10.2.1 Why Buy Capabilities?

A capability refers to the organization's ability to perform a specific security function, such as detecting and responding to threats, protecting sensitive data, or managing identities. These capabilities are built upon a foundation of people, processes, and technology working in concert to achieve a desired security outcome.

A capabilities-based approach offers several key benefits. First, it aligns security efforts with business needs. With a focus on capabilities, CISOs ensure that security investments target the organization's most critical risks and support its strategic objectives. Second, it optimizes resource allocation, making sure that tools and resources are used efficiently without duplication or waste. Third, this approach enhances effectiveness by creating a stronger, more resilient security posture through a well-integrated set of capabilities. Finally, it promotes adaptability, as capabilities can evolve more easily than individual tools, allowing the organization to respond to new challenges more effectively.

10.2.1.1 Example

Instead of focusing on procuring a specific vulnerability scanning tool, a CISO 3.0 might identify the need for a "vulnerability management capability". This capability encompasses not only the technology for scanning and identifying vulnerabilities but also the processes for prioritizing and remediating them, as well as the people and skills required to manage the program effectively.

I often tell CISOs looking to purchase vulnerability scanners that there are better starting points for building a vulnerability management program. Instead, they should have their penetration testing provider perform a single vulnerability scan. That scan will likely uncover hundreds of vulnerabilities that need remediation. Some will be solved by patching, but others, like outdated cipher suites or expired certificates, will require more bespoke, hands-on interventions. Even for organizations with a robust patching program, these remediations will take time to set up and deliver.

Most of the remediation will be performed by other teams and will require ticketing and tracking. Figuring out how to orchestrate this process often takes months, and there's no point in having a scanning tool or a person performing scans every month while remediations aren't yet underway. The scan would be a waste of effort and resources, returning the same or similar results each time. So, beginning with a single point-in-time scan saves on cost and effort while giving you time to build your remediation process.

After you've closed the majority of the vulnerabilities, you can pay for the penetration testing provider to come back and rescan to validate your fixes. When your remediation program runs smoothly, you can move to a managed service that scans your environment regularly. Once scanning is occurring on a cadence, you can start figuring out how to prioritize remediations. This will likely take several more months.

While formalizing the vulnerability management process, hiring a threat and vulnerability management person will eventually make sense. Once onboard, that person will likely have a preference for specific scanning tools. Since most scanning tools have feature parity, you can let them pick the tool they like.

In this way, you've built out a capability and avoided a protracted solution selection process for a tool that would have been underutilized for months or even years.

10.2.2 How to Buy Capabilities

Shifting to a capabilities-based procurement model requires a well-thought-out and methodical approach. It starts with a comprehensive evaluation of the organization's existing security capabilities. This assessment should go beyond identifying general strengths and weaknesses: It should focus on understanding where the current security measures are effective and where they fall short in addressing potential threats. Engaging stakeholders across various departments can provide a clearer view of how security integrates with day-to-day operations and where critical improvements are needed.

Once the assessment is complete, the next step is to develop a capabilities matrix. This tool maps the organization's current security capabilities against the desired level of maturity for each. The matrix serves as a visual guide, offering a clear snapshot of the organization's security posture. It helps prioritize areas for improvement by highlighting the most significant gaps, making it easier to focus investment where it will have the greatest impact.

It's not simply about filling gaps: It's about aligning new capabilities with the organization's broader strategic goals and risk appetite. Any new security capabilities you bring on board must contribute directly to mitigating risks that could threaten the organization's success. It's essential to ensure that these investments not only address immediate security concerns but also support long-term business objectives, reinforcing the idea that cybersecurity is a strategic enabler, not just a cost center. This approach ensures that security investments are both meaningful and measurable, driving business value while safeguarding critical assets.

10.3 CAPABILITY ASSESSMENTS

A capability assessment is the first step in understanding your organization's security posture and identifying areas for improvement. It provides an overview of your current capabilities, which will help inform your decisions about resource allocation, tool selection, and strategic planning. Without a clear understanding of your existing capabilities, it's impossible to identify gaps and prioritize investments effectively.

10.3.1 Assessing Your Team's Skills

People are the backbone of any cybersecurity program. Therefore, assessing your team's skills is just as important as evaluating your technology and processes. This involves understanding their technical expertise, soft skills, and areas of interest.

Here are some ways to assess your team's skills:

Self-Assessments: Have team members evaluate their own skills and knowledge using questionnaires or surveys.

Skills Matrices: Create a matrix that maps team members' skills against the required competencies for their roles.

Performance Reviews: Use performance reviews to assess individual contributions and identify areas for development.

Third-Party Assessments: Engage external providers to conduct skills assessments or training needs analysis.

10.3.1.1 Examples

Technical Skills: Proficiency in specific programming languages, operating systems, or security tools.

Soft Skills: Communication, problem-solving, teamwork, and leadership abilities.

Areas of Interest: Identify areas where team members have a passion or desire to learn more, which can guide future training and development opportunities.

10.3.2 Creating a Skills Gap Roadmap

Once you've assessed your team's skills, you can create a roadmap to address any identified gaps. This roadmap should prioritize training and development opportunities based on the organization's most critical needs and the individual's career aspirations.

10.3.2.1 Example Skills Gap Roadmap

Short-Term (0–6 Months): Provide immediate training on critical skills gaps, such as incident response or cloud security.

Mid-Term (6–12 Months): Offer opportunities for certifications or advanced training programs to enhance expertise in key areas.

Long-Term (12+ Months): Encourage professional development through conferences, mentorship programs, and opportunities for career advancement within the organization.

10.3.3 Program and Individual Capabilities Assessments

Both program and individual capabilities assessments are essential components of a capability assessment.

Program Capabilities Assessment: Evaluates the organization's overall security capabilities across key areas, like vulnerability management, threat intelligence, access management, incident response, and data protection.

Individual Capabilities Assessment: Focuses on the skills and knowledge of individual team members, identifying areas where training and development are needed.

10.3.4 Example Assessment Questions

10.3.4.1 Program Capabilities Assessment

Threat Intelligence

- How effective are our current threat intelligence capabilities in identifying and proactively addressing emerging threats? Can we collect, analyze, and disseminate relevant threat information to inform our security decisions?
- Vulnerability Management: What is our process for managing vulnerabilities? How efficient is it in identifying, prioritizing, and remediating vulnerabilities across our IT environment? Do we have the necessary tools and resources to address vulnerabilities in a timely manner?
- Incident Response: How well-prepared are we to respond to and recover from a cyber incident? Do we have a well-defined incident response plan, trained personnel, and the necessary tools and technologies to contain the breach, minimize damage, and restore normal operations?
- Identity and Access Management: Do we have robust identity and access management controls in place to protect sensitive data and prevent unauthorized access? Are we effectively managing user identities, authentication, and authorization throughout the organization?
- Data Protection: How do we ensure the confidentiality, integrity, and availability of our critical data, both at rest and in transit? Do we have appropriate encryption, data loss prevention, and backup solutions in place?
- Third-Party Risk Management: What is our approach to third-party risk management? How do we assess and mitigate the risks associated with vendors and partners? Do we have clear contractual agreements and ongoing monitoring processes in place?
- Security Awareness Training: How effective is our security awareness training program in educating employees about potential threats and fostering a security-conscious culture? Do we regularly assess and update our training program to address emerging threats and risks?

10.3.4.2 Individual Capabilities Assessment

Technical Expertise

- What are your areas of expertise and certifications in cybersecurity (e.g., network security, cloud security, incident response, or penetration testing)?
- What specific technical skills do you possess, and how proficient are you in them (e.g., programming languages, operating systems, or security tools)?

Role-Specific Skills

- How comfortable are you with performing your current role's responsibilities?
- What areas within your role do you need more training or support?
- What are your strengths and weaknesses in relation to your current role?

Professional Development

- Which cybersecurity domains or specializations are you most interested in developing further?
- What specific training programs, certifications, or conferences would you like to attend to enhance your skills and knowledge?
- What are your career aspirations within the field of cybersecurity?

Security Awareness

- How do you stay updated on the latest cybersecurity trends and threats?
- What resources do you use to stay informed about security best practices and emerging risks?
- How do you apply your security knowledge in your day-to-day work?

Consider your current team's skillset. Are there any gaps that must be addressed to manage your organization's security capabilities effectively? How could you leverage self-assessments, skills matrices, or performance reviews to identify and bridge these gaps?

10.4 SKILLS MATRIX

A skills matrix, also known as a competency or proficiency matrix, is a powerful tool for evaluating and visualizing the expertise of your cybersecurity team. It's essentially a table that lists the necessary skills for each role and assesses each team member's proficiency in those areas. This simple format helps you identify skill gaps, track progress, and better allocate training and development.

The primary benefit of a skills matrix is its ability to highlight gaps in your team's expertise. It offers a clear snapshot of current skills and areas needing improvement. You can also easily identify where additional training or development is required.

Additionally, it serves as a tool for tracking progress. Regular updates allow you to monitor individual growth and identify where further support might be needed.

This proactive approach ensures your team continues to develop the skills necessary to support the organization's security objectives.

When it comes to training and development decisions, a skills matrix offers valuable insight. It helps you prioritize resources, focusing on closing the most critical skill gaps. This ensures your team is equipped to excel in their roles and strengthen your security program overall.

A skills matrix is also useful for succession planning. Use it to identify high-potential team members and the specific skills required for future roles. Then, you can strategically plan for leadership transitions and maintain expertise within the team, minimizing disruption.

The skills matrix can also facilitate performance reviews. It provides a clear and objective assessment of each individual's skills and contributions, making it easier to provide constructive feedback and set goals for future development. This contributes to a culture of continuous improvement and helps team members understand their role in the bigger picture of the cybersecurity program.

10.4.1 Example

Here's a simplified example (Table 10.1) presenting a skills matrix for a cybersecurity team:

TABLE 10.1
Team Skills Matrix

Skill/Competency	Security Analyst	Security Engineer	Incident Responder
Network Security	Proficient	Expert	Intermediate
Cloud Security	Intermediate	Proficient	Basic
Incident Response	Basic	Intermediate	Expert
Vulnerability Management	Proficient	Proficient	Intermediate
Risk Assessment	Intermediate	Basic	Intermediate
Communication Skills	Proficient	Intermediate	Proficient
Problem-Solving Skills	Expert	Proficient	Expert

This matrix quickly identifies areas where individuals excel and where they may need additional training or support. For example, the security analyst might benefit from further training in cloud security, while the security engineer could enhance their risk assessment skills.

A more complex example of a skills matrix for a cybersecurity team is given in Figure 10.1.

A skills matrix is a living document that should be regularly updated to keep pace with changes in your team's skills and the evolving needs of your cybersecurity program. It helps ensure that your team has the right capabilities to manage cyber risks and protect your organization effectively. For a practical example, you can find a sample skills matrix in the downloadable resources section at CISO30.com.

Skills Matrix

Category	Practice	Prioritized Offerings	Score (Name)	Interest Area	Legend
Practice Areas	IAM	Overall Practice	4 - Advanced Ability	Yes	0- No Ability
		IAM Assesment	3 - Strong Skills		1 - Limited Interest or Ability
		Access Mgmt	4 - Advanced Ability		2 - Intermediate or Apprentice Level Skill
		ID Governance	2 - Intermediate or Apprentice Level Skill		3 - Strong Skills
		PIM/PAM Archetecture & Strategy	5 - Subject Mater Expert		4 - Advanced Ability
		Identity for Zero Trust	2 - Intermediate or Apprentice Level Skill		5 - Subject Mater Expert
	Cloud Security	Overall Practice			This section is about your knowledge of service delivery practice area's and their prioritized service offerings.
		Cloud Migration Strategy			
		Cloud Security Architecture			
		Cloud Security Assessment			
		CNAPP- Cloud Threat/ Posture MGMT			
Practice Areas	Threat / Vulnerability	Overall Practice			
		Vulnerability Management			
		Patch Management			
		Penetration Testing			
		Purple Teaming			
		Breach Simulation			
	Strategy & Risk	Overall Practice			
		Integrated Risk Management			
		Risk Quantification			
		GRC Platforms			
		Security Program Initiation and Remediation			
		SAP Security			
		Internal Audit			

FIGURE 10.1 Skills Matrix.

10.5 CAPABILITIES MATRIX

The capabilities matrix presented here offers a valuable framework for evaluating an organization's security capabilities by examining the relationship between staffing levels, process maturity, and tool deployment. Each combination of these three factors results in a distinct "Control Capability" outcome, highlighting the potential effectiveness of a particular security control (Table 10.2).

10.5.1 Understanding the Matrix

- **Optimized Control Capability:** This ideal scenario occurs when a control is fully staffed, supported by a formal process, and fully deployed. This combination ensures efficient and effective operation, maximizing the control's potential to mitigate risks.
- **Low Adoption Control Capabilities:** Despite having formal processes and fully deployed tools, understaffed controls suffer from low adoption. This indicates that the control might not be utilized to its full potential due to a lack of personnel to manage and operate it effectively. Similarly, controls without any staffing or process, even if fully deployed, also fall into the low adoption category.
- **Inconsistent Operation:** Several combinations lead to inconsistent operation. This includes fully staffed and fully deployed controls with informal processes, indicating a lack of standardized procedures, as well as scenarios where there's staffing but no established process. These situations can result in unpredictable and unreliable control performance.
- **Shelfware:** This unfortunate outcome occurs when tools are fully or partially deployed but lack the necessary staffing and processes to support them. In these cases, the tools are essentially unused and provide little to no value to the organization.
- **Limited Ability to Execute:** This occurs when there's partial staffing and either informal processes or no process at all, even with fully deployed tools. It signifies a situation where the control might be functional but lacks the necessary resources and structure for consistent execution.
- **Low Return on Investment (ROI)/Wasted Effort:** These outcomes highlight scenarios where investments in tools or processes are not yielding the desired return. This can occur when controls are partially deployed or not deployed at all, coupled with inadequate staffing or processes. It indicates a misalignment between resources and objectives, resulting in inefficient or ineffective security measures.
- **Unscalable:** This outcome arises when a control is fully staffed but lacks formal processes or tool deployment. It signifies a situation where people are the control, and without a tool or process, the control might be effective on a small scale but cannot be scaled to meet the organization's growing needs.
- **No Control:** The worst-case scenario, where there is no staffing, no process, and no tool deployment, resulting in a complete absence of control and represents a vulnerability for the organization.

TABLE 10.2
Capabilities Matrix

Staffing	Process	Tooling	Control Capability
Fully staffed	Formal process	Fully deployed	.Optimized
Partially staffed	Formal process	Fully deployed	Low adoption
Unstaffed	Formal process	Fully deployed	Low adoption
Fully staffed	Informal process	Fully deployed	Inconsistent operation
Partially staffed	Informal process	Fully deployed	Inconsistent operation
Unstaffed	Informal process	Fully deployed	Shelfware
Fully staffed	No process	Fully deployed	Inconsistent operation
Partially staffed	No process	Fully deployed	Limited ability to execute
Unstaffed	No process	Fully deployed	Shelfware
Fully staffed	Formal process	Partially deployed	Low ROI
Partially staffed	Formal process	Partially deployed	Low ROI
Unstaffed	Formal process	Partially deployed	Wasted effort
Fully staffed	Informal process	Partially deployed	Underutilized
Partially staffed	Informal process	Partially deployed	Low maturity
Unstaffed	Informal process	Partially deployed	Low ROI
Fully staffed	No process	Partially deployed	Limited ability to execute
Partially staffed	No process	Partially deployed	Shelfware
Unstaffed	No process	Partially deployed	Shelfware
Fully staffed	Formal process	Not deployed	Unscalable
Partially staffed	Formal process	Not deployed	Wasted effort
Unstaffed	Formal process	Not deployed	Wasted effort
Fully staffed	Informal process	Not deployed	Wasted effort
Partially staffed	Informal process	Not deployed	Unscalable
Unstaffed	Informal process	Not deployed	Wasted effort
Fully staffed	No process	Not deployed	Limited ability to execute
Partially staffed	No process	Not deployed	Unscalable
Unstaffed	No process	Not deployed	No control

You've likely heard the phrase "People, Process, and Technology" in the context of security. This matrix highlights just how important it is to align your staffing, processes, and tools to achieve the best security outcomes. A CISO 3.0 knows that buying tools alone isn't enough: You also need to invest in the people and processes that support those tools and ensure they function effectively. By assessing each control within this framework, CISOs can pinpoint areas for improvement and make smarter decisions about where to allocate resources, develop processes, and choose the right tools to build a strong, resilient cybersecurity program. A downloadable capability matrix template is available at CISO30.com.

10.6 THE PROBLEM WITH RFPS

Another major pitfall I see CISOs 2.0s and their programs fall into is the RFP process. RFPs can be a nightmare. They're often driven by procurement teams who are focused on price, not on the actual capabilities the organization needs. I've seen RFPs that were so poorly written that they attracted responses from completely unqualified vendors. It's a waste of time and resources. Honestly, the RFP is probably the worst possible way to buy capabilities instead of tools and avoid shelfware. The CISO 3.0 needs to take control of the procurement process and ensure that it's aligned with the security strategy.

Here's why RFPs usually fail. The RFP is designed to ask vendors for a response to questions about their solution's functionality. However, vendors are incentivized to say "yes" to everything and make their solution fit within the box you've drawn, regardless of whether it's a good match.

Here's what typically happens: imagine you're creating an RFP for a vehicle. You research common vehicle features and consult your architects and engineers on what's necessary and desirable. Let's say you end up with a list that includes four wheels, a dashboard with gauges, gas-powered, leather seats, a heads-up display, windows, doors, etc.

First, how do you distinguish between must-haves like wheels and a dashboard and nice-to-haves like leather seats and a heads-up display? Also, even though it seems obvious that something like four wheels is so standard that there would be no reason to request it, you still have to include that to avoid getting responses for skid steer equipment with tracks or two-wheeled motorcycles. You send the RFP to vendors and receive several positive responses, all claiming to meet your criteria. One is for a Ferrari, one is for a Honda Accord, one is for a school bus, one is for an airplane, and one is for a front-end loader. Your team reviews the demos and narrows it down to three: the Ferrari, the school bus, and the airplane. I see this scenario play out all the time. Then, you consult your advisors, asking which one to choose.

At this point, I must ask questions about the vehicle's intended use. Are they going to race, transport dozens of kids to school, or move earth? The answer is usually no, as most people need a sedan, SUV, or pickup truck. We could have narrowed down the options in advance.

Now, you're left with absurd choices because all the practical options, like Hondas and Toyotas, were eliminated for lacking flashy extras like flight capability or massive passenger capacity. Often, by the time I'm involved, we have to either start from scratch or settle for the least bad option.

This could all be avoided by starting with the business problem: What are we trying to accomplish? I know the vehicle analogy seems extreme, but it's not. Countless times, I've seen RFPs for security solutions result in responses for completely different types of tools. This happens because vendors try to shoehorn their solution into every use case.

One of the main reasons organizations still use RFPs is the belief that it fosters competition, leading to a price advantage. Consequently, the procurement team, rather than the security team, often drives the RFP process. Unfortunately, there aren't substantial savings to be had by leveraging RFPs. Even disregarding the low

ROI and underutilized shelfware, most organizations don't get the best pricing this way on a deal-by-deal basis.

Only the largest global Fortune 50 type organizations have enough negotiating power to make it worthwhile to go directly to the manufacturer. It's more beneficial for most organizations to leverage a value-added reseller (VAR) and engage them early in the process. VARs handle such large volumes that they have deeply negotiated discounts, often exceeding 60% off list price. Even with the VAR's 10–15% markup, it's still generally cheaper.

Engaging the VAR early allows them to assist in determining the business case, selecting appropriate selection criteria, and sizing and scoping the solution. Many CISOs delegate these tasks to their engineers and architects, tasks which VARs will usually perform for "free" as part of their value-add. There's no reason to burden your team with tasks you can outsource at no cost.

CISOs often mistakenly believe they can get better deals by pitting VARs against each other. However, the best way to obtain tailored advisory services from a VAR is to cultivate a deep, ongoing relationship where they intimately understand your environment and business. Moreover, the incentive structure doesn't favor competing VARs.

For instance, the VAR who initially sold the solution typically has an incumbent discount advantage of 20–25% off list price, a portion of which is passed on to you. I've witnessed many CISO 2.0s request vendors to remove this discount to create a "level playing field" for VARs to bid against each other. This well-intentioned move backfires, as no one receives the discount, and the price increase is passed on to the organization, resulting in paying 10–20% more for the illusion of competition.

10.7 HOW TO SHOP FOR CAPABILITIES INSTEAD

The CISO 3.0 takes a more strategic and results-focused approach when it comes to acquiring security solutions. Rather than getting lost in the details of technical specifications, they concentrate on finding solutions that provide the specific capabilities necessary to tackle the organization's unique risks and align with its overall security strategy.

This approach starts by understanding your "why". Before evaluating tools or solutions, it is critical to ask foundational questions about the current state, desired future state, and expected outcomes. By clearly defining what you are trying to achieve, how it aligns with business goals, and what success will look like, you set the stage for acquiring solutions that will actually build capabilities and deliver value.

10.7.1 Start with the Current State

Begin by assessing where you are today. Identify the pain points and challenges you are trying to solve. Consider your current environment, including the tools and resources already in place, the staffing situation, and any formal policies, processes, or procedures that support existing operations. Evaluate the steps you've already taken and the gaps that remain. This baseline understanding provides a clearer picture of the areas that require attention and ensures that you avoid redundancy when acquiring new solutions.

10.7.2 Define the Desired Future State

Next, determine what "good" looks like for your organization. What are the specific functions or capabilities that your future state should include? Will achieving this vision require new tools, additional staffing, or outsourcing? Consider how the solution will be managed and maintained over time, including whether new policies or processes will be needed to support it. This vision helps align your team's efforts and ensures that every step you take moves the organization closer to its goals.

10.7.3 Focus on Positive Business Outcomes

A successful future state is not only about solving technical problems but also about delivering measurable business outcomes. Consider how your desired state will impact broader business goals or initiatives. Will it reduce risk, cut costs, or improve operational efficiency? For example, if the primary business driver is risk reduction, identify how the new solution will help mitigate key risks. If the goal is cost reduction, quantify how the solution will save money or optimize resource allocation.

10.7.4 Establish Clear Requirements

With your goals in mind, translate them into actionable business and functional requirements. Define the capabilities you need, the key risk indicators (KRIs) you aim to improve, and any compliance or regulatory requirements the solution must meet. Identify the users who will interact with the tool and their specific needs. Additionally, consider scalability, integration, and performance requirements to ensure the solution can grow with your organization and align with existing systems.

10.7.5 Define Success Metrics

Success must be measurable. Determine how you will evaluate potential solutions against each other (selection metrics), assess the success of the implementation (project metrics), and monitor long-term performance (program metrics). Establishing these metrics upfront provides a clear framework for decision-making and accountability.

10.7.6 Address Current Capabilities and Deficiencies

Finally, examine any internal deficiencies that could hinder your ability to achieve the desired future state. For example, do you need additional skills, training, or stakeholder buy-in? Are there risks or potential roadblocks, such as interdepartmental communication challenges or a lack of end-user participation? Identifying these issues early allows you to develop mitigation strategies and ensure a smoother path to success.

By systematically addressing these areas, the CISO 3.0 ensures that every investment in a new tool or solution is purposeful, aligned with strategic goals, and capable of delivering measurable value. This approach shifts the focus from acquiring technology for its own sake to building capabilities that directly support the organization's security and business objectives.

Once you understand your "why", your starting point, and your end goals, you can begin to shop around for solutions.

Because you understand your success criteria, you can now shop for solutions based on capabilities instead of merely listing features. For example, instead of saying a solution must include "X number of threat feeds" or "Y type of encryption", you should express the need for "advanced threat detection and response" or "robust data protection". Even better, try to define the business purpose behind these needs and focus on the desired outcomes. For instance, if your goal is to transport kids to school, you wouldn't waste time evaluating all types of vehicles: You would simply look for school buses. Similarly, if you need to enhance your cybersecurity, instead of examining every security information and event management (SIEM) on the market, you would determine if the function could be outsourced and then focus on solutions that provide capabilities that cover and complement your engineering and analyst skills gaps. This approach opens up a more relevant range of solutions that meet your capability needs, even if they don't check every box on a traditional RFP.

You may remember the example from the beginning of this chapter about buying a wrench. Some might think security tools are different, viewing them more like utility tools with lots of attachments for specialized tasks. Perhaps you feel like your program has a "specialty bolt" that requires a unique attachment, and you're trying to find the perfect tool that fits this specific need. I often see folks get bogged down by these corner cases, leading to analysis paralysis, where they're stuck comparing tools that are largely the same, but each has some minor shortcomings.

When this happens, it's important to take a step back and consider the bigger picture. Ask yourself: Does this shortcoming prevent you from delivering the outcome, or is it just a "nice to have" feature? How often does this corner case actually come up? Is it a daily need or an occasional inconvenience? Keep in mind that your goal isn't to buy the best or most perfect tool in existence: It's to build a capability that achieves a specific outcome. Remember, the wrench analogy still applies, whether or not you have the specialty attachment, just buying the tool doesn't make you good at, or even capable of, completing the job. Technology is only one-third of the answer, and tool selection isn't the best predictor of success. People and process are what determine if you can create a capability and convert the tool into an outcome. Focus on which tool you can operationalize most effectively and efficiently, not which tool has the most features.

Before investing in new capabilities, you need to understand your team's skills, talents, and interests. This means identifying any overall skills gaps and creating a roadmap to address them, similar to how you would assess gaps in controls and tools. Many providers offer security skills training and can help with these assessments; some, like CDW, even provide this service for free. When you get a clear picture of your team's capabilities, you can ensure that any new solutions you acquire can be seamlessly integrated and effectively operationalized within your security program.

Next, conduct proof of concept (POC) evaluations to test potential solutions in a real-world environment. The most important aspect of any POC is the testing plan. You need to approach POCs with a clear agenda and a set of outcomes you are looking to achieve. A common mistake organizations make is performing unstructured POCs that drag on for months. There is generally no reason for this; a couple of weeks is usually sufficient if you have a planned agenda. Prolonged buying cycles only add confusion and can lead to people forgetting the initial goals or the situation changing, undermining the entire project. I see many teams tinker around with no clear objective, just to "see if they like it". Worse yet, POCs often result in a list of dislikes or a feature wish list without determining if the tool can be utilized as-is for its stated purpose.

The POC should validate that the tool aligns with your business goals, team skills, and day-to-day workflow, confirming it's a good fit. The POC can also be an opportunity to deploy the tool and set up the initial configuration. While reducing professional services hours and cost is admirable, don't aim to eliminate professional installation and setup services entirely.

A common misconception among CISOs is that if their engineers handle the installation and initial configuration, they will inherently learn how the tool works and be able to operationalize it. This rarely works for most organizations. Imagine you own a restaurant and need a new commercial oven. Would you have your executive chef install it? Of course not. While they are experts in using the oven, they aren't installation experts, and the installation process has little to do with daily oven use. Your chef is a valuable resource who should focus on their primary role. The same applies to security tools.

There are multiple roles associated with security tools. The engineer provides daily care and feeding of the solution and might benefit from observing the initial configuration but likely won't gain much from the installation itself. The analyst consumes the tool's output and can't perform their tasks until it's up, running, and tuned. In some organizations, these roles might be combined, but even then, the individual needs the skills and functions of both roles to transform the tool into a capability.

Another mistake CISOs often make is budgeting only for the tool cost, neglecting professional services, managed services, or training, which are the people and process components of the solution. This leads to partially or unstaffed tools with informal or no processes, resulting in shelfware and limited ROI.

When evaluating security solutions, take the total cost of ownership (TCO) into account and make sure to prioritize integration and interoperability with your existing tools. It's also important to assess the vendor's reputation and track record.

10.8 CONCLUSION

In this chapter, we've challenged the traditional "tool-first" mindset and focused on a more strategic approach to security. We've talked about the importance of conducting capability assessments to identify gaps in your defenses and prioritizing investments that align with your overall goals. We've also questioned the effectiveness of the traditional RFP process, which often leads to expensive and underutilized tools. Instead, we've advocated for a more strategic approach to solution shopping, one that emphasizes POC evaluations, understanding your team's skills, and considering the TOC.

The truth is, real security isn't about having the most tools; it's about having the right capabilities. It's about understanding your organization's unique risks and building a security program that can actually address them. That's where the CISO 3.0 comes in. They're not just tool collectors: They're strategic thinkers who prioritize building capabilities over simply acquiring the latest gadgets.

Remember, your people are your most valuable asset. Investing in their skills and development is just as important as investing in technology. Building a team with the right capabilities and fostering a culture of continuous learning can help ensure that new security solutions are not only implemented effectively but also fully integrated into your overall security program.

This chapter also stressed the importance of approaching solution shopping with the business problem in mind. It's not about checking off a list of features; it's about finding solutions that deliver the capabilities you need to address your organization's specific risks. The most important aspects are focusing on solving the business problem, conducting targeted POCs, considering the TOC, and prioritizing integration. In the end, it's about building a security program that's more than a collection of tools but a powerful set of capabilities you can use to protect your company's assets.

10.8.1 Key Takeaways

1. **The Shift from Tools to Capabilities**
    - The CISO 3.0 emphasizes building comprehensive security capabilities instead of focusing solely on acquiring tools.
    - Capabilities combine people, processes, and technology to address organizational risks effectively and align with strategic objectives.
2. **Avoiding "Widgetism"**
    - Overreliance on tools without proper integration and skilled personnel leads to inefficiencies such as "shelfware" and diminished program effectiveness.
    - Success lies in deploying tools that align with a well-defined capability strategy rather than amassing technology for its own sake.
3. **Conducting Capability Assessments**
    - Assessments help identify gaps in existing security measures, focusing on areas like team skills, program effectiveness, and individual competencies.
    - This process ensures resource allocation targets critical vulnerabilities and aligns with business goals.
4. **Skills and Capabilities Matrices**
    - **Skills Matrix:** Evaluates team expertise, identifies gaps, tracks progress, and informs training and development priorities.
    - **Capabilities Matrix:** Examines the relationship between staffing levels, process maturity, and tool deployment to measure the effectiveness of security controls.

5. **Rethinking the RFP Process**
   - The traditional RFP process often leads to tool-centric decisions, resulting in suboptimal solutions and wasted resources.
   - A strategic approach to procurement prioritizes capability development and alignment with organizational needs.
6. **Strategic Solution Shopping**
   - Define requirements based on desired capabilities and outcomes rather than specific features.
   - Conduct focused POC evaluations to validate alignment with business objectives, team skills, and workflows.
7. **Integration and TOC**
   - Evaluate potential solutions based on their ability to integrate with existing security systems and their overall cost, including deployment, training, and ongoing maintenance.
   - Prioritize tools that maximize interoperability and align with long-term goals.
8. **People and Processes as Foundational Elements**
   - Security capabilities are built on skilled teams and robust processes supported by appropriate technology.
   - Investing in training and fostering a culture of continuous learning enhances the effectiveness and adaptability of security programs.
9. **The Role of the CISO 3.0**
   - The modern CISO is a strategic thinker who prioritizes capability building, ensuring that cybersecurity initiatives align with business objectives and effectively address evolving threats.

10.8.2 Reflection and Exploration Questions

Understanding the Concepts

1. What are the key differences between a tool-centric and a capability-focused approach to cybersecurity?
2. How do skills and capabilities matrices contribute to identifying and addressing gaps in a security program?

Connecting the Dots

3. How does the concept of avoiding "widgetism" connect to earlier discussions on risk quantification and strategic alignment?
4. In what ways do capability assessments enhance the alignment of security programs with organizational objectives?

Real-World Application

5. Can you think of examples where a tool-centric approach led to inefficiencies in a security program? How could a capabilities-based approach have improved the outcome?
6. How might a comprehensive POC evaluation prevent common pitfalls in tool acquisition?

Personal and Professional Growth

7. How can you apply the principles of capability assessments and matrices to enhance your role or organization's security program?
8. What steps can you take to improve your understanding of the TOC for security solutions?

Further Exploration

9. What additional tools or frameworks could support the transition to a capabilities-based approach in your organization?
10. How might emerging trends, such as AI and automation, influence the evolution of security capabilities?
11. How can you begin building out capabilities and integrating people, processes, and technology into a cohesive cybersecurity strategy?

11 Leading Effective Teams

11.1 INTRODUCTION

At the heart of any successful cybersecurity program is a team of dedicated and skilled professionals. The CISO 3.0 knows that it's the people on the front lines who make all the difference. Building a high-performing security team isn't just about hiring the best and brightest: It's also about creating an environment where they can thrive. It's about fostering a culture of continuous learning, where everyone is encouraged to grow and develop their skills. It's about promoting collaboration, where team members work together seamlessly, sharing knowledge and supporting each other.

This dream team needs to be more than just a collection of individuals: It needs to be a well-oiled machine ready to tackle the challenges of the modern threat landscape. They need the skills and knowledge to implement the latest security tactics, from protecting user identities to securing cloud environments. And they need to be able to work together like a symphony orchestra, each member playing their part to create a harmonious and effective defense.

Because, in the end, it's not only about technology but also about people. It's about building a team that's passionate about protecting the organization's assets and enabling its success. It's about creating a security culture where everyone feels empowered and invested in the mission. That's the true key to building a cybersecurity program that can withstand the test of time.

11.2 UNDERSTANDING SECURITY ROLES

The CISO 3.0 knows that a successful defense depends on having the right people in the right roles. Building an effective team means carefully considering key roles and ensuring the right people are positioned to contribute to the organization's security efforts. Let's explore some of the essential roles that should be part of every CISO's ideal security team.

CISO: The strategic leader and visionary responsible for setting the overall direction and vision for the security program. They align security initiatives with business objectives, manage risk, and communicate effectively with stakeholders at all levels. The CISO is the ultimate owner of the security program and is accountable for its success.

Director/Manager: The operational manager is responsible for overseeing the security team's day-to-day activities. They manage personnel, allocate resources, and ensure that projects are completed on time and within budget. The director/manager acts as a bridge between the CISO and the rest of the security team, translating strategic goals into actionable plans.

DOI: 10.1201/9781003510789-16

Architect: The designer and planner responsible for creating and implementing secure IT architectures and solutions. They select and evaluate security technologies, design network and system configurations, and ensure security is built into every aspect of the IT environment. The architect is a key contributor to the organization's overall security posture.

Engineer: The hands-on implementer responsible for deploying, configuring, and maintaining security tools and technologies. They troubleshoot issues, respond to incidents, and ensure that the security infrastructure operates effectively. The engineer is the technical backbone of the security team.

Analyst: The investigator and problem-solver responsible for analyzing security data, identifying potential threats, and escalating incidents as needed. They conduct forensic investigations, perform vulnerability assessments, and develop remediation plans. The analyst's primary role is to detect and respond to cyberattacks.

Project Manager: The coordinator and organizer responsible for tracking and managing security projects, ensuring they are completed on time and within budget. They manage resources, track progress, and communicate with stakeholders. The project manager ensures that security initiatives are executed efficiently and effectively.

These roles can often specialize in expertise areas like the following:

Risk, Compliance, and Governance: Focuses on identifying, assessing, and mitigating risks, ensuring compliance with regulations and industry standards, and establishing effective governance frameworks.

Identity and Access Management (IAM): Specializes in implementing and managing IAM controls, such as multifactor authentication (MFA), privileged access management (PAM), and identity governance and administration (IGA), to secure identities and access rights.

Data Protection: Implements and manages data protection measures, such as encryption, data loss prevention (DLP), and data access governance, to safeguard sensitive data throughout its lifecycle.

Security Monitoring and Operations: Monitors and analyzes security data, detects and responds to threats, and manages security incidents.

Threat Intelligence: Gathers and analyzes threat information to identify emerging threats and vulnerabilities.

Cloud Security: Ensures the security of cloud environments and workloads, leveraging tools like cloud security posture management (CSPM) and cloud-native application protection platform (CNAPP).

Incident Response: Specializes in responding to and recovering from security incidents, minimizing their impact on the organization.

DevSecOps: Integrates security into the software development lifecycle, ensuring security is built into applications from the ground up.

Infrastructure Security: Focuses on securing the organization's IT infrastructure, including networks, servers, and endpoints.

Vulnerability and Threat Management: Identifies, assesses, and prioritizes vulnerabilities in systems and applications and develops remediation plans.

It's common to have roles like a vulnerability and threat management engineer, a data protection engineer, or an incident response analyst, reflecting the specialized expertise required in these areas.

It is important to understand that these roles represent functional areas of expertise rather than rigid job titles. In smaller organizations, a single individual might fulfill multiple roles, while larger organizations might have dedicated teams for each function. The key is to ensure that the necessary capabilities are present within the team, regardless of how they are structured or titled. The CISO 3.0 understands the importance of flexibility and adaptability in team structure, recognizing that the ideal configuration will evolve with the threats and the organization's needs.

You need to inventory your talent and figure out which roles you already have on staff and which are gaps. This should align with the capabilities we defined in Chapter 10. You need to determine which roles and capabilities you need for your team and want to staff and develop in-house. You can use the assessment methodology from Chapter 10 to determine which team members are suited for which roles.

In addition to building out your direct team, you also need to develop a network of subject matter experts who can provide specialized knowledge when needed. Start by identifying any gaps in your team's skills, particularly in areas that are vulnerable to emerging risks or shifting regulations. Addressing these gaps doesn't always require new hires. Sometimes, the right expertise can be found within other departments in your organization, so reaching out to colleagues with alternative insights and skillsets can often be a great help. Additionally, don't hesitate to bring in external consultants or specialists on a case-by-case basis for those situations where deeper expertise is necessary.

By combining internal talent with strategic external support, you ensure your team remains flexible, knowledgeable, and prepared to handle changes and challenges. This approach not only enhances your team's capabilities but also fosters collaboration across the organization, ensuring that no skill gap becomes a critical vulnerability.

Take a moment to mentally evaluate your current team structure. Are the essential roles clearly defined and filled?

11.2.1 The NIST NICE Framework

The National Institute of Standards and Technology (NIST) National Initiative for Cybersecurity Education (NICE) Framework, which can be found at https://www.nist.gov/itl/applied-cybersecurity/nice, can be a valuable tool in defining roles and identifying the necessary skills for each role, enabling you to assess and categorize your team's current capabilities accurately. The NICE Framework also facilitates mapping roles to actual job titles and creating detailed job descriptions, ensuring clarity and consistency in your staffing strategy.

The NICE Framework categorizes cybersecurity work into seven categories: securely provision, operate and maintain, protect and defend, investigate, collect and operate, analyze, and oversee and govern. Each category encompasses various specialty areas and work roles, each with its own set of knowledge, skills, and abilities (KSAs). Utilizing the NICE Framework can help you gain an understanding of the diverse roles within cybersecurity and the specific skills required for each.

NICE Seven Categories

1. **Securely Provision:** Focuses on designing and building secure IT systems from the ground up, ensuring security is embedded at every stage of development.
2. **Operate and Maintain:** Manages the administration and ongoing support needed to keep IT systems performing securely and efficiently.
3. **Protect and Defend:** Actively identifies, analyzes, and mitigates threats to the organization's internal IT systems and networks.
4. **Investigate:** Examines cybersecurity incidents, extracting key insights to prevent future occurrences and strengthen defense strategies.
5. **Collect and Operate:** Specializes in conducting denial and deception operations while gathering intelligence to support broader cybersecurity initiatives.
6. **Analyze:** Evaluates cybersecurity data and vulnerability reports to detect, assess, and neutralize potential risks.
7. **Oversee and Govern:** Provides leadership and direction, ensuring the organization's cybersecurity initiatives align with strategic objectives and regulatory requirements.

To leverage the NICE Framework effectively, start by identifying the key cybersecurity functions that your organization needs to perform. Then, map these functions to the relevant NICE categories and specialty areas. This will help you identify the specific work roles that are most critical for your organization's security program. Once you have identified the necessary work roles, you can use the NICE Framework to define the KSAs required for each role. This information can then be used to assess your team's current capabilities, identify skills gaps, and develop targeted training and development programs.

For instance, if your organization is focused on safeguarding its IT systems and networks, you might identify the need for roles like security analyst, incident responder, and vulnerability management specialist. The NICE Framework can help you define the specific knowledge, skills, and abilities (KSAs) needed for each role, such as understanding network security protocols, incident response procedures, and vulnerability assessment tools. Comparing your team's existing skills against the KSAs outlined in the NICE Framework can pinpoint where additional training or development might be necessary.

The NICE Framework also helps translate roles into actual job titles, which is particularly useful for crafting job descriptions or recruiting new talent. Aligning your job titles and descriptions with the NICE Framework can help attract candidates who have the right skills and experience for each position.

For example, the NICE Framework work role of "vulnerability assessment analyst" could correspond to job titles like "security analyst" or "penetration tester" in your organization. Using the NICE Framework as a guide allows you to create job descriptions that accurately reflect the required KSAs, helping you attract qualified candidates.

Overall, the NICE Framework is a valuable resource for defining roles, identifying skills gaps, and developing job descriptions within your cybersecurity team. By leveraging this framework, you can ensure your team has the capabilities needed to effectively implement and manage your security program.

11.2.1.1 Example

When it comes to staffing a security operations center (SOC), for instance, the jobs primarily fit into two roles: the engineer role and the analyst role. The engineer manages the inputs and handles the care and feeding of the solutions. Their responsibilities include onboarding log sources, creating rules and correlations, setting up alerts, tuning rules, creating dashboards, building reports, integrating threat intelligence, and developing automation and orchestration workflows. The analyst, on the other hand, manages the outputs of the solutions. SOC analysts generally work in tiers:

Tier 1: Handles triage and enrichment of security alerts, filtering out false positives and escalating potential threats to tier 2 analysts.
Tier 2: Conducts in-depth analysis and investigation of potential threats, determines the scope and impact of incidents, and initiates containment and remediation efforts.
Tier 3: Performs proactive threat hunting, searching for signs of malicious activity that may have evaded detection by traditional security tools.
Tier 4: Handles advanced forensic investigations and malware reverse engineering, requiring specialized skills and expertise.

It generally takes at least one full-time employee (FTE) to handle the engineering role, but this FTE often comprises several partial FTEs since it's difficult for one person to be an expert on every tool in the SOC. On the other hand, analysts need to be ready for alerts 24 x 7 x 365, which would require a minimum of five to six FTEs to staff in-house. However, many organizations opt for alternative staffing models, such as an 8 am–5 pm workday with 5 pm–8 am on-call support, or outsourcing some or all of the analyst functions to a managed security service provider (MSSP).

11.3 INSOURCING VERSUS OUTSOURCING: A STRATEGIC CHOICE

The insourcing versus outsourcing debate is a classic one in cybersecurity. There's no easy answer, and the right choice depends on the organization's specific needs and resources. I've seen both approaches work well, but the key is to be strategic and intentional in your decision-making. Don't just follow the crowd; do what's right for your team and your organization.

The decision to insource or outsource security functions depends on various factors, such as your organization's size, resources, risk appetite, and strategic objectives. Insourcing involves building and maintaining an in-house security team, while outsourcing entails contracting with external providers to deliver specific security services or functions.

Insourcing offers greater control over the security program and allows for closer alignment with the organization's culture and values. It also enables the development of specialized expertise and institutional knowledge within the organization. However, it can be costly and resource-intensive, requiring a high level of personnel, training, and technology investments.

On the other hand, outsourcing can provide access to specialized skills and expertise that may not be readily available in-house. It can also offer cost savings and scalability, allowing organizations to adjust their security resources based on their needs. Outsourcing also involves relinquishing some control over the security program and relying on external providers to meet the organization's security requirements.

The CISO 3.0 carefully evaluates the pros and cons of each approach, considering the organization's specific needs and risk profile. They might choose to insource core security functions that require deep integration with the business while outsourcing specialized services or functions that can be delivered more efficiently by external providers. The key is to strike the right balance between control, cost-effectiveness, and access to expertise.

Once you have determined the roles you need for your program, you must decide how you want to staff those roles. Do you want to insource, hire staff, and build those capabilities in-house, or would you rather outsource some or all of those roles to another party? This choice is remarkably similar to the risk investment mix choice of risk transfer versus risk mitigation. You have to decide "what business you want to be in". Assessing your team's skills will help a great deal in determining which skills and roles you want to keep in-house and which would be better performed by someone else.

You need to ask: Where does your team currently have talent, experience, and desire? Is it in the following?

Program Improvement: Planning, design, and control maintenance

or

Program Maintenance: Execution, deployment, and control monitoring

Do you want to outsource the parts that are boring, repetitive, and time-consuming so your team can focus on advanced tasks like threat hunting and forensic investigation? Or does your team lack those advanced skills, and you need to farm out those functions while your team focuses on basic blocking and tackling?

Once you've determined what you want to outsource, you can start to figure out how you want to outsource it. You have a lot of options for the method and duration of outsourcing, and each has its own pros and cons. Some options are as follows:

Staff Augmentation/Temporary Staffing: This involves hiring temporary staff to fill specific roles or skill gaps within your team. It's a flexible option that allows you to scale your resources up or down as needed. However, it can be more expensive than other outsourcing options, and temporary staff may not be as invested in the organization's long-term success. *Example: Hiring a temporary penetration tester to conduct an annual assessment.*

Strategic Staffing/Temp-to-Perm: This is similar to staff augmentation but with the intention of potentially converting the temporary staff into permanent employees. It allows you to evaluate a candidate's performance before making a long-term commitment. *Example: Hiring a security analyst on a temp-to-perm basis to assess their fit within the team and culture.*

Professional Services/Contract/SOW Bound: This involves engaging external consultants or service providers to deliver specific projects or tasks. It's a good option for specialized expertise or one-time projects. However, it can be expensive, and you may have less control over the project's execution. *Example: Hiring a consulting firm to conduct a risk assessment or implement a new security tool.*

Managed Services: This involves outsourcing the management and operation of specific security functions or technologies to a third-party provider. It can free up internal resources and provide access to specialized expertise. It also involves relinquishing some control and relying on the provider to meet your security requirements. *Example: Outsourcing your security monitoring and incident response to an MSSP.*

As-a-Service: This is a service-based delivery model where you pay for security services on a subscription basis. As-a-service abstracts you from the tools and processes and simply delivers an outcome. It offers flexibility, scalability, and often lower upfront costs. However, it can also lead to vendor lock-in and potential concerns about data security and privacy. *Example: Subscribing to a cloud-based email security solution.*

It's also possible to outsource all of the roles we mentioned earlier. There are many managed service providers that will take on the bulk of your program, but you can also outsource the individual roles across different services and service types. Analysts can be outsourced to managed services, engineers can be outsourced to professional or managed services, architects can be outsourced to staff augmentation (and your in-house architects can be augmented by architects from your VAR as discussed in Chapter 10), managers and directors can be outsourced via staff augmentation and strategic staffing, and even the top CISO job can be outsourced via virtual CISO (vCISO) services. Outsourcing SOC functions to MSSP or managed detection and response (MDR) providers is extremely common, but it's also possible to outsource nearly every individual function of your security program. Some functions commonly managed by third parties include

- endpoint security, detection, and response;
- network anomaly detection and response;
- incident response and forensics;
- threat hunting;
- vulnerability management;
- patch management;
- security awareness programs;

- cyber threat intelligence;
- identity governance;
- data protection;
- frontline support;
- platform management;
- third-party risk management;
- attack surface management;
- application security scanning;
- web application firewall monitoring;
- digital asset monitoring.

The choice of outsourcing model depends on your organization's specific needs, budget, and risk tolerance. It's important to carefully evaluate each option and select the one that best aligns with your overall security strategy and business objectives.

When evaluating what to insource or outsource, consider what has become commoditized. For instance, SOC outsourcing is highly commoditized, and services like Arctic Wolf have built world-class SOCs that will likely perform better than an in-house SOC ever could. Even if the costs were equal, you would be getting a more sophisticated and feature-rich SOC. All you need to do is tune and integrate it into your specific environment. Because SOC has reached this level of commoditization, it rarely makes sense to build your own, that is, unless you're a massive multinational corporation with dozens or hundreds of IT security FTEs or if you can leverage the SOC as a revenue-generating service by reselling SOC services to customers or cross-charging business units. Even in such cases, there are still aspects of the SOC that can be outsourced, such as endpoint triage through solutions like Falcon Complete from CrowdStrike.

Once you've decided what functions to insource and which to outsource, only then can you begin to determine the best options for related products and services. Often, it makes more sense to allow MSSPs to handle the commoditized tasks like detection and response while focusing internal efforts on SOC-related services that are not easily outsourced, such as risk management, insider threat detection, and advanced, continuous threat and vulnerability management. In the next chapter, we will explore building out a risk operations center (ROC), which focuses on these more strategic, high-impact functions, rather than the commoditized SOC tasks.

Reevaluating how you balance onshore and offshore staffing can significantly reduce costs while increasing efficiency. Instead of keeping high-cost talent on standby, you could focus on scaling specific roles, like vulnerability assessors or incident response team members, in regions with lower operational costs. By equipping these offshore teams with the right tools, training, and oversight, you can centralize critical functions while maintaining high standards.

For example, imagine outsourcing routine security monitoring tasks to a trusted partner in a lower-cost region, freeing up your in-house team to focus on higher-value activities such as advanced threat analysis or strategic planning. This way, you not only save money but also create flexibility in your budget to invest in specialized roles or emerging technologies that are crucial for staying ahead of evolving threats.

This approach helps you make the most of your available resources, ensuring your security program is not only cost-effective but also agile enough to adapt to future challenges. By reallocating these savings to more strategic areas, you're setting up your organization to handle whatever comes next.

11.4 INSOURCING

The decision to outsource doesn't absolve you of the need for internal expertise. Even if you entrust your entire security program to external providers, you'll still require a core team in-house to act as the bridge between your organization and those providers. Think of it as having a dedicated liaison who understands your unique needs and can effectively communicate them to the outsourced team. If you outsource engineering roles, for instance, you'll need someone internally who can raise the flag when changes occur in your environment, like new systems being added or existing ones being modified. Similarly, on the analyst side, you'll need a point person who can receive and interpret communications from the outsourced team, ensuring that you're kept in the loop about any actions taken or incidents detected. The bottom line is that even in a heavily outsourced model, certain roles and skills must remain in-house to ensure seamless integration and effective oversight of your security program.

I often hear complaints about MSSPs, where CISOs believe the services are lacking. While some services may indeed underperform, more often than not, the problem stems from the organization's inability to properly consume the service. Inadequate structuring to support the outsourced model leads to dissatisfaction, and ironically, this drives some organizations to insource these functions. However, this can result in costly failures when they couldn't even handle the outsourced service, let alone manage the entire function internally.

Building a strong security team is like assembling a championship sports team. You need to find the right players, train them, and create a culture of collaboration and trust. It's not easy, but it's incredibly rewarding. I've been fortunate to work with some amazing security professionals over the years, and I've learned that the best teams are built on a foundation of shared passion and commitment.

When it comes to insourcing, there's a four-step path to team building:

1. **Hiring:** This requires finding talent, determining market value, selecting "A" players, and attracting talent.
2. **Training:** This involves determining skills gaps, ensuring relevance to the role, and timing and budgeting training efforts.
3. **Retention:** This necessitates up-leveled skills migration, challenge and goal setting, managing stress levels, and providing development and support.
4. **Career Pathing and Succession Planning:** This includes offering promotion opportunities, establishing a clear title structure, providing coaching and development, and implementing metrics and management.

Let's explore each of these steps in more depth.

11.4.1 Hiring

The first step in building a high-performing security team is to hire the right people. This involves several key activities:

Finding Talent: The cybersecurity talent pool is highly competitive, so it's essential to be proactive in your search. Expand your search by leveraging multiple avenues, including online job platforms, professional networks like LinkedIn, industry conferences, and employee referral programs. Consider partnering with specialized recruiters who have expertise in the cybersecurity field.

Determining Market Value: Cybersecurity salaries can vary widely depending on location, experience, and skillset. It's important to understand the market value for the roles you're hiring for to ensure that your compensation packages are competitive and attract top talent. Conduct salary surveys, consult with industry experts, and leverage online resources to gather information about current market trends.

Attracting Talent: Once you've identified top candidates, it's important to create a compelling offer that entices them to join your team. This goes beyond just salary and benefits; it's about showcasing your organization's culture, values, and opportunities for growth and development. Highlight the impact that security professionals can make within your organization and the potential for career advancement.

Selecting Top Talent: Hiring the right people is not only about finding individuals with the right technical skills, but it's also about identifying those who possess the drive, passion, and cultural fit to thrive in your organization. Look for candidates who are eager to learn, adaptable to change, and committed to the organization's mission and values. Conduct thorough interviews, reference checks, and assessments to evaluate candidates' skills, experience, and personality. Identifying and attracting top performers, those who consistently exceed expectations and drive innovation, is a hallmark of the CISO 3.0. For a deeper dive into how to identify and recruit "A" players, read Matt Sharp and Rock Lambros' book, *The CISO Evolution: Business Knowledge for Cybersecurity Executives*. Chapter 10 of that book offers great guidance on how to hire a high-performing security team that aligns with the strategic vision of the CISO 3.0.

11.4.2 Training

Once you've hired the right people, investing in their ongoing training and development is an essential next step. This ensures that they have the skills and knowledge to deliver capabilities that contribute effectively to the security program.

Determining Skills Gaps: Conduct regular skills assessments to identify areas where your team members might need additional training or development. This could involve self-assessments, skills matrices, performance reviews, or even engaging external providers to conduct training needs analysis. The NIST NICE Framework can be a valuable resource in this process, providing an overview of the KSAs required for different cybersecurity roles.

Relevance to Role: Ensure that the training and development opportunities you provide are relevant to the individual's current role and future career aspirations. This helps keep employees engaged and motivated, and it ensures that they are acquiring the skills that are most valuable to the organization.

Timing and Budgeting: Develop a training and development plan that outlines the specific training programs, certifications, or conferences that each team member will participate in. Allocate the necessary budget and time for these activities, ensuring that they are scheduled at appropriate intervals to minimize disruption to their day-to-day work.

11.4.3 Retention

Retaining top talent is possibly more important than recruiting it. The CISO 3.0 creates a work environment where employees feel valued, challenged, and supported, which builds loyalty and commitment. They also recognize that cybersecurity professionals are motivated by opportunities to learn and grow. Providing clear paths for career advancement and skill development is key to keeping the team engaged and motivated.

Skill Application and Growth: As team members gain new skills and expertise, it's important to offer opportunities for them to use those skills in their current roles or transition into new positions where their knowledge can shine. This keeps employees challenged and ensures the organization is getting the most out of its talent.

Setting Goals and Providing Challenges: Give each team member clear goals and expectations, along with opportunities to tackle challenges that help them grow. Regularly check in to review progress, offer feedback, and celebrate successes while addressing areas that need improvement.

Managing Stress: Cybersecurity can be a high-stress field, so it's essential to create a supportive atmosphere where employees feel comfortable asking for help or taking a break when needed. Foster open communication and provide resources for stress management and mental health support.

Investing in Development: Support your team's professional development by offering mentorship, opportunities to attend conferences or training programs, and access to online learning tools. Investing in their growth shows that you're committed to their long-term success.

Career Pathing: One of the best ways to retain talent is by offering clear, rewarding career paths. Recognize and celebrate skill advancement through visible acknowledgments like badges or other forms of recognition. This demonstrates your investment in your team's growth and helps reinforce their commitment to the organization.

One of the biggest challenges in retaining talent comes after employees have been reskilled or upskilled. Upskilling often leads to higher employee churn for two main reasons. The first is that with new skills, workers become more valuable in the market and may seek new opportunities elsewhere. To deal with this challenge, we need to be sure to advocate for and allocate budget specifically to compensate employees for their enhanced skills, ensuring that their salaries reflect current market rates.

To justify pay increases for upskilled employees, CISOs need to present a compelling business case to leadership. Start by outlining the tangible value the new skills bring to the organization. For example, if an employee gains advanced threat detection skills, this could reduce response times, mitigate breaches more effectively, and lower overall cybersecurity risk, saving the company potential costs in incident response or fines from regulatory breaches. Quantify these benefits wherever possible and show how reducing the average cost per incident or improving compliance can directly impact the bottom line.

Additionally, highlight the cost of replacing that employee if they leave. Recruiting, hiring, and onboarding new talent typically cost far more than retaining and developing existing staff. Factor in these costs along with the time it takes for a new hire to reach the same productivity level as the upskilled worker, and compare that to the investment in a salary adjustment. Presenting these numbers clearly helps justify salary increases as a cost-saving measure in the long run.

For instance, if a security analyst gains expertise in cloud security, which is an area where outsourcing would typically cost the company an additional $100,000 annually, then proposing a $20,000 salary increase is easily justified by the cost savings in-house talent can bring. Framing these conversations around cost–benefit analysis helps leadership understand the value of paying employees for their new skills.

The second major reason upskilled workers leave is tied to how their new skills are utilized. Too often, employees gain new expertise but continue to perform the same tasks as before or, worse, are asked to take on additional responsibilities without any increase in compensation. This can lead to burnout, frustration, and, ultimately, employee departure.

To avoid this, the CISO 3.0 should adopt the capability acquisition model, ensuring that tools, processes, and roles are aligned with the employee's new skills. This means integrating those skills into the workflow in a way that not only benefits the security program but also allows the employee to apply their training in meaningful ways. For example, if a team member gains new skills in automation and orchestration, they should be moved into a role where they can actively design and implement automated security processes rather than continuing their previous duties.

Additionally, setting clear boundaries between old and new responsibilities helps prevent employees from becoming oversubscribed. Without properly addressing this, you risk overloading them with tasks from both roles, leading to disengagement, especially if no extra compensation accompanies the added work.

Incorporating these skills into a well-structured security program turns training into actionable capabilities, ensuring that employees feel valued, their growth is recognized, and the organization fully benefits from the investment in their development. This strategy not only boosts retention but also strengthens the overall effectiveness of the cybersecurity team.

Your retention capabilities should be factored into your decision to insource or outsource. If you choose to insource and don't have strong retention practices, you will probably struggle to deliver consistent services that rival outsourced service-level agreements (SLAs). Worse yet, you may end up paying to train talent for other local companies. If you decide to build in-house expertise, you need to make sure that you have the budget and executive buy-in to keep your team incentivized, compensated fairly, and motivated to stay.

11.4.4 Career Pathing and Succession Planning

The CISO 3.0 takes an active role in career development and succession planning, ensuring there's a pipeline of skilled individuals ready to step into key positions when needed. This involves identifying high-potential team members, offering them the right training and growth opportunities, and laying out a clear path for career progression.

Promotion Opportunities: Establish a transparent promotion process that rewards strong performance and recognizes future potential. Give employees the chance to take on new responsibilities and challenges, and provide them with the support and guidance they need to thrive in these roles.

Title Structure: Create a clear, well-defined title structure that outlines various levels of responsibility and expertise within the security team. This clarity helps employees see their potential growth paths and also makes it easier to benchmark compensation and benefits.

Coaching and Development: Offer regular coaching and mentorship to help employees enhance their skills and reach their full potential. This can include one-on-one meetings, group training, or pairing team members with mentors, either from inside or outside the organization.

Metrics and Management: Use performance metrics and management tools to track individual and team progress, pinpoint areas for improvement, and provide constructive feedback. This ensures accountability and promotes continuous improvement across the security program.

Milestones and Achievements: Consider adding an element of fun to skill development with gamification, such as points, leaderboards, challenge coins, or even martial arts-style belts to celebrate achievements. By offering a clear path for advancement and recognizing accomplishments, you show your team that you value their efforts and are committed to their growth within the organization.

11.4.5 The CISO as a Talent Magnet

The CISO 3.0 is more than just a leader: They serve as a guiding light, attracting top-tier cybersecurity talent through a blend of strategic vision, a vibrant organizational culture, and a genuine commitment to their team's growth and development. In today's competitive cybersecurity landscape, they know that attracting and retaining the best talent require more than just a good salary. It's about fostering an environment where individuals feel not only employed but also truly valued, intellectually engaged, and empowered to reach their full potential.

To create this environment, the CISO 3.0 champions a culture of continuous learning, recognizing that change is the only constant in cybersecurity. They offer numerous professional development opportunities, encouraging team members to stay updated on the latest security trends and technologies through training programs, certifications, and mentorship. They also promote a collaborative and inclusive atmosphere where diverse perspectives are not only welcomed but also actively sought after. This openness cultivates innovation, allowing team members to feel comfortable sharing ideas, questioning assumptions, and contributing to the collective strength of the security program.

The CISO 3.0's ability to attract talent stems from their ability to create a sense of purpose and belonging. They articulate a clear and inspiring vision for the security program, emphasizing its role in the organization's success and its contribution to a safer digital world. They empower their team members to take ownership of their work by providing the autonomy and resources necessary for success. Additionally, they celebrate achievements, fostering an environment of appreciation and motivation that drives the team forward.

Here are some specific examples of how a CISO 3.0 can attract and retain top talent:

Champion a Culture of Learning and Growth: Offer generous professional development budgets, encourage participation in industry conferences and training programs, and create internal knowledge-sharing platforms like lunch-and-learns or brown bag sessions. I understand that when times are tight, training budgets are frequently the first thing to go, but continuous training doesn't have to cost a lot. Training credits are one of the easiest things to get vendors to "throw in" when you're closing a deal on a new product. Beyond that, many vendors and VARs will provide complimentary passes for conferences and events. You can leverage those same vendors and VARs to put on executive briefings (EBCs) where they will educate you on any number of topics. Most will even cover travel and accommodations for EBCs. You can also take advantage of local groups and organizations like the Information Systems Security Association (ISSA) and the Information Systems Audit and Control Association (ISACA), which usually put on educational events monthly. There are a ton of cheap and free educational options, so don't let budget constraints be an excuse for not building a culture of continuous learning.

Foster a Collaborative and Inclusive Environment: Encourage open dialog within the team, create opportunities for cross-functional collaboration, and celebrate the diversity and unique perspectives each team member brings.

Provide Mentorship and Guidance: Set up formal mentorship programs where junior team members are paired with experienced mentors. Offer regular feedback and coaching to help everyone continue growing and developing their skills.

Recognize and Reward Accomplishments: Celebrate both individual and team successes, no matter how big or small. Make it a point to acknowledge contributions and show appreciation for the hard work being done.

Offer Flexible Work Arrangements: Provide options for remote work and flexible schedules to meet the varying needs of your team. Supporting work–life balance not only benefits your team but also strengthens their overall productivity.

Showcase the Impact of Their Work: Regularly highlight how the security team's efforts protect the organization and its stakeholders. Emphasize the positive, real-world impact their work has on the company and society at large.

Lead by Example: Show your commitment to continuous learning, personal growth, and ethical leadership. Set a strong example for your team, inspiring them to pursue excellence and integrity in their work.

11.4.5.1 Example

One of my clients, a mid-sized manufacturing company nestled in a small Midwest town, was facing a persistent challenge: Attracting and retaining top cybersecurity talent. The allure of their industry and location was, to put it mildly, underwhelming. The security team was chronically understaffed and overworked, and morale was at an all-time low. The newly appointed CISO, recognizing the urgent need for change, embarked on a mission to revitalize the security culture and build a high-performing team despite the geographical and industry-specific hurdles.

She began by addressing the elephant in the room: The perceived "boringness" of the manufacturing sector and the remoteness of the company's location. To combat this, she boldly opened up roles to remote work, casting a wider net to attract top talent regardless of their geographical location. This strategic move instantly expanded the pool of potential candidates, allowing her to tap into a national, even global, talent market.

Next, she turned her existing team members into advocates and champions for the security program. She recognized that passionate and engaged employees are the best recruiters, so she empowered them to share their experiences and enthusiasm with potential candidates. This created a sense of authenticity and camaraderie that resonated with prospective hires, showcasing the positive culture and growth opportunities within the team.

Simultaneously, she invested heavily in her existing team, providing them with ample opportunities for training and development. She implemented a mentorship program, pairing junior analysts with seasoned security professionals to foster knowledge sharing and career growth. She also cultivated a more collaborative and inclusive environment, encouraging open communication, recognizing individual contributions, and celebrating successes.

She revamped the company's recruiting strategy, shifting the focus from simply filling vacancies to attracting "A" players who were not only technically skilled but also passionate about cybersecurity and aligned with the company's values. She actively sought out individuals who were eager to learn, adaptable to change, and excited about the opportunity to make a meaningful impact in a revitalized security program.

Within a year, the transformation was nothing short of extraordinary. The once disheartened security team was now energized and motivated, and the company had become a sought-after destination for top cybersecurity talent. The CISO's leadership, vision, and strategic initiatives had not only overcome the initial challenges but had also created a thriving security ecosystem where individuals felt valued, challenged, and empowered to make a difference.

What strategies do you currently employ to attract and retain top cybersecurity talent? How can you enhance your organization's appeal as a talent magnet and create an environment where individuals feel valued, challenged, and empowered to grow?

11.5 LEADING THROUGH CHANGE

Change is a constant in cybersecurity, and as CISOs, we must be prepared to lead our teams through it. Whether it's the introduction of new technology, a reorganization, or a shift in strategy, change can be challenging. What I've learned is that successful change management hinges on communication, transparency, and empathy. Involving your team in the process, listening to their concerns, and making them feel part of the solution can turn what feels like disruption into an opportunity for growth and innovation. The way you lead through these transitions can be the difference between a well-defended organization and one that's vulnerable.

Change in cybersecurity is often driven by the need to address outdated systems, implement new technologies, or realign with strategic goals. Staying ahead of cyber threats, maintaining compliance, improving efficiency, reducing costs, and enhancing customer trust all depend on your ability to guide your team through these shifts. Whether it's rolling out a Zero Trust Architecture, adapting to cloud security solutions, or responding to a major cyberattack, change leadership is the key. Beyond managing the technical side, you're also dealing with the psychological impact on your team and the organization, rebuilding trust, and ensuring everyone stays engaged in the security mission.

As a CISO, you are naturally a change agent. Your role involves not only introducing new security technologies but also adapting to emerging threats and fostering a security-conscious culture across the organization. The responsibility extends beyond your security team: Your influence must permeate every department, encouraging a mindset where security is everyone's responsibility. Leading this charge means shaping behaviors, influencing attitudes, and cultivating a shared sense of ownership when it comes to protecting the company's critical assets.

Several key strategies are essential to lead through change effectively. Start with a clear vision, communicating not only the "what" but also the "why" behind the change. Address concerns openly and maintain ongoing dialog so that your team feels informed and heard throughout the process. Involve your team in the journey, empowering them by soliciting their input and giving them ownership of the changes being implemented.

Be flexible, understanding that change is rarely a smooth, linear process. Adjust when necessary without losing sight of your long-term goals. Lastly, show resilience. Change often brings resistance and setbacks, but your ability to stay the course will inspire confidence and encourage your team to remain committed, even when the path is difficult.

Leading through change is a skill that requires both strategic thinking and emotional intelligence. Done well, it strengthens not only the organization's security posture but also its resilience in the face of future challenges.

11.5.1 Key Tenets of Change Leadership

Transparent Communication: The foundation of effective leadership during change is open and honest communication. The CISO 3.0 must articulate the reasons for change, painting a vivid picture of the expected benefits and acknowledging the potential challenges. They must create a safe space for dialog and feedback, actively listening to their team's concerns and anxieties. This transparency builds trust and fosters a sense of shared purpose, even in the face of uncertainty.

Empowerment and Ownership: Change is not something that is done *to* people; it's something that is done *with* them. The CISO 3.0 understands this and actively involves team members in the change process. They solicit input and ideas, encouraging everyone to contribute to the solution. They also empower their team members with the autonomy to implement solutions, fostering a sense of ownership and accountability. This approach not only generates valuable insights but also cultivates a sense of shared responsibility for the success of the change initiative.

Lead by Example: The CISO 3.0 embodies the change they wish to see in their team. They demonstrate a personal commitment to learning and adaptation, actively seeking out new knowledge and skills. They model the behaviors they expect from their team, inspiring them to embrace change and strive for continuous improvement.

Celebrate Success: Recognizing and rewarding accomplishments, both big and small, is important for maintaining morale and motivation during times of change. The CISO 3.0 celebrates successes, creating a positive and supportive environment where everyone feels valued and appreciated. This reinforces positive behaviors, encourages further innovation, and builds a sense of camaraderie within the team.

Provide Support and Resources: Change can be challenging, and the CISO 3.0 ensures that their team has the necessary resources and support to navigate it successfully. This could include providing training programs to equip team members with new skills, offering mentorship opportunities to guide their development, or even providing access to mental health resources to support their well-being during times of stress. The CISO 3.0 is not only a leader but also a coach and a confidant, providing the support and encouragement their team needs to thrive amidst change.

To be an effective change leader, lead by example, embrace change yourself, and model the behaviors you expect from others. Build strong relationships with key stakeholders across the organization and develop emotional intelligence to understand and manage your own emotions while empathizing with others. Celebrate success, acknowledging and rewarding progress to maintain momentum and motivation.

11.5.2 Change Management Models

Several tools and techniques can support your change leadership efforts. Kotter's eight-step change model provides a structured framework for planning and implementing change. Lewin's change management model outlines three stages of change: unfreezing, changing, and refreezing. The ADKAR Model focuses on the individual change process, addressing awareness, desire, knowledge, ability, and reinforcement.

Kotter's Eight-Step Change Model

Pros: Provides a clear, step-by-step roadmap for large-scale organizational change, emphasizing the importance of building a sense of urgency and creating a guiding coalition.

Cons: This approach can be time-consuming and might not be the best fit for smaller, fast-moving changes.

When and How to Use: Best suited for major transformations, such as implementing a company-wide cybersecurity awareness program or transitioning to a new security framework.

Cybersecurity Application

Step 1: *Create a sense of urgency* by presenting compelling data on the current threat landscape and the potential consequences of inaction.

Step 2: *Build a guiding coalition* by enlisting influential stakeholders from across the organization, including IT, HR, and legal.

Step 3: *Develop a clear vision* by outlining the desired cybersecurity posture and the key steps to achieve it.

Step 4: *Rally employee involvement* by engaging employees at all levels in the change process through training, communication, and recognition.

Step 5: *Enable action by removing obstacles* by addressing any obstacles that might hinder the implementation of new security measures.

Step 6: *Generate short-term wins* by celebrating early successes and showcasing the positive impact of the change.

Step 7: *Sustain acceleration* by building on the momentum and continuously improving cybersecurity practices.

Step 8: *Institute change* by embedding the new security culture into the organization's DNA.

2. ADKAR Model

Pros: Focuses on the individual's experience of change, ensuring that employees have the awareness, desire, knowledge, ability, and reinforcement to tackle the change successfully.

Cons: May require significant investment in training and support and can be challenging to measure progress.

When and How to Use: Particularly useful when introducing new technologies or processes that require employees to learn new skills or adapt their behaviors.

Cybersecurity Application

Awareness: Communicate the need for change, highlighting the risks associated with current practices and the benefits of the new approach.

Desire: Foster a desire for change by demonstrating how the new security measures will improve employees' work experience, protect their personal data, and contribute to the company's success.

Knowledge: Provide training and resources to equip employees with the necessary skills and understanding.

Ability: Offer hands-on opportunities for employees to practice the new behaviors and receive feedback.

Reinforcement: Recognize and reward positive behaviors and address any challenges or resistance promptly.

3. Lewin's Change Management Model

Pros: Simple and easy to understand, providing a basic framework for managing the transition from the current state to the desired state.

Cons: May oversimplify the complexities of change and neglect the emotional and psychological aspects.

When and How to Use: Suitable for relatively straightforward changes or as a starting point for change management initiatives.

Cybersecurity Application

Unfreezing: Create awareness of the need for change by highlighting vulnerabilities in the existing security posture.

Changing: Implement the new security measures, providing clear guidance and support to employees.

Refreezing: Reinforce the new behaviors and processes, ensuring that they become embedded in the organizational culture.

Leading through change is an ongoing process requiring continuous learning and adaptation. Embrace your role as a change agent, leverage best practices, and utilize available tools and techniques to ensure your organization's resilience. Remember, change is about people as much as technology. Foster a culture of security awareness and empower your team to create a lasting foundation for a secure future.

11.5.3 Behavioral Change Management

Behavioral change management is a systematic approach to understanding and influencing human behavior to achieve desired outcomes. It recognizes that change is not just about implementing new technologies or processes: It's also about shifting through processes and behaviors. The CISO 3.0 can leverage behavioral change management principles to create a security-conscious culture where employees actively participate in protecting the organization from cyber threats.

The FOGG Behavior Model, developed by Dr. BJ Fogg, provides a useful framework for understanding and influencing behavior change. It states that for a behavior to occur, three elements must converge at the same time: Motivation, ability, and a prompt.

Increase Motivation: Instead of simply highlighting the risks of weak passwords, incentivize the behavior you want. You could provide rewards for early adopters or high complexity scores.

Increase Ability: While providing a password manager is helpful, go a step further by offering training sessions on how to use it effectively and securely. Consider integrating the password manager with single sign-on solutions to streamline the login process and reduce friction.

Provide Prompts: Rather than relying solely on reminders, use creative prompts to reinforce good password hygiene. For instance, display eye-catching posters with password tips in high-traffic areas or incorporate password-themed quizzes into company newsletters. Offer small incentives, like coffee vouchers or extra paid time off (PTO), for employees who consistently demonstrate strong password practices.

The goal is to shift the perception of security from a burden to a shared responsibility, where employees are actively engaged and motivated to contribute to the organization's cyber resilience.

11.5.4 Influence and Persuasion

The CISO 3.0 must also be skilled in the art of influence and persuasion. They need to be able to effectively communicate the importance of security to stakeholders at all levels, from the boardroom to the front lines. Books like *Influencer: The Power to Change Anything* and *Nudge: Improving Decisions About Health, Wealth, and Happiness* offer valuable insights into the psychology of influence and how to nudge people toward desired behaviors.

Some strategies for improving influence and persuasion include the following:

Building Relationships and Trust: Establish strong relationships with key stakeholders across the organization, based on mutual respect and trust.

Framing the Message: Tailor your communication to resonate with your audience, highlighting the benefits of security in a way that aligns with their priorities and concerns.

Leveraging Social Proof: Highlight the positive actions of others to encourage desired behaviors.

Making It Easy: Remove barriers and simplify processes to make it easier for people to adopt secure behaviors.

Providing Incentives: Offer rewards or recognition for individuals or teams who demonstrate good security practices.

11.5.5 Gamification

In an era where human error remains a leading cause of security breaches, the CISO 3.0 recognizes that traditional, compliance-driven security awareness training often misses the mark. What's needed is a fresh approach, one that taps into the innate human desire for engagement, achievement, and recognition. Enter gamification: The practice of applying game-design elements and principles to everyday, nongame situations.

Gamification works by introducing game-like elements, points, leaderboards, badges, and simulation into training, turning it from a task to an experience. When training feels more like a game, employees are more likely to pay attention and engage actively rather than simply checking a box. This interactive approach not only captures interest but also improves knowledge retention by making the learning process more enjoyable and memorable. It goes beyond simply teaching security best practices. Gamification encourages employees to apply these principles in real-world scenarios, rewarding positive behaviors and providing immediate feedback. It also helps foster a sense of community and shared responsibility, building a proactive security culture.

While gamification is a powerful tool, it's not the be-all, end-all solution. It works best when the goal is to drive behavior change, such as encouraging employees to use stronger passwords, report phishing attempts, or follow data protection procedures. However, it's important to consider your audience. Gamification tends to resonate most with employees who are comfortable with technology and enjoy game-like experiences. Additionally, not all security topics lend themselves to this approach, more complex or sensitive issues may require traditional methods of training, as gamification could trivialize critical information.

When implementing gamification, it is essential to align these initiatives with your business goals. The aim should always be to support the organization's security goals, not just provide entertainment. Measuring success is also key to tracking participation, engagement, and knowledge retention and gauging the effectiveness of your efforts. Gathering feedback and refining your approach over time will help maximize the impact of your gamification strategy. Also, remember that gamification should be enjoyable and engaging, not a cutthroat competition or another mandatory training exercise. When done right, it can transform how employees interact with security initiatives, making them more invested in protecting the organization.

11.5.5.1 Security Gamification Ideas

Gamification isn't just about using external apps: It's also about infusing game-like elements into your security initiatives. Here are some ideas to spark your creativity, along with examples of apps that can help you bring them to life:

Points, Leaderboards, and Rewards

Points System: Award points for completing security training modules, reporting phishing emails, or demonstrating secure behaviors.

Leaderboards: Create leaderboards to track progress and foster friendly competition among employees. Publicly recognize top performers to incentivize participation.

Rewards: Offer tangible or intangible rewards for achieving milestones or demonstrating exceptional security awareness. Consider gift cards, extra time off, or company-branded merchandise.

App Examples: KnowBe4, Bunchball Nitro.

Badges and Achievements

Skill-Based Badges: Award badges for mastering specific security skills, such as identifying phishing attacks, using encryption tools, or securely configuring devices.

Knowledge-Based Badges: Recognize employees for completing security awareness training modules or passing security quizzes.

Behavior-Based Badges: Reward employees for consistently demonstrating secure behaviors, such as reporting suspicious activity or using strong passwords.

App Examples: Badgeville, Credly.

Simulations and Challenges

Phishing Simulations: Add a program like "catch of the day" to your regular simulated phishing campaigns to test employees' ability to identify and report them.

Capture the Flag (CTF) Challenges: Organize CTF events where teams compete to solve security challenges and exploit vulnerabilities in a controlled environment.

Escape Room-Style Challenges: Build interactive scenarios where employees must use their security knowledge and skills to solve puzzles and escape a "locked room".

Quiz Show: Host a live or virtual quiz show where employees can test their security knowledge in a fun and competitive setting, with prizes for the winners.

App Examples: KnowBe4, Kahoot!, Immersive Labs.

Other Gamification Ideas

Storytelling and Narratives: Weave security awareness messages into engaging stories and narratives that resonate with employees.

Microlearning: Break down complex security topics into bite-sized, gamified modules that employees can complete in short bursts.

Social Elements: Encourage social interaction and collaboration around security topics through online forums, discussion boards, or team-based challenges.

Personalized Learning Paths: Tailor security awareness training to individual roles and skill levels, providing a more engaging and relevant experience.

Mobile-First Design: Make security awareness training and resources easily accessible with apps like Centrical on mobile devices to accommodate employees on the go.

Remember, successful gamification is about more than just adding game elements to your security program. It's about creating a fun, engaging, and rewarding experience that motivates employees to take ownership of their security responsibilities and contribute to a strong security culture.

11.5.6 Building a Positive Security Culture

One of the most common missteps I observe among CISO 2.0s is the reliance on fear and punishment to enforce security compliance. It's understandable why the idea of setting strict thresholds for phishing failures and resorting to termination for repeat offenders might seem appealing. After all, these individuals are potentially jeopardizing the organization and increasing the workload for the security team. However, the "stick" approach rarely yields the desired results and can actually backfire. It can create a culture of fear, where employees become hesitant to interact with the security team, ultimately hindering your overall security goals.

People often feel like they are being scolded or are "in trouble" when they have to report to the security team. We want to encourage people to report security incidents, even if they've made mistakes. Our aim should be to encourage open communication and prompt reporting, even if it means admitting to clicking on a phishing link. We don't want people to be afraid and cover up that type of thing. When people feel safe and supported, they are more likely to be transparent and honest with the security team, which is fundamental but often requires a culture shift.

Sometimes, a little empathy and understanding can make a world of difference. I recall a client who discovered that one of their frequent phishing victims had dyslexia. By collaborating to implement accessibility features, that employee was empowered to improve their security awareness and quickly ceased to be a repeat offender. This experience underscores the power of compassion, understanding, and open dialog in building a security culture. Remember, a little grace can go a long way in achieving lasting security compliance.

11.5.6.1 Example

One of my clients implemented a "Catch of the Day" program to gamify security reporting. Employees who reported phishing emails were entered into a daily drawing for gift cards. This initiative leveraged both gamification and positive reinforcement to not only increase phishing reporting but also foster a sense of shared responsibility for security. Employees began to see themselves as part of the solution rather than a potential problem for the security team.

11.6 THE CISO 3.0 AS A LEADER

Leadership is about more than just managing: It is the art of guiding, motivating, and inspiring others toward a common goal. It's the key difference between a team that's engaged, resilient, and ready to take on challenges and one that's stagnant, disengaged, and vulnerable. Effective leadership fosters an environment where people feel valued and supported, empowering them to perform at their best.

Consider two CISOs responding to a major security incident. One remains calm and composed, communicates clearly, and empowers their team to take charge of the situation. The other panics, micromanages, and looks for others to blame. In the first scenario, the response is swift and organized, with minimal disruption and a boost in team morale. In the second, the result is confusion, delays, and a demoralized workforce. The difference? Leadership.

For the CISO 3.0, leadership skills aren't just an added bonus; for the CISO 3.0, they are fundamental to their success. Beyond technical know-how, CISOs need executive-level skills to navigate complex business landscapes. What separates a great leader from the rest is a blend of vision, integrity, empathy, decisiveness, and accountability.

Vision provides a clear direction that motivates the team. Integrity, built on honesty and transparency, earns the trust and respect of colleagues. Empathy creates a supportive and inclusive atmosphere. Decisiveness enables leaders to make timely, informed choices, even in the face of uncertainty. And accountability means owning both successes and failures while empowering others to take responsibility as well.

In the end, strong leadership is what drives teams to rise above challenges and achieve their goals, turning obstacles into opportunities for growth and success.

Some essential executive-level skills include the following:

Visionary Leadership: Articulating a clear and compelling vision for the future of the security program, inspiring and motivating others to work toward that vision.

Strategic Thinking: Seeing the big picture, anticipating future trends, and aligning with the organization's overall strategy.

Change Management: Leading and managing change effectively, ensuring that security initiatives are embraced and implemented successfully.

Financial Acumen: Understanding and managing budgets, demonstrating the value of security investments to the organization.

Relationship Building: Cultivating strong relationships with key stakeholders across the organization, including the C-suite, the board, and other departments.

Communication and Collaboration: Communicating effectively with both technical and nontechnical stakeholders, building strong relationships, and fostering collaboration across the organization.

Decision-Making and Problem-Solving: Making sound decisions under pressure, solving complex problems, and adapting to changing circumstances.

Emotional Intelligence: Understanding and managing one's own emotions and the emotions of others, building trust and empathy within the team.

Influence and Persuasion: Influencing and persuading others, gaining buy-in for security initiatives, and fostering a culture of security awareness.

Mentorship and Coaching: Guiding and developing team members, providing them with the support and resources they need to succeed.

Take a moment to reflect on that list. Where do your skills align? Where do you have opportunities for growth?

CISO 3.0s should consistently look for ways to strengthen their leadership abilities, recognizing that becoming an effective leader is a continuous endeavor that requires dedication and self-awareness. There are several paths CISOs can take to develop these valuable skills.

Participating in *formal leadership training* programs or workshops can provide practical tools and fresh perspectives on leading teams. *Mentorship and coaching* from seasoned leaders or executive coaches also offer invaluable guidance, helping CISOs refine their leadership style and work through areas for improvement.

Additionally, *self-reflection* plays a key role. Regularly evaluating your own leadership style and seeking honest feedback from colleagues can offer insights into areas where growth is needed. Engaging in *continuous learning*, whether through reading leadership books, attending industry conferences, or staying current with evolving leadership practices, ensures that CISOs are always honing their skills and adapting to the demands of their role.

11.6.1 The CISO's Self-Assessment

The CISO 3.0, while adept at assessing their team's capabilities, must also turn the lens inward. Self-awareness is a cornerstone of effective leadership, and CISOs must be acutely conscious of their own strengths and weaknesses. Just as they meticulously evaluate their team's skills using tools like the skills matrix discussed in previous chapters, CISOs should maintain a similar matrix for themselves. This self-assessment is a mirror, reflecting their current capabilities and highlighting areas for growth and development.

To facilitate this introspective process, we've created the downloadable CISO Skills Assessment & Matrix, available in the resources at CISO30.com. This tool offers a structured framework for self-evaluation, divided into three sections:

Executive Skills: This section evaluates your leadership, communication, strategic planning, decision-making, and business acumen capabilities.

CISO Career Development Skills: This section focuses on skills like networking, personal brand building, and career development.

Technical Skills: This section assesses your technical proficiency and understanding of cybersecurity concepts.

Each tab is further divided into subsections that categorize the different skill areas. Within each skill area is a list of questions designed to help you assess your prowess. You do this by rating your skill level, its importance, and the relevance or frequency you engage in each activity for each question on a scale of 1–5.

By completing this assessment, you can gain valuable insights into your strengths and weaknesses, enabling you to identify areas for targeted improvement. The completed assessment can also function as an ongoing skills matrix, allowing you to track your progress over time and ensure continuous growth and development.

The CISO 3.0 must possess a balanced blend of technical and executive skills. While technical expertise remains foundational, it's the executive skills that truly distinguish the CISO 3.0 as a strategic leader.

The executive skills laid out in the downloadable skills assessment and matrix template include the following:

Leadership: Inspiring, motivating, and guiding others toward a shared vision.

Team Development: The capacity to build and nurture a high-performing team, fostering collaboration, growth, and a sense of shared purpose.

Talent Management: Attracting, developing, and retaining top talent, creating an environment where individuals thrive and contribute their best.

Self-Awareness/Personal Growth: The willingness to introspect, recognize one's own strengths and weaknesses, and actively seek opportunities for personal and professional development.

Ethical Leadership: The commitment to integrity, transparency, and ethical decision-making, setting a positive example for the entire organization.

Conflict Management: The ability to effectively manage and address disputes, promoting a cooperative and productive workplace atmosphere.

Effective Communication: Communicating clearly and persuasively with diverse audiences, both technical and nontechnical.

Strategic Planning: The ability to anticipate future challenges and opportunities, craft forward-thinking strategies, and create actionable plans to achieve long-term objectives.

Decision-Making: Making sound and timely decisions, even in the face of uncertainty and pressure.

Relationship Building: The skill to cultivate strong and meaningful relationships with stakeholders across the organization, fostering trust and collaboration.

Crisis Management: Leading effectively during times of crisis, maintaining composure, making critical decisions, and guiding the organization toward recovery.

Business Acumen: The understanding of the organization's business objectives, financial metrics, and risk appetite, enabling the alignment of security initiatives with broader business goals.

Self-awareness is not just about recognizing areas for improvement: it's also about actively working toward growth and development. Good leaders understand that leadership is a journey, not a destination. They are constantly seeking to refine their skills, expand their knowledge, and evolve their leadership style to meet the ever-changing demands of their role. CISOs can become more effective leaders by investing in their own development and inspiring their teams, building trust with stakeholders, and driving the organization toward a secure and successful future.

11.6.2 Assessments and the CISO Wheel

Two more powerful tools that CISOs can leverage to enhance their leadership skills are 360 assessments and the CISO Wheel. These tools can help CISOs gain a deeper understanding of their leadership style and identify areas for growth.

360 Assessments: A 360 assessment provides comprehensive feedback on an individual's leadership abilities from a wide range of perspectives. This process involves collecting confidential input from colleagues at different levels, peers, direct reports, and supervisors who interact with the CISO in various settings. By gathering feedback from multiple sources, this tool offers a well-rounded view of how a leader is perceived in terms of key competencies like communication, decision-making, collaboration, and strategic thinking.

For CISOs, a 360 assessment can be eye-opening. It provides a chance to see yourself through the eyes of others, revealing strengths you may not have recognized and uncovering blind spots that need attention. This kind of feedback can be invaluable for personal growth, offering a clear roadmap for honing leadership skills and improving your impact on the team and organization. By actively seeking feedback from different stakeholders, CISOs can develop a deeper understanding of their leadership style and take focused action to enhance their effectiveness.

If you're interested in conducting a 360 assessment, you can take advantage of a platform like EchoSpan. You can set up a free account at www.echospan.com/360-degree-feedback-process.asp. The process typically includes selecting a group of raters, distributing a structured questionnaire, and analyzing the feedback. The results are presented in a confidential report, highlighting key themes and areas where growth is needed.

The CISO Wheel: A model that assesses across the 12 key dimensions of executive leadership listed above. The CISO Wheel helps create a roadmap for leadership development, highlighting specific areas where CISOs can focus their efforts to become more effective and impactful leaders (Figure 11.1). It encourages stronger leadership, recognizing that effective leaders excel in multiple dimensions, not just one or two.

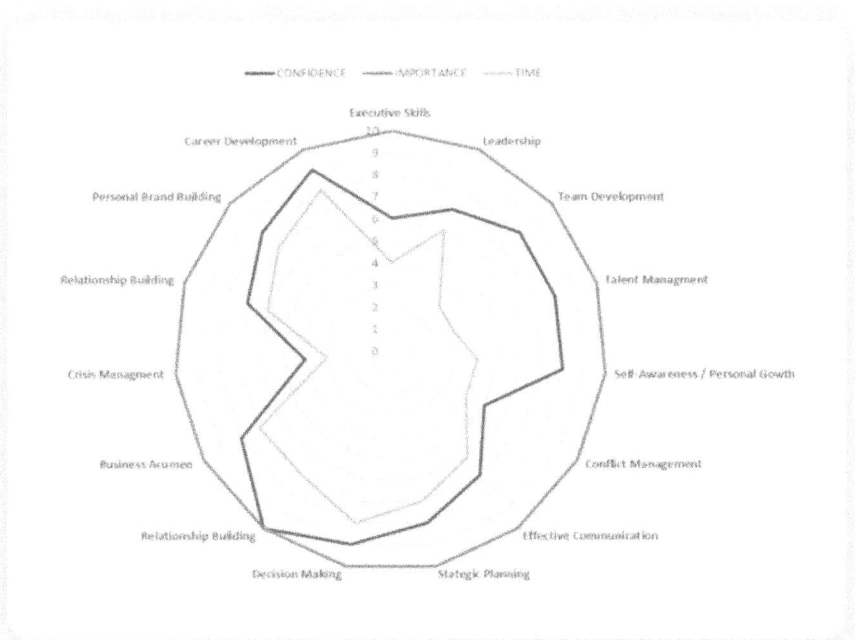

FIGURE 11.1 Radar Graph.

The CISO Strengths Wheel, available for download at CISO30.com, is a practical tool designed to facilitate this self-assessment process. Similar to the self-assessment and skills matrix, the CISO Wheel prompts you to rate your confidence and competence in each of the 12 leadership dimensions, as well as their importance to your specific role and the amount of time you currently invest in them. The tool then generates a visual radar map, allowing you to see how your skills align with the requirements of your job and how your time allocation matches up with your role's priorities. It also allows for an assessment of technical skills across 22 domains, providing an overview of your overall capabilities as a CISO.

By actively seeking out opportunities for development and leveraging tools like 360 assessments and the CISO Strengths Wheel, CISOs can continuously enhance their leadership skills and become the transformative leaders their organizations need. They can build high-performing teams, foster a culture of security awareness, and effectively communicate with stakeholders at all levels to drive the organization to even greater success.

11.7 CONCLUSION

Great cybersecurity leadership isn't just about having the shiniest tech or the most complex strategies. It's also about building a team that's firing on all cylinders, that is, technically skilled, strategically minded, and ready to adapt to whatever comes their way. In this chapter, we've shone a light on the importance of assembling and nurturing such a high-performing security team.

We've taken a closer look at the diverse roles that make up a strong security team, from the eagle-eyed analysts who spot threats to the visionary architects who design defenses. We've discussed the pros and cons of building your team in-house versus bringing in outside help, recognizing that there's no one right answer. And we've dug into the NIST NICE Framework, a valuable tool for figuring out which roles you need and the skills those roles require.

But cybersecurity isn't a static field. It's constantly moving, and your team needs to move along with it. That's where the CISO 3.0 steps up, not only managing technology and risk but also guiding and inspiring their team through the inevitable changes. Their influence reaches beyond the security department, creating a culture where everyone in the company understands the importance of security and plays a part in protecting the organization.

Building a top-notch security team isn't easy. It takes a strategic approach that includes finding the right people, training them, keeping them engaged, and giving them room to grow. The CISO 3.0 needs to be a talent magnet, creating an environment where people are excited to learn, collaborate, and contribute their best. It's also about finding the right balance between people, processes, and technology, making sure they all work together seamlessly.

Attracting and keeping great talent isn't a one-and-done: It's an ongoing commitment. The CISO 3.0 needs to create a workplace where people feel valued, challenged, and supported. Through fostering a culture of learning and growth, they'll build a team that's not only technically skilled but also strategically aligned, adaptable, and resilient.

Think of the CISO 3.0 as the architect of a human firewall, a living, breathing defense system made up of individuals who are not only experts in their field but also passionate about protecting the company's digital future. The success of your security program rests on the shoulders of this team, and your leadership will shape their destiny and, ultimately, the long-term security and success of the entire organization.

11.7.1 Key Takeaways

1. **Building a High-Performing Team**
   - Effective cybersecurity programs depend on skilled, engaged teams working cohesively toward common goals.
   - A high-performing team integrates technical expertise, strategic alignment, and a collaborative culture to navigate evolving cybersecurity challenges.
2. **Defining Roles and Responsibilities**
   - Key roles include CISO, director/manager, architect, engineer, analyst, project manager, and specialized functions such as risk management and cloud security.
   - Clear delineation of roles and responsibilities, guided by frameworks like NIST NICE, ensures teams are aligned and capable of addressing organizational needs.
3. **Balancing Insourcing and Outsourcing**
   - Strategic decisions on insourcing versus outsourcing depend on factors like budget, organizational needs, and risk tolerance.
   - Core functions often benefit from in-house expertise, while commoditized tasks (e.g., SOC monitoring) are frequently outsourced.
4. **Talent Acquisition and Retention**
   - Hiring the right talent requires understanding market dynamics, offering competitive compensation, and fostering an attractive work environment.
   - Retention is enhanced through skill development opportunities, clear career progression, and a culture of recognition and inclusion.
5. **Continuous Learning and Development**
   - Investing in ongoing training ensures team members stay ahead of emerging threats and technologies.
   - Tools like the NIST NICE Framework and structured training plans help identify and address skills gaps effectively.
6. **Leading through Change**
   - Change management is critical for cybersecurity success, requiring communication, transparency, and team involvement.
   - Frameworks like Kotter's eight-step model and ADKAR provide structured approaches to navigating organizational shifts.
7. **Behavioral Change and Culture Building**
   - Fostering a security-conscious culture involves understanding human behavior and leveraging tools like the FOGG Behavior Model.
   - Gamification and positive reinforcement can drive engagement and encourage proactive security practices.

8. **The CISO 3.0 as a Leader**
   - Modern CISOs combine technical expertise with leadership skills, guiding their teams through strategic decision-making and operational challenges.
   - Self-assessment tools like the CISO Wheel help identify strengths and areas for improvement in leadership capabilities.

11.7.2 Reflection and Exploration Questions

Understanding the Concepts

1. What are the main roles of leadership in building effective security teams?
2. How do frameworks like NIST NICE help in defining roles and addressing skills gaps?

Connecting the Dots

3. How does the concept of insourcing versus outsourcing relate to earlier discussions about capability-building and strategic alignment?
4. In what ways do behavioral change strategies enhance the effectiveness of security programs?

Real-World Application

5. Can you identify examples where strong leadership transformed a struggling security team into a high-performing one?
6. How might gamification be applied in your organization to promote security awareness and engagement?

Personal and Professional Growth

7. How can you use tools like the CISO Skills Assessment & Matrix to evaluate and improve your leadership capabilities?
8. What steps can you take to foster a culture of continuous learning and collaboration within your team?

Further Exploration

9. What additional tools or strategies could enhance team performance and engagement in your organization?
10. What could you do to begin growing your influence?
11. How could you start to build an organization-wide culture of security?

12 Security Tactics

12.1 INTRODUCTION

The chief information security officer (CISO) 3.0's path is one of growth, a strategic climb from hands-on security to the big-picture view of how security fits into the business. But don't get it twisted, this doesn't mean leaving the trenches behind. Even as a master strategist, the modern CISO needs to stay sharp, keeping up with the latest security tactics and knowing when to use them.

In this chapter, we're going to unpack the essential tactics that every CISO 3.0 needs in their toolkit. We'll explore the power of putting identity at the heart of your security, the importance of always keeping an eye on things, the wisdom of good data governance, the resilience of a zero-trust approach, and the flexibility of cloud-native protection. They're cutting-edge tactics that will help you keep your organization's assets safe in the years to come.

12.2 IDENTITY CENTRICITY

The old days of building a fortress around your network and calling it a day are over. With cloud computing, remote work, and cyberattacks getting smarter every day, it's clear that the traditional perimeter is crumbling. Today's security landscape is more like a sprawling kingdom, with data and resources scattered across countless locations.

In this new world, identity is your castle wall. It's not only about protecting the network itself anymore but also about safeguarding the keys that unlock the gates. The CISO 3.0 understands this shift. They know that in a world where data is accessed from everywhere, controlling who has access to what is paramount.

The old model of trusting everything inside the network and blocking everything outside just doesn't cut it anymore. We've got people working from home, apps running in the cloud, and a whole zoo of devices connecting from who-knows-where. It's a complex and ever-changing environment, and the traditional perimeter is no longer a moat you can rely on.

I'm trying not to use the old and wildly overused trope "identity is the new perimeter", but the fact is the focus has shifted from securing the network to securing the individuals and devices that traverse it. Organizations can ensure that only authorized entities gain access to critical and sensitive resources, regardless of their location or network connection, by verifying and fortifying user identities. In the digital realm, identity is the ultimate gatekeeper, the guardian of the organization's most valuable assets.

Identity governance, which encompasses the policies, processes, and technologies for managing and controlling user identities and access rights, is a fundamental prerequisite for both zero trust and AI security.

DOI: 10.1201/9781003510789-17

Zero trust is a security model that operates on the principle of "never trust, always verify". It assumes that every user, device, and application is a potential threat and must be continuously authenticated and authorized before accessing any resources. Effective identity governance is essential for implementing zero trust, as it provides the foundation for verifying and authorizing identities, ensuring that only legitimate users have access to sensitive data and systems.

Similarly, AI security relies heavily on identity governance. AI systems often process and analyze vast amounts of data, including sensitive information. Robust identity governance ensures that AI systems have access only to the data they need to perform their functions, minimizing the risk of unauthorized access or data breaches. Identity governance helps track and audit AI system activity, ensuring transparency and accountability in their use.

Shifting to an identity-centric security model isn't a simple switch; it requires a multipronged approach that involves your people, your processes, and your technology. It all starts with a crystal-clear understanding of how identity works within your organization. Who are your users? What devices are they using? Which applications are they accessing? And what kind of data are they handling? It's about taking a complete inventory of your identity landscape and pinpointing any potential weak spots.

Once you've mapped out your identity model, it's time to put some serious safeguards in place. We're talking about robust identity and access management (IAM) controls like multifactor authentication (MFA) and privileged access management (PAM). But it's not just about the tech: You also need clear policies and procedures for managing identities. Think of it as creating a rulebook for who can access what, when, and how.

And remember, the process isn't complete once you've implemented these controls. Identity-centric security is an ongoing process. You need to constantly monitor and audit identity-related activities, looking for any signs of suspicious behavior. It's like being a vigilant guardian, always on the lookout for potential threats before they can breach your defenses.

Identity isn't just a buzzword: It should be the guiding principle behind your entire security strategy. From risk assessments to vulnerability management, incident response, and even security awareness training, everything should revolve around protecting those identities.

When you're evaluating risks, think about the potential fallout if someone's credentials are compromised or an insider decides to go rogue. Make smart investments in IAM controls and processes that proactively address these threats. And when the inevitable incident happens, make sure your response plan is ready to handle identity-related breaches. That means revoking access, resetting passwords, and conducting thorough investigations to get to the root of the problem and prevent it from happening again.

Don't forget your employees are your first line of defense. Educate them about the importance of protecting their credentials, spotting phishing scams, and speaking up if they see something suspicious. You can empower your entire workforce to become active participants in protecting your organization's most valuable assets by fostering a culture of security awareness.

Identity-centric security isn't just a tactical change: It's a fundamental shift in how we approach security in today's digital world. It's about recognizing that the old perimeter is no longer enough and that identity is the new cornerstone of a resilient organization. Prioritizing identity security allows you to not only mitigate risks but also enable your business to thrive in the new, fully distributed world.

12.2.1 Insider Threats

The concept of identity-centric security extends beyond external threats; it's equally important for mitigating insider threats, which can be just as damaging, if not more so, than external attacks. Insider threats encompass a range of risks posed by individuals within an organization who misuse their legitimate access to sensitive data or systems. These individuals could be current or former employees, contractors, or even trusted partners. Their actions can be intentional, driven by malicious intent or personal gain, or unintentional, resulting from negligence or lack of awareness. The consequences of insider threats can be severe, including data breaches, financial losses, and even operational disruptions.

The CISO 3.0 understands that insider threats pose a significant, yet often underestimated, risk. Even with strong perimeter defenses and external threat detection, the biggest threat can come from within the organization. The key challenge is balancing employee access needed to perform their jobs with preventing misuse of those privileges.

An aspect of insider threats that often goes unnoticed is the transformation of external threats into insider threats. When attackers successfully compromise user credentials through phishing attacks, social engineering, or other tactics, they essentially gain insider access. This allows them to operate within the network, often undetected, and potentially cause considerable damage. The stolen credentials might belong to a regular employee, granting the attacker access to sensitive data or systems they wouldn't normally have access to. Or, even more alarmingly, the compromised credentials could belong to a privileged user, such as a system administrator or executive, giving the attacker elevated access and the ability to inflict even greater harm.

This external-to-insider threat transformation highlights the critical importance of identity-centric security. Implementing robust IAM controls, such as MFA and PAM, organizations can reduce the risk of external attackers gaining insider access, even if they manage to compromise user credentials.

Identity centricity is the key to mitigating insider threats by focusing on verifying and securing user identities and access rights. With robust IAM controls, organizations can limit the potential damage that insiders can cause.

Key IAM controls for mitigating insider threats include the following:

- **Least Privilege Access:** Granting users only the minimum level of access necessary to perform their job functions reduces the potential impact of a compromised or misused account.
- **Role-Based Access Control (RBAC):** Assigning permissions and access rights based on predefined roles ensures that users have access only to the resources they need for their specific job functions.

- **User and Entity Behavior Analytics (UEBA):** Monitoring user and entity behavior patterns can help identify anomalies and potential insider threats, such as unusual access patterns, data exfiltration attempts, or attempts to bypass security controls.
- **Privileged Access Management:** Implementing strict controls and monitoring for privileged accounts, which have elevated access to critical systems and data, helps prevent misuse and abuse of these powerful accounts.

To create a secure environment that mitigates insider threats, organizations must implement comprehensive IAM controls alongside a broader insider threat program. An effective program requires a multifaceted approach that integrates technology, processes, and people. It begins by assessing the organization's risk landscape, identifying critical assets, and understanding potential vulnerabilities that could be exploited by internal actors.

The first step in building this program is conducting thorough risk assessments to uncover potential insider threats. This involves analyzing both the organization's assets and possible motivations for malicious behavior. Once the risks are clearly defined, preventive measures can be implemented. These include security awareness training, establishing clear policies, and deploying technical controls like access management and data loss prevention (DLP) tools.

Employee education plays a vital role in mitigating insider threats. A well-informed workforce is more likely to recognize suspicious behavior and understand the potential consequences of violating security policies. The insider threat program should promote a culture of vigilance, ensuring employees feel comfortable reporting unusual activity without fear of retaliation.

While many activities within an insider threat program resemble those of a security operations center (SOC), they differ in focus. SOCs primarily deal with external threats, whereas insider threat programs concentrate on risks posed by individuals inside the organization. Effective collaboration across departments such as HR, legal, and compliance is critical to ensure the program aligns with the organization's risk management strategy and complies with relevant laws.

The CISO 3.0 must champion the development of a robust insider threat program by fostering a culture of security awareness and trust. This includes encouraging employees to report suspicious behavior and ensuring they understand their responsibility in protecting the organization's assets.

By prioritizing identity-centric security measures and embedding them into the overall security strategy, the CISO 3.0 can significantly reduce insider threats. This proactive approach not only strengthens the organization's defenses but also demonstrates the CISO's leadership in safeguarding critical assets while enabling business growth. With these efforts, CISOs position themselves as strategic leaders who contribute to the organization's long-term success.

12.3 ZERO TRUST

The concept of Zero Trust has become a focal point in modern cybersecurity, yet its implementation often presents challenges for CISOs. The promise of Zero Trust lies in its ability to mitigate risks like data breaches, ransomware, and insider threats, but its complexity and the multitude of approaches can lead to confusion and stalled progress. Zero Trust is the solution that became its own problem. The CISO 3.0, however, sees Zero Trust not as a problem but as a strategic opportunity to enhance the organization's security posture and align it with business objectives.

At its core, Zero Trust can be viewed as "next-generation access control". It's about automating just-in-time access control, making it granular, and contextualizing it with real-time risk information from the SOC. This approach ensures that users have access only to the resources they need when they need them and from authorized devices and locations, reducing the attack surface and minimizing the potential impact of a breach.

The National Security Agency (NSA) defines Zero Trust as follows:

> Zero Trust is a security model, a set of system design principles, and a coordinated cybersecurity and system management strategy based on an acknowledgment that threats exist both inside and outside traditional network boundaries.
>
> **– NSA | Embracing a Zero Trust Security Model**

The business problem that Zero Trust addresses was succinctly captured by a CISO I work with:

> We need to give users access to what they need, when they need it, from anywhere, on any device, without compromising security.

Zero Trust is often perceived as a technology issue, but in reality, it's a business problem disguised as a technical challenge. Many organizations struggle to understand what Zero Trust truly means and how to implement it effectively.

Several organizations offer guidance and frameworks for implementing Zero Trust. Each offers similar guidance, definitions, and pillars but with subtle differences in terminology and focus areas. Some prominent examples are as follows:

- **National Institute of Standards and Technology (NIST) SP 800–207 Zero Trust Architecture (August 2020):** This framework provides a detailed overview of Zero Trust principles and components, offering guidance on designing and implementing a Zero Trust architecture.
- **Department of Defense (DOD) Zero Trust Reference Architecture (February 2021):** This architecture outlines the key components and considerations for implementing Zero Trust within the DOD.
- **NSA Embracing a Zero Trust Security Model (February 2021):** This document provides practical guidance on adopting a Zero Trust security model, emphasizing the importance of continuous verification and least privilege access.
- **Cybersecurity and Infrastructure Security Agency (CISA) Zero Trust Maturity Model (June 2021, Pre-decisional Draft):** This model offers a staged approach to Zero Trust implementation that helps organizations assess their maturity level and plot a course toward improvement.

- **Office of Management and Budget (OMB) Federal Zero Trust Strategy: Moving the US Government toward Zero Trust Cybersecurity Principles (January 2022):** This strategy outlines the federal government's approach to adopting Zero Trust principles, providing guidance for agencies and departments.

NIST's guidance on Zero Trust is anchored in seven core tenets that provide a foundational understanding of the model's principles:

1. **All data sources and computing services are considered resources.** This tenet emphasizes that all assets within an organization, whether on-premises or in the cloud, are potential targets and must be protected.
2. **All communication is secured regardless of network location.** This underscores the need to encrypt all data in transit, regardless of whether it's within the internal network or traversing the Internet.
3. **Access to individual enterprise resources is granted on a per-session basis.** This principle advocates for dynamic, just-in-time access control, where access is granted only for the duration of a specific task or session, minimizing the window of opportunity for attackers.
4. **Access to resources is determined by dynamic policy.** This emphasizes the need for adaptive and context-aware access control policies that consider various factors, such as user identity, device posture, location, and real-time risk assessments.
5. **The enterprise monitors and measures the integrity and security posture of all owned and associated assets.** This highlights the importance of continuous monitoring and visibility into the security posture of all assets, enabling proactive threat detection and response.
6. **All resource authentication and authorization are dynamic and strictly enforced before access is allowed.** This emphasizes the "never trust, always verify" approach, ensuring that every access request undergoes strict authentication and authorization before being granted.
7. **The enterprise collects as much information as possible about the current state of assets, network infrastructure, and communications and uses it to improve its security posture.** This underscores the value of data-driven decision-making in Zero Trust, leveraging information about the environment to inform security policies and controls.

These seven tenets can be distilled into four primary activities that form the foundation of a Zero Trust implementation:

1. **Establish identity-based access controls (identity centricity)**
2. **Establish a micro-segmented perimeter**
3. **Contextualize access controls with additional security verification**
4. **Automate and continuously monitor secure access connections.**

The CISA Zero Trust Maturity Model offers a structured approach to Zero Trust adoption, guiding organizations through a phased implementation. It outlines five pillars that encompass the key components of a Zero Trust architecture (Figure 12.1):

Identity: This pillar focuses on establishing strong identity and access management controls, including MFA, PAM, and identity governance and administration (IGA). It stresses the necessity of verifying and authenticating user identities before granting access to resources.

Device: This pillar prioritizes device security and posture assessment, ensuring that only trusted and compliant devices can connect to the network to access resources.

Network/Environment: This pillar focuses on network segmentation and micro-segmentation, creating isolated zones to limit lateral movement in case of a breach. It also spotlights the necessity for continuous monitoring and visibility into network traffic and activity.

Application Workload: This pillar centers on securing applications and services, including vulnerability management, application security testing, and secure coding practices.

Data: This pillar emphasizes data protection, including encryption, DLP, and access control. It focuses on classifying data based on its sensitivity and implementing appropriate security measures to protect the data throughout its lifecycle.

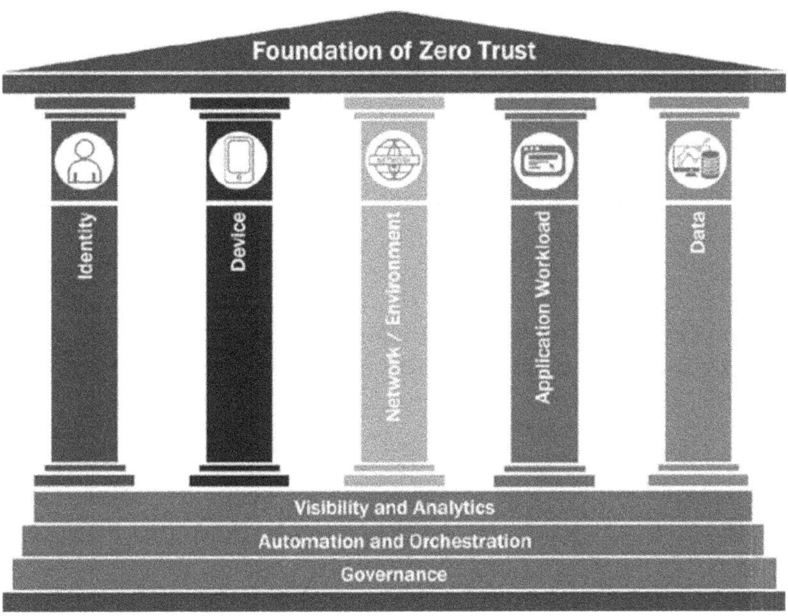

FIGURE 12.1 CISA Zero Trust Pillars.

Successfully implementing a mature Zero Trust architecture depends on mastering four key capabilities: Visibility, analytics, automation, and governance.

1. *Visibility* is the foundation, providing a real-time understanding of users, devices, networks, applications, and data. This deep insight helps make informed decisions about access control and risk management.
2. *Analytics* plays a crucial role by processing data from various sources, such as security logs, user behavior patterns, and threat intelligence. By leveraging advanced analytics, organizations can proactively detect and respond to potential threats, strengthening their security posture.
3. *Automation* simplifies routine tasks like access provisioning, de-provisioning, and incident response workflows. This not only reduces the workload for security teams but also minimizes human error, increasing overall efficiency.
4. *Governance* is essential, ensuring clear policies, procedures, and standards for IAM. Strong governance ensures that security practices remain consistent and compliant across the organization.

Together, these four capabilities create a robust backbone for a strong Zero Trust architecture.

Implementing Zero Trust should be seen as a gradual process, not something to tackle all at once. Start by laying the groundwork with core security controls and building from there.

In the initial phase, focus on establishing foundational security controls like governance frameworks that define roles, set policies, and measure success. Begin with identity and access management (IAM) controls, such as multi-factor authentication (MFA), privileged access management (PAM), and identity governance and administration (IGA). Data security should also be a priority, involving data classification and appropriate security measures like encryption and access control. The network should be fortified through segmentation, limiting lateral movement in case of a breach, while privacy processes and technologies should be set up to protect personal data and ensure regulatory compliance.

In the second phase, integrate these core controls into your broader identity program. This enables more granular, context-aware access control policies that consider factors like user identity, device health, and location. Adaptive authentication adds an extra layer of protection by adjusting based on real-time risk assessments. Strengthen network segmentation through micro-segmentation and implement software-defined perimeters (SDPs) for secure, on-demand access to resources.

The third phase involves making access controls dynamic and adaptive. Use risk-based authentication that changes requirements based on contextual information like location, device health, and behavior patterns. Automate access decisions based on real-time assessments to reduce the load on security teams and use behavioral analytics to continuously monitor user and entity activity, catching anomalies early. Automate the rotation of passwords, keys, and other credentials to reduce the risk of theft and unauthorized access and ensure data access is granted based on the context of each request.

Before jumping into Zero Trust, establish a strong foundation with solid governance and visibility. Effective governance means setting clear policies for identity management, device standards, network security, application deployment, and data management. These policies should define who can access what, when, and how, including standards for device procurement and security configurations.

Visibility means having a clear, real-time understanding of your entire IT environment. This includes keeping an up-to-date inventory of all assets – hardware, software, and data – and using monitoring tools to track network traffic and spot potential threats. By leveraging tools like User and Entity Behavior Analytics (UEBA), you can better understand user and device behavior, enabling the proactive detection of unusual activity. Regular vulnerability scanning and timely remediation are also critical to maintaining a strong security posture.

The CISA Zero Trust Maturity Model provides a useful framework for assessing how far an organization has progressed in implementing Zero Trust. It evaluates five key areas – identity, device, network, application, and data – across three maturity levels:

Traditional: This baseline level focuses on perimeter-based security, with controls that often rely on implicit trust.

Advanced: At this stage, the organization has implemented more proactive controls, such as MFA, network segmentation, and data classification.

Optimal: The highest level of maturity, where Zero Trust principles are fully integrated. Continuous verification, least privilege access, and real-time risk-based decisions are routine.

While frameworks like NIST and CISA are great guides for implementing Zero Trust, they present a few challenges. One common issue is the heavy focus on "users" without adequately considering usability and potential friction. If security measures make it difficult for employees to do their jobs, productivity can suffer. You have to find the right balance between security and ease of use.

The following table provides a more detailed breakdown of what each maturity level might look like for each pillar (Figure 12.2).

Another challenge lies in managing machine identities. Many models overlook the complexities of machine identifiers versus human identities, not to mention the unique security needs of different types of machines within an organization. Additionally, these frameworks sometimes fail to provide clear guidance on how to apply Zero Trust principles to specific workloads, particularly in dynamic and cloud-native environments.

Data is another significant hurdle. The amount of data that needs to be collected and analyzed to support Zero Trust can be overwhelming. Organizations must invest in robust infrastructure and processes to handle this volume of information efficiently. Addressing this data challenge is essential for successfully implementing and sustaining a Zero Trust architecture.

As Zero Trust strategies continue to advance, we can expect major developments in the near future. Automation will play a key role, with automated workload discovery, profiling, and modeling becoming the norm. This will help organizations keep pace with the increasingly agile and dynamic nature of modern IT environments. Data analytics will also become more advanced, providing predictive insights and higher-confidence intelligence to support automated security processes. We may

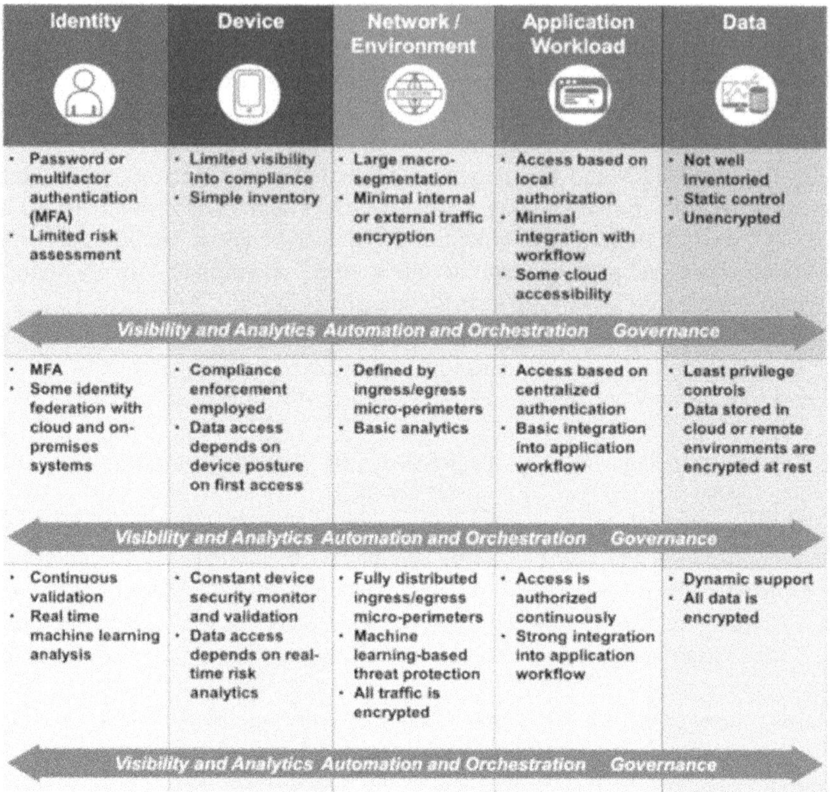

CISA – Zero Trust Maturity Model
https://www.cisa.gov/zero-trust-maturity-model

FIGURE 12.2 CISA Zero Trust Maturity.

even see a shift toward limited autonomous protection, where AI and machine learning (ML) drive real-time access decisions and enforce security policies, especially in highly regulated sectors.

For the CISO 3.0, the journey involves moving from a reactive, compliance-based mindset to a proactive, strategic, and business-aligned approach. Zero Trust is an important part of this shift, helping CISOs protect their organizations' most critical assets. The CISO 3.0 doesn't view Zero Trust as just another security model because it's a strategic necessity that aligns with business goals, risk management, and data-driven decision-making.

By embracing Zero Trust, CISOs can significantly reduce the attack surface, making it much harder for attackers to gain unauthorized access. This approach also helps defend against a variety of threats, such as data breaches, ransomware, and insider attacks, by enforcing least privilege access and continuously monitoring for suspicious activity. Additionally, Zero Trust supports the flexibility of remote work and cloud adoption – providing secure access to resources from any location or device is an essential factor in modern business operations. The core principle of "never trust, always verify" fosters a security-conscious culture across the organization, encouraging everyone to take responsibility for safeguarding company assets.

However, implementing Zero Trust comes with challenges. CISOs need to navigate the complexity of the rollout, minimize potential disruptions for users, and ensure the solution fits the organization's specific needs and risk profile. But the rewards, a stronger security posture, improved productivity, and greater business agility, are well worth the effort. The CISO 3.0 is equipped to lead this transformation, aligning Zero Trust with the company's strategic goals and ensuring the organization is prepared.

12.4 CONTINUOUS MONITORING

Not all that long ago, we used to run vulnerability scans once a quarter or even once a year. But in today's fast-paced threat landscape, that's simply not enough. Continuous monitoring is essential for detecting and responding to threats in real-time. I've seen organizations that have moved to continuous monitoring significantly reduce their risk of breach and improve their overall security posture.

Threats and vulnerabilities can emerge at any moment, this new normal demands a shift toward continuous monitoring. The CISO 3.0 model embraces this shift, promoting a proactive approach that provides real-time visibility into the security environment. This continuous monitoring enables organizations to detect and address threats before they cause significant harm.

The shortcomings of traditional point-in-time monitoring are clear. It only offers a limited snapshot of security at a given moment, leaving organizations vulnerable to threats that surface between assessments. The gap between these checks gives attackers an opportunity to exploit vulnerabilities and go undetected. Additionally, point-in-time monitoring often targets known threats, potentially missing zero-day attacks and new vulnerabilities that may arise.

In contrast, continuous monitoring delivers a steady stream of security data and insights, enabling organizations to respond in real time. This proactive strategy leads to quicker incident response, minimizing the impact of breaches and reducing system downtime. Continuous monitoring also helps maintain regulatory compliance by offering ongoing proof of security and risk management efforts.

Some key benefits of continuous monitoring include the following:

Early Threat Detection: By constantly analyzing security data, organizations can spot potential threats and vulnerabilities before they're exploited, allowing for swift mitigation and reducing the risk of successful attacks.

Faster Incident Response: Real-time visibility means security teams can detect and respond to incidents more quickly, minimizing downtime and damage.

Improved Compliance: Continuous monitoring provides ongoing documentation of security practices, making it easier to meet regulatory requirements and demonstrate diligent risk management.

Stronger Security Posture: Continuously addressing vulnerabilities helps organizations strengthen their defenses and reduce their overall attack surface.

To implement continuous monitoring effectively, organizations must leverage the right mix of technology and processes. This might include deploying security information and event management (SIEM) systems to collect and analyze data from multiple sources, using intrusion detection and prevention systems (IDPS) to monitor network traffic for suspicious activity, and utilizing vulnerability scanning tools to identify weaknesses in systems and applications.

Clear, well-defined incident response processes are essential. This means ensuring security teams can act quickly and efficiently when threats are detected. It's important to establish roles and responsibilities, create communication channels, and develop playbooks that guide the response to common incidents.

12.4.1 Continuous Threat Exposure Management

Gartner defined continuous threat exposure management (CTEM) as "a proactive and continuous approach to identifying and prioritizing an organization's external attack surface, vulnerabilities, and exposures to threats". It involves the use of various tools and capabilities to gain real-time visibility into the organization's security posture and identify potential risks before attackers can exploit them.

Key tools and capabilities included in CTEM are as follows:

- **External Attack Surface Management (EASM):** This involves continuously discovering, inventorying, and assessing the organization's Internet-facing assets, such as websites, applications, and cloud services, to identify potential vulnerabilities and exposures.
- **Cyber Asset Attack Surface Management (CAASM):** This focuses on identifying and managing the attack surface associated with an organization's internal assets, such as devices, applications, and data, to proactively address potential vulnerabilities.
- **Breach and Attack Simulation (BAS):** This involves simulating real-world attacks to test the effectiveness of security controls and identify weaknesses in the organization's defenses.
- **Penetration Testing as a Service (PTaaS):** This provides on-demand access to penetration testing expertise to identify and exploit vulnerabilities in the organization's systems and applications.
- **Extended Security Posture Management (xSPM):** This technology extends traditional cloud security posture management (CSPM) capabilities to cover a broader range of assets, including on-premises infrastructure, endpoints, and cloud environments. It provides a unified view of the organization's security posture across its entire IT estate.

CTEM is more than a tool or set of tools: It is a program that involves simultaneous, repeatable cycles of discovery, assessment, prioritization, and remediation. It covers exposures that are both patchable risks, like software vulnerabilities, and those that can't be fixed with patches, such as misconfigurations or human error, giving organizations a fuller understanding of their risk landscape.

One of CTEM's key strengths is its ability to validate priorities by quantifying the potential impact of different risks. This helps organizations focus their resources on the most critical areas. Instead of waiting for issues to escalate, CTEM offers real-time insights into emerging threats, empowering teams to take proactive measures to protect valuable assets before a problem occurs.

CTEM is not only about identifying known vulnerabilities, but it's also about uncovering the "unknown unknowns" – the risks we're not even aware exist. In the context of cybersecurity, unknown unknowns represent the hidden risks and exposures that organizations are not even aware of. These could be vulnerabilities in unmanaged or forgotten assets, shadow IT resources, or even misconfigurations in cloud environments. CTEM, with its focus on continuous discovery and assessment of the attack surface, helps identify these unknown unknowns and bring them to light, enabling organizations to proactively address them before attackers can exploit them. Unknown exposures equal unmanaged risks.

Traditional threat and vulnerability management (TVM) tends to focus on identifying known vulnerabilities through periodic assessments. But with the growing complexity of today's IT environments, a more proactive and continuous approach is essential. CTEM broadens the scope of traditional TVM by continuously monitoring the entire attack surface, identifying both known and unknown vulnerabilities, and prioritizing remediation based on real-time risk assessments. This shift helps organizations stay ahead of emerging threats and reduces overall risk exposure.

For the CISO 3.0, it's important to advocate for the expansion of TVM and secure the necessary resources to make it happen. Continuous monitoring should become a core part of the SOC, allowing analysts to detect and respond to threats more effectively. This involves integrating SIEM systems to collect and analyze data from multiple sources, using IDPS to monitor network traffic, and employing vulnerability scanning tools to pinpoint weaknesses.

Additionally, establishing clear processes for incident response and communication ensures that the SOC can react quickly and efficiently when threats are detected, helping minimize damage and maintain security.

12.4.1.1 Recommendations

To successfully implement *CTEM* and continuous monitoring, organizations should take several key steps to ensure the program's effectiveness:

Allocate Resources: Make sure at least 20% of your SOC and engineering teams are dedicated to CTEM. This ensures you have enough people to manage the continuous stream of security data and make informed decisions quickly.

Enhance Skills: Invest in ongoing training for your security team. Equip them with the skills they need to use CTEM tools effectively and navigate processes that extend beyond traditional security roles. This means developing a deep understanding of new technologies and cross-functional workflows.

Expand the Scope: CTEM shouldn't just cover IT assets – extend its reach to all digital assets. This helps give your team a full view of your organization's attack surface, enabling more proactive risk management.

Adopt New Technologies: Stay on the cutting edge by embracing tools that support continuous monitoring and CTEM. The right technology can make all the difference in quickly identifying and mitigating threats.

Establish Clear Processes: Develop clear processes for acting on CTEM insights. Define roles, set up communication channels, and create playbooks to ensure quick and effective responses to any identified risks or vulnerabilities.

Identify and Prioritize Risks: Make it a habit to continuously evaluate risks, ranking them based on their potential impact and likelihood. This ensures your team focuses remediation efforts where they're needed most.

Regularly Assess Tools: Don't settle for tools that aren't delivering. Regularly review your threat and vulnerability management tools and consider upgrading or replacing those that fall short. The *CISO 3.0* should always be on the lookout for new technologies to strengthen the security program.

12.4.2 CTEM and AI

A major driver behind the growing need for CTEM is the adversarial use of artificial intelligence (AI) by threat actors. As AI technologies become more accessible and advanced, cybercriminals are leveraging them to automate and scale their attacks, making traditional, reactive security strategies insufficient. The use of AI by attackers has fundamentally changed the threat landscape, driving organizations to adopt a proactive, continuous approach like CTEM to stay ahead.

Threat actors are increasingly using AI to enhance the speed, precision, and complexity of their attacks. For example, AI-powered tools can rapidly scan networks for vulnerabilities, automatically exploit weak points, and evade detection by adapting to security measures in real-time. One notable example is the use of AI in phishing campaigns, where ML algorithms can craft highly personalized phishing emails by analyzing large datasets of potential targets. These AI-enhanced attacks bypass traditional defenses by mimicking legitimate communications more convincingly than ever before.

AI is also being used to create sophisticated malware that can autonomously adapt its behavior based on the environment it's deployed in. For instance, AI-enhanced ransomware can analyze a victim's network, identify high-value assets, and adjust its encryption strategy to maximize damage. In these scenarios, AI isn't just automating known attack vectors – it's enabling attackers to explore new methods and adapt faster than human defenders can respond. We'll discuss AI security in depth in Chapter 14.

The more AI accelerates the threat lifecycle, the harder it is for traditional security programs to keep pace. While many security measures are designed to detect known threats, AI-driven attacks evolve too quickly for static defenses. This is where CTEM becomes crucial – it provides continuous visibility into an organization's security posture, helping teams stay one step ahead by identifying potential vulnerabilities before AI-powered threats can exploit them.

CTEM's focus on continuous discovery and real-time risk assessment is vital in countering AI-driven attacks. For example, External Attack Surface Management

(EASM) and Cyber Asset Attack Surface Management (CAASM) tools can identify and prioritize potential weak points that might be targeted by automated AI scanning tools. Breach and Attack Simulation (BAS) can simulate AI-driven attack scenarios, allowing organizations to test their defenses against these sophisticated threats. This constant cycle of discovery, assessment, and remediation ensures that security teams are always aware of their current vulnerabilities and can address them before adversarial AI tools can exploit them.

12.4.3 CTEM and the Development of Meaningful KPIs

The data generated by CTEM tools can be invaluable in developing meaningful Key Performance Indicators (KPIs) that accurately reflect the organization's security posture and progress toward its goals. The reports from these tools offer a wealth of information on vulnerabilities, exposures, and potential threats, which can be leveraged to create KPIs that track the effectiveness of security controls and identify areas for improvement.

For instance, EASM tools can provide data on the number and severity of vulnerabilities discovered on external-facing assets. This data can be used to create KPIs that track the organization's progress in reducing its external attack surface and improving its vulnerability management processes. Similarly, CAASM tools can offer insights into the internal attack surface, allowing for the development of KPIs that measure the effectiveness of internal security controls and identify potential insider threats.

BAS and PTaaS tools, by simulating real-world attacks, can help identify weaknesses in the organization's defenses. The results of these simulations can be used to create KPIs that track the organization's ability to detect, respond to, and recover from attacks, providing a measure of its overall security resilience. We'll discuss the strengths of breach and attack simulation (BAS) and ways to leverage it to drive your strategy in the next chapter.

xSPM tools, with their unified view of the security posture, can provide data on various security controls and their effectiveness. This data can be used to create KPIs that track the performance of individual controls, identify areas where improvements are needed, and measure the overall effectiveness of the security program.

By leveraging the data and insights from CTEM tools, CISOs can develop KPIs that are not only relevant and actionable but also aligned with the organization's strategic objectives and risk appetite. These KPIs can then be used to track progress, identify trends, and help right-size resource allocation and security investments, ultimately contributing to a more proactive and effective security posture.

Continuous monitoring and CTEM are essential for CISO 3.0s as they enable a proactive and data-driven approach to security. This proactive approach aligns with the CISO 3.0's focus on business enablement and risk management, ensuring that security efforts are not only reactive but also contribute to the organization's overall success.

The shift from CISO 2.0 to 3.0 is marked by a transition from tactical, compliance-driven tasks to a strategic, business-aligned perspective. Continuous monitoring and CTEM can help with this transformation by providing the CISO 3.0 with the data and insights needed balance security with business objectives. The CISO 3.0, by proactively identifying and addressing risks, can demonstrate their value as a strategic partner who contributes to the organization's overall success.

12.5 CLOUD-NATIVE PROTECTION

The migration of workloads and data to the cloud has been an ongoing trend for years, driven by the promise of scalability, flexibility, and cost savings. However, while the cloud isn't new, and migrating workloads to it isn't a novel idea, it continues to be a major undertaking for most organizations. The recent acquisition of VMware by Broadcom, leading to changes in licensing and increased costs for many, has further fueled this renewed interest in cloud migration. This might involve simply lifting and shifting existing applications or, more strategically, re-architecting them to a service-oriented model that supports containerization or serverless deployment.

The cloud model today is diverse and complex. The initial euphoria surrounding the cloud, with its promises of reduced costs and increased agility, has been tempered by the reality that these benefits aren't always guaranteed. Many applications migrated to the cloud in the early wave have since been repatriated, creating hybrid environments where workloads are distributed across on-premises and cloud infrastructure. Organizations have discovered various use cases for building their own private clouds, further adding to the complexity. The result is a heterogeneous environment where securing workloads and their runtime environments can be a daunting task.

12.5.1 Cloud Security Challenges

Governance and compliance are a primary challenge in cloud security. Cloud environments, despite their inherent differences, must adhere to the same regulatory and industry standards as on-premises infrastructure. This includes compliance with regulations such as HIPAA, PCI-DSS, SOX, and various data privacy laws such as the California Consumer Privacy Act (CCPA), the California Privacy Rights Act (CPRA), and the General Data Protection Regulation (GDPR). Organizations must also ensure that their cloud environments meet the requirements of industry benchmarks and frameworks like CIS Benchmarks, NIST SP 800–53, FedRAMP, ISO 27001 & 27002, and SOC 2 & 3.

Maintaining consistent security policies and controls across hybrid environments can be complex, requiring careful planning and coordination. Organizations need to ensure that their cloud providers adhere to their security standards and that data is protected throughout its lifecycle, regardless of its location.

Cloud environments face a range of threats and vulnerabilities. Cryptocurrency mining is one such threat, where attackers exploit vulnerabilities in cloud infrastructure to hijack computing resources for their own gain. Accidental data disclosure from misconfigured resources, like unsecured S3 buckets, is another concern, potentially leading to data breaches and compliance violations. Host malware can infect cloud servers and virtual machines, compromising data and spreading to other parts of the network. Compromised cloud servers can serve as pivot points for launching further attacks, such as port scans, DDoS attacks, or spam campaigns, targeting both internal and external systems.

Cloud servers, along with phishing and Internet-facing remote access servers, are among the most common entry points for threat actors to infiltrate an organization's network. Misconfigurations remain a top concern for cybersecurity professionals, as they can create vulnerabilities that are easily exploited by attackers.

12.5.2 Cloud-Native Application Protection Platforms

To address the unique security challenges of cloud environments, cloud-native application protection platforms (CNAPPs) have emerged as a novel solution. CNAPPs consolidate various security capabilities into a single platform, providing end-to-end protection for cloud-native applications and infrastructure.

The key components of a CNAPP platform include the following:

CSPM: CSPM provides visibility and control over how your cloud infrastructure is configured across multiple cloud providers. It helps identify security misconfigurations and ensures your setup aligns with both internal security policies and industry standards, reducing the risk of vulnerabilities.

Cloud Workload Protection (CWP): CWP offers flexible, adaptive protection for workloads running in the cloud, such as virtual machines, containers, and serverless functions. It protects these environments from threats throughout their lifecycle, helping keep the cloud operations secure.

Cloud Identity Entitlement Management (CIEM): CIEM tackles the risks tied to overly permissive roles, poor credential management, and accidental public exposure of sensitive data. It enforces least-privilege access, ensuring that identities are properly managed and access is limited to what's absolutely necessary.

Cloud Network Security: This focuses on mapping out how applications interact and enforcing micro-segmentation to isolate workloads. By limiting lateral movement in the event of a breach, you can build a more secure and resilient cloud network architecture.

Infrastructure as Code (IaC) Security: By integrating security checks early into the DevOps workflow, IaC security scans and secures infrastructure code before deployment. This proactive approach helps prevent vulnerabilities and misconfigurations from entering your cloud environment in the first place.

CNAPPs offer comprehensive cloud protection by consolidating multiple security functions, helping CISOs manage risks, optimize investments, and enhance the organization's security posture in cloud and hybrid environments.

It's important to understand that in addition to the *CSPM* found in CNAPP solutions, there are also other security posture management (SPM) tools. *Data security posture management* (*DSPM*) ensures data is secured and regulated, providing visibility into data risks and compliance across cloud environments. *Application security posture management* (*ASPM*) strengthens application security by ensuring adherence to security standards. *Artificial intelligence security posture management (AI-SPM)* addresses the unique risks of AI assets and models, safeguarding these emerging technologies from exploitation.

These tools provide real-time visibility into vulnerabilities, misconfigurations, and compliance gaps across cloud environments, applications, data, and AI systems. For example, CSPM ensures cloud environments are properly configured, while DSPM helps monitor sensitive data and ensure compliance with regulations like GDPR. By automating the detection and remediation of potential issues, SPM tools minimize security risks before they can be exploited. Beyond risk mitigation, SPM tools also simplify regulatory compliance. They continuously monitor systems for adherence to industry standards and legal requirements, helping organizations avoid costly fines and reputational damage. Ultimately, SPM tools provide organizations with the insight and control needed to maintain a robust security posture and stay ahead of emerging threats.

While the different SPM tools can be used individually, leveraging them together will produce the best results and is the future of cyber threat management. The problem is that each solution requires significant administrative overhead and specialized expertise. As organizations adopt CNAPPs and SPM tools, SOC operations will have to be expanded to accommodate these new functions. However, the ability to integrate SPM tools with SOC operations enhances incident response, ensuring quicker reactions to threats while reducing overall operational overhead.

12.5.3 Managing CNAPP

While CNAPPs offer a powerful solution for cloud security, managing all the interconnected tools effectively can be a complex undertaking. Integrating a CNAPP platform into your existing security ecosystem requires careful planning and execution. It's essential to ensure seamless integration with your SIEM; security orchestration, automation, and response (SOAR); and other security tools to enable efficient data sharing, correlation, and incident response.

CNAPPs generate a vast amount of security data and alerts, requiring skilled analysts to interpret and prioritize them effectively. This often necessitates upskilling your existing security team or hiring additional personnel with expertise in cloud security and CNAPP technologies.

Ultimately, the goal is to transform the CNAPP toolset into a real capability that enhances your organization's security posture. This requires not only technical expertise but also a clear understanding of the organization's cloud strategy, risk appetite, and business objectives.

Given the complexity and resource requirements of managing CNAPPs, many organizations are turning to managed service providers (MSPs) for assistance. Just as outsourcing traditional SOC functionalities has become common, outsourcing CNAPP management and monitoring is likely to become increasingly prevalent in the coming years. This allows organizations to leverage the expertise of specialized providers while freeing up internal resources to focus on other strategic initiatives.

12.5.4 The CISO 3.0 and Cloud-Native Security

Cloud-native security and Cloud-Native Application Protection Platforms (CNAPP) should play a central role in the CISO 3.0's strategy for aligning security with business objectives. With these technologies, CISOs can facilitate cloud adoption, which is essential for fostering business agility and innovation. Implementing robust cloud security measures not only supports the secure integration of cloud technologies but also aligns with the organization's broader digital transformation efforts.

The cloud brings unique security challenges that require a proactive approach from CISOs to safeguard the organization's assets and data. CNAPPs offer a comprehensive solution for managing these risks, allowing CISOs to maintain a strong security posture in cloud environments. CNAPPs consolidate multiple security functions into one platform, which can lead to cost savings and increased efficiency compared to juggling various point solutions. This enables CISOs to optimize their security investments and demonstrate tangible value to the business.

CNAPPs provide visibility into the cloud environment, enabling CISOs to identify and address potential risks before they can be exploited. This proactive approach enhances the organization's security posture and reduces its attack surface.

In conclusion, cloud-native protection and CNAPPs are essential components of a modern cybersecurity strategy. The CISO 3.0's focus on business enablement, risk management, and data-driven decision-making makes them well-equipped to navigate the complexities of cloud security and leverage these technologies to protect their organizations.

12.6 DATA GOVERNANCE

12.6.1 The Data Governance Lifecycle

The data governance lifecycle outlines the key stages involved in managing data throughout its lifecycle. It begins with information lifecycle governance, which involves managing the flow of data from creation and acquisition through archival and deletion, based on the organization's policies and legal requirements. Data stewardship/council is integral to ensuring that data governance policies are implemented and followed, acting as a bridge between IT and business functions. Data quality management ensures that data is accurate, consistent, and complete across all systems, while metadata management provides a structured approach to defining and documenting data assets. Master and reference data management focuses on the structure, integration, and flow of data across the organization, including data dictionaries and catalogs.

Data security management, a fundamental component of the data governance lifecycle, protects sensitive data from unauthorized access or breaches. It involves enforcing data access controls, data masking, encryption, and other security measures. Information privacy/compliance ensures that data management practices comply with relevant laws, regulations, and standards.

Good data governance requires a cultural shift within the organization. It involves collaboration across teams to develop a common data literacy and identify gaps in current data management practices. This collaborative approach ensures that all stakeholders understand the importance of data governance and are committed to its success.

Key questions that good data governance addresses include the following:

Stewardship Concerns: Who owns data at our organization? Who can make decisions for data-related matters?

Data Lineage Concerns: What data assets do we have across our organization? Where are they? Where did they come from?

Quality Concerns: Why is this value on my presentation wrong? Why do I have two different values for the same metric on these two reports?

Privacy Concerns: Can we handle requirements from new privacy laws? Can we provide or delete a customer's data if they request it?

Lifecycle Concerns: Should we keep or delete the data in this retired system?

12.6.2 Modern Data Platforms

Good data governance often necessitates the adoption of modern data platforms. These next-generation solutions leverage advanced technologies and architectures to store, process, and analyze large volumes of diverse data in real-time or near-real-time. They can handle structured, semi-structured, and unstructured data, and they are designed to be scalable, distributed, and flexible, often leveraging cloud technologies. These platforms support a wide variety of use cases, from business intelligence and analytics to AI and ML workloads.

12.6.3 Data Lakehouse

A *data lakehouse* blends the strengths of both data lakes and data warehouses. A data lake stores vast amounts of raw data in its original form, while a data warehouse organizes processed data for reporting and analysis. The data lakehouse brings these two together, offering a single platform for storing both raw and processed data, making it easier to analyze everything in one place.

Benefits of a Data Lakehouse

Flexibility: It supports a variety of data types and formats, from structured to unstructured.

Scalability: Designed to handle massive amounts of data, making it ideal for organizations with growing data needs.

Cost-Effectiveness: By using cloud storage and computing power, it reduces the reliance on expensive on-premises infrastructure.

Versatility: It caters to a wide range of use cases, including data science, ML, business intelligence, and reporting.

A data lakehouse typically connects to a range of tools, from data ingestion systems that pull in data from different sources to transformation tools that clean and prepare data for analysis. It can also work with data processing technologies like Databricks and AI/ML models to turn raw data into actionable insights.

12.6.4 Data Security and Protection

Embedded within the data governance lifecycle is the data protection lifecycle, which covers various aspects such as data lifecycle management, resiliency and recovery, identification and classification, secure processing, storage and retention, security, and privacy. Simply put, data protection ensures that crucial information is shielded from corruption, compromise, or loss. As organizations generate and store increasing amounts of data, a robust data protection strategy becomes essential, especially since executives, boards, and customers have little patience for downtime or data breaches.

12.6.4.1 Data Security versus Data Privacy

Data security is all about using technical solutions to safeguard information from unauthorized access, cyberattacks, and tampering. These protections include everything from DLP to encryption, access controls, SIEM, and auditing user activity.

On the other hand, data privacy focuses on the legal and ethical handling of data. This includes managing how sensitive data is collected, processed, and stored, as well as giving individuals the right to decide how their data is used. Data privacy requires organizations to respond to requests from individuals to access, delete, or control their data – a task that can be overwhelming due to the vast amount of dispersed data within most companies.

12.6.4.2 Data Mapping

This task of locating and understanding all instances of a user's data can be nearly impossible for most organizations due to the sheer volume and dispersion of data. Data mapping can help address this challenge. Data mapping involves creating a visual diagram that illustrates how data moves and transforms within an organization's IT infrastructure. It helps identify data sources, data transformations, and data dependencies, providing an overview of the data.

Data mapping tools can automatically synthesize data across multiple databases and sources, uncovering relationships and patterns that would take humans years of work to discover manually. This enables organizations to gain a deeper understanding of their data, and identify potential risks and compliance issues.

12.6.4.3 Building Data Resiliency

Data resiliency refers to an organization's ability to continue operating despite disruptions like cyberattacks or natural disasters. To establish strong data resiliency, you need a tailored approach based on your organization's needs and risk profile. Key steps include the following:

Defining Resiliency Goals: Determine recovery time objectives (RTOs) and recovery point objectives (RPOs) for critical data.

Inventory and Classification: Identify and prioritize data based on its importance and sensitivity.

Implementing Regular Backups: Follow the 3-2-1 backup rule: three copies of your data on two different storage media, with one stored offsite.
Testing Recovery Plans: Regularly test your data recovery plan to ensure critical information can be restored quickly.
Continuous Monitoring: Keep a constant eye on your data systems to detect any potential issues that could affect availability.
Using Cloud Solutions: Take advantage of cloud-based backup and disaster recovery solutions for scalability and geographic diversity.

You've probably heard the phrase "data is the new oil". I don't know if I'd invest all my money in data governance tools in the hopes of moving to Beverly Hills, but the environment is definitely shifting. It is possible today to mine sales data, sorting by product line, cost, size, and quantity, and then map it to regional sellers and their customer spending ratios. That's obviously incredibly powerful, but would you confidently present that data to your board or share it with investment analysts, knowing they'll hold you accountable to those numbers? Probably not, because many organizations still don't fully trust the quality of their data or the accuracy of their models.

That kind of high-stakes data-driven decision-making will only become more prevalent in the next 5 years, and stakeholders will begin to expect that level of transparency and insight. To meet these future expectations, we need to start by cleaning and governing our data today. For any data protection strategy to succeed, it needs the full backing of executives. For a long time, data governance was viewed as a pure cost center, and traditional DLP tools often fell short of expectations. However, as data becomes not only a commodity but also a core product, addressing the "strategic debt" around data governance and protection will become the pathway to unlocking future data-driven revenue streams. So, the first step of a successful data protection initiative is securing executive sponsorship. Once you have that, it's important to establish a data governance steering committee to define objectives, policies, and oversight mechanisms. Conducting a thorough data risk management assessment to identify threats to data integrity, confidentiality, and availability is also essential. This assessment will guide the creation of a detailed roadmap to mitigate those risks.

By implementing initiatives such as data access governance, data discovery, and privacy operations, organizations can not only reduce their exposure to data-related risks but also position themselves to capitalize on data as a strategic asset, paving the way for future revenue opportunities. Today's modern data security posture management (DSPM) platforms, powered by AI and business intelligence, are helping turn data governance into a value-add, making it easier for CISOs to secure buy-in from leadership.

12.6.5 Data Security Posture Management (DSPM) Platforms

Data security posture management platforms (DSPMs) represent a progression in data protection, offering a unified approach to safeguarding sensitive information throughout its lifecycle. These platforms bring together various security functions under one umbrella, making it easier for companies to discover, classify, protect, and monitor their data throughout its lifecycle.

At the core of a DSPM is its ability to automatically locate and classify sensitive data across multiple environments, whether on-premises or in the cloud. In today's complex IT landscape, where data is often scattered across many systems and applications, this feature is critical. DSPMs utilize advanced technologies like ML and data fingerprinting to identify important data, like personally identifiable information (PII), financial records, or intellectual property. This enables organizations to focus their protection efforts where they matter most.

Once the sensitive data is identified, DSPMs implement detailed access controls and data masking policies. These measures ensure that only authorized personnel can view or interact with the data, and even then, in ways that protect privacy. This reduces the risk of data breaches and helps safeguard an organization's most valuable information.

DSPMs offer a range of integrated tools to help secure data throughout its lifecycle. Some of the primary features include the following:

Data Discovery and Classification: Automatically find and categorize sensitive data across various storage locations, both on-site and in the cloud. This helps organizations identify their most critical assets and apply the right security measures.

Data Access Governance: Enforce strict access controls and data masking to ensure that only authorized users can access sensitive data, and that it's displayed in ways that maintain privacy.

Data Loss Prevention (DLP): Monitor and prevent unauthorized sharing or leaks of sensitive information, whether through email, network traffic, or endpoint activities.

Encryption: Protect data both at rest and in transit by encrypting it, ensuring that even if it's intercepted or stolen, it remains secure.

Data Activity Monitoring: Track and log access to sensitive data, providing a comprehensive audit trail for compliance and investigations.

By consolidating these functions into one platform, DSPMs help organizations manage data security more effectively and reduce the complexity of juggling multiple, disconnected tools.

One of the biggest strengths of DSPMs is the use of advanced technologies like ML, which automates essential tasks like data discovery, classification, and protection. This automation drastically reduces the need for manual processes, which can be time-consuming and prone to error. With DSPMs, organizations can scale their data protection efforts more efficiently and with fewer resources.

Another key benefit is the clear visibility DSPMs provide into how data is being used and where it moves within an organization. This transparency helps security teams assess risks more accurately and direct resources where they're needed most.

Unlike older DLP solutions, which were often seen as disruptive or difficult to use, DSPMs are designed with the user experience in mind. They allow employees to work smoothly and productively while still maintaining robust security controls.

DSPMs offer a modern, automated, and user-friendly approach to data protection. By using these platforms, organizations can more effectively manage and secure their data, reduce risks, stay compliant with regulations, and get the most value from their data. It's a solution built for today's fast-moving, data-driven world.

12.6.6 The CISO 3.0 and Data Governance

Data governance and data protection are nonnegotiable for the modern CISO. For the CISO 3.0, data controls aren't just a checklist: They're the foundation for aligning security with business objectives. Treating data as a strategic asset, which it absolutely is, means understanding that protecting it is mission-critical to achieving business outcomes.

Think of it this way: Effective data governance and protection are two sides of the same coin. They ensure responsible and secure data usage, fostering business growth and innovation. At the same time, they mitigate the risks associated with data breaches and other data-related incidents that can have a devastating impact on your organization's finances. Proactive data risk management, through robust governance and protection, allows you to minimize the potential fallout from such incidents, safeguarding your organization's bottom line.

But here's the thing: Data governance has often been seen as a cost center, a necessary evil with little tangible return on investment (ROI). That perception is rapidly changing. Thanks to the rise of generative AI and advanced data analytics, data itself is becoming a valuable commodity. If your organization wants to play in the future data economy, then data governance is no longer optional: It's a prerequisite. You need clean, high-quality data that take data lineage, privacy, and ethical considerations into account.

This shift is a game-changer, tying your data governance activities to a potential revenue stream can make the ROI conversation much easier. Suddenly, investing in data hygiene isn't just about mitigating risk; it's about unlocking new business opportunities.

Of course, technology is only part of the equation. You also need to foster a culture of security. Data governance and protection require buy-in and active participation from every level of your organization. Cultivating a culture of security and data awareness helps empower your employees to become guardians of your organization's data assets.

Ultimately, data governance and data protection are not just technical necessities: They're strategic imperatives. The CISO 3.0 recognizes the inherent value of data and the critical importance of protecting it. They also understand that data governance and data protection measures can facilitate business growth and help prove their value as a strategic leader.

12.6.7 Data Privacy

Data privacy regulations are becoming increasingly central to organizational strategy, driven by heightened consumer expectations and regulatory scrutiny. Major privacy laws like the GDPR in the EU and the CCPA in the United States impose stringent requirements on how personal data is collected, processed, and stored. However, while GDPR provides a centralized regulatory framework in Europe, the

United States has taken a decentralized approach, leaving it up to individual states to craft their own privacy laws. This patchwork of regulations makes compliance especially challenging for organizations operating across multiple states, each with its own nuances, exceptions, and enforcement mechanisms. The result is an environment where complete compliance feels unattainable, forcing organizations to strike a delicate balance between adhering to the strictest regulations and maintaining a standard that is both reasonable and defensible.

For CISOs, this regulatory complexity represents a growing responsibility. They are increasingly tasked not only with ensuring data security but also with managing privacy compliance. The CISO 3.0 is uniquely positioned to help shape privacy strategies by applying the same mindset they use to determine "reasonableness" in security controls. CISOs must balance legal, technical, and operational demands to craft privacy programs that stand up to scrutiny without hindering business operations. This is particularly important as privacy regulations evolve and enforcement actions increase. Organizations that fail to prioritize privacy face significant consequences, from hefty fines to brand damage.

Privacy operations are deeply interconnected with cybersecurity, as they govern how personal data is managed, requiring processes for access control, consent management, and data retention. Data mapping is essential for both privacy and security. It helps organizations identify where personal data is located, which is crucial for both protecting it and responding to privacy requests, such as an individual's right to be forgotten. This need for clear visibility into data flows underscores the importance of DSPMs, which automate the discovery, classification, and protection of sensitive data. DSPMs ensure that privacy regulations are met by controlling access, monitoring data usage, and securing data throughout its lifecycle.

The role of the CISO is expanding to encompass privacy responsibilities. Even in organizations with a chief privacy officer (CPO) or data privacy officer (DPO), CISOs are increasingly collaborating with privacy teams, as privacy and security become more intertwined. In some cases, CISOs are directly responsible for privacy operations, managing both cyber risks and compliance with privacy laws. With advancements in technology, such as AI adoption, the CISO's role in safeguarding personal data is growing, as these technologies introduce new privacy concerns that require oversight. AI systems, for example, can inadvertently expose sensitive data, making it essential for CISOs to implement robust privacy safeguards.

Traditionally, privacy and security were seen as distinct domains, where privacy was handled by legal or compliance teams, and security focused on protecting against cyber threats. However, these lines are blurring. The CISO 3.0 must expand their knowledge of privacy laws and regulations while working closely with legal teams to ensure both compliance and security measures are aligned. This convergence requires collaboration across departments to successfully manage the dual responsibilities of privacy and security. Most privacy programs address compliance through three core pillars: The first is legal, where the focus is on regulatory adherence, ensuring proper consent mechanisms, and opt-ins. The second pillar is security, which deals with conducting data risk assessments, implementing access controls, and maintaining robust monitoring systems. The third pillar is IT, which is tasked with delivering privacy operations. For example, most regulations grant data subjects the right to request access to, modify, or delete their personal data. While creating web pages and portals for these requests may seem straightforward, the real challenge lies in mapping out all the locations where a user's data reside and ensuring the necessary changes or deletions are executed across the board.

Privacy operations directly impact cybersecurity because they govern how personal data is managed, requiring processes for access control, consent management, and data retention. Data mapping, for example, is essential for both privacy and security. It helps identify where personal data is located across the organization, which is key to both protecting it and responding to privacy requests from individuals, such as the right to be forgotten. This need for clear visibility into data flows also underscores the role of DSPMs, which are already being explored by security teams. These tools are capable of automatically locating and classifying sensitive data across various environments, making it much easier for IT teams to fulfill these privacy requests. However, there's no simple solution here. Organizations are grappling with the enormity of the task, from managing vast amounts of data to untangling the legal implications of privacy breaches. DSPM solutions may help, but they don't erase the reality that privacy compliance will remain an ongoing challenge that will only intensify as consumer awareness and regulatory oversight grow.

CISOs are well-positioned to lead the privacy charge because they already possess the technical expertise needed to implement technical solutions. Many privacy teams lack this know-how that CISOs bring to the table. As CISOs gain more insight into privacy requirements, they can leverage their experience in deploying technical solutions to improve how privacy is managed across the organization. For instance, many privacy programs rely on outdated tools and processes, creating gaps in compliance. By applying their knowledge of security operations, CISOs can help streamline privacy efforts and close these gaps, reducing risk and ensuring compliance.

Looking ahead, the role of the CISO will increasingly overlap with privacy functions as regulatory demands and consumer expectations evolve. Privacy and security are not separate issues, they are two sides of the same coin, with privacy focusing on protecting individual rights and security ensuring confidentiality. The CISO 3.0, equipped with technical expertise and an understanding of privacy frameworks, is well-positioned to lead privacy efforts. By leveraging their experience in deploying technical solutions, CISOs can streamline privacy operations, closing compliance gaps and reducing risk.

In today's privacy-first environment, CISOs must also anticipate future trends. Governments and regulators are increasingly prioritizing privacy enforcement, and consumers are becoming more serious about protecting their personal information. This shift places added pressure on CISOs to not only protect data but also ensure that privacy practices are transparent, robust, and aligned with regulatory requirements. As privacy laws evolve and enforcement actions increase, organizations that fail to prioritize privacy will face significant consequences.

Ultimately, CISOs must be prepared to take on privacy responsibilities, even in organizations with dedicated privacy officers, shaping a privacy strategy that aligns with both regulatory demands and consumer expectations. By embracing this reality, CISOs can help create a secure, privacy-first culture that ensures long-term success in an increasingly privacy-conscious world.

12.6.7.1 CISO versus DPO

The role of a data protection officer (DPO) is to ensure that personal data is handled in compliance with privacy laws, such as GDPR, and to guide organizations on proper data processing practices. In some regions, appointing a DPO is legally required, especially for organizations processing large volumes of sensitive data or regularly monitoring individuals. Even when not mandated, it is highly recommended that privacy responsibilities are assigned to someone who can act independently, whether it's a DPO, chief privacy officer (CPO), or another privacy-focused leader.

A DPO's key responsibility is ensuring compliance with the GDPR, acting as the organization's custodian of data privacy practices. The DPO's main job is to act as the organization's custodian of data privacy practices. Their primary mission is to align data processing with legal and ethical standards, serving as the point of contact for individuals with concerns about how their personal data is handled. Beyond regulatory enforcement, the DPO integrates privacy considerations across all aspects of the organization, from managing data breaches to conducting privacy impact assessments.

The DPO must possess strong legal and regulatory expertise, particularly around GDPR, and have the ability to translate complex privacy requirements into practical, organization-wide actions. Effective communication and up-to-date knowledge of evolving privacy frameworks are essential for the role.

A DPO's responsibilities include conducting data protection impact assessments (DPIAs) to identify and mitigate risks associated with new technologies or processes. They also educate employees on data protection best practices, embedding privacy considerations into the organization's culture. In the event of a data breach, the DPO leads the response, ensuring swift communication with regulators and affected individuals while maintaining compliance with legal obligations.

The DPO must be able to advise on privacy compliance without taking direction from other departments, ensuring they remain independent in their decision-making. In many organizations where a formal DPO is not appointed, legal, compliance, or security teams may take on this role. However, where required, failure to appoint an independent DPO can result in significant fines, such as those under GDPR, which could reach up to 10 million euros or 2% of global revenue, depending on the circumstances.

In cases where a DPO is not required by law, organizations often assign privacy-related tasks to the CISO, chief risk officer (CRO), or legal teams. While this may be voluntary, it is still essential to maintain checks and balances to ensure privacy compliance is not compromised. The DPO's independence is vital in this context, as they should not take direction from the CISO, reinforcing the need for separation of duties.

As organizations face tighter regulations and are expected to do more with fewer resources, there is increasing pressure on CISOs to take on privacy-related duties. However, combining the roles of CISO and DPO presents a conflict of interest. A DPO should report to the highest levels of management, ideally the board, to maintain their independent oversight. In contrast, the current CISO 2.0s typically report to the C-suite, and even as CISOs advance toward 3.0, their focus should be on enterprise security and addressing broader risk management challenges. Additionally, DPOs bear legal accountability for privacy breaches. Under GDPR, they can be held personally responsible for compliance failures, making the role particularly high-stakes. This is why CISOs should be cautious about taking on informal DPO responsibilities. If a CISO is going to act as a DPO, they should ensure they are included under the company's directors and officers (D&O) insurance policy, and they may also need errors and omissions (E&O) coverage to mitigate personal risk.

In organizations where a CISO is expected to manage both privacy and security, it's important to maintain clear boundaries. If a CISO is tasked with DPO responsibilities, they must avoid overriding privacy compliance needs with security objectives. If this conflict becomes problematic, it's the CISO's responsibility to alert executive leadership and recommend assigning DPO duties to an independent role.

Due to the significant responsibilities and potential legal consequences, many organizations opt to outsource the DPO role. "DPO-for-hire" services are becoming increasingly common, where external professionals take on the role for multiple organizations in a fractional capacity, much like a virtual CISO. These arrangements ensure independence and transfer part of the legal accountability to the outsourced DPO, which can be advantageous for organizations needing the expertise without the potential conflict of interest.

To summarize, while privacy and security must collaborate closely, they should generally not be managed by the same individual. The DPO must remain independent from the CISO to effectively oversee privacy compliance. If a CISO does take on privacy tasks, they should focus on technical and operational controls, while leaving strategic privacy decisions to other areas, such as legal or compliance. This separation is especially important in organizations where appointing a DPO is legally required. That said, privacy will be part of what a CISO 3.0 must tackle in one form or another, and the privacy requirements for CISOs are only going to increase going forward. This is why CISO 3.0s must familiarize themselves with privacy regulations and prepare their programs for privacy integration.

12.7 BRINGING THE CAPABILITIES TOGETHER

As we explore the modern tactics essential for a CISO 3.0, a common thread emerges: Each of these capabilities, identity centricity, CTEM, Zero Trust, cloud-native protection, data governance, and privacy, has the potential to enhance the Security Operations Center (SOC). However, this integration presents challenges, primarily due to the labor-intensive nature of these tactics. Gartner's suggestion that 20% of SOC resources should be dedicated to CTEM alone is telling. When we consider the additional demands of identity, Zero Trust, CNAPP, data protection, and privacy, it's clear that the size and skillset of the SOC will need to be expanded.

Each of these tactics alone is powerful and vital to the future of cybersecurity, but the real focus of this chapter is that these capabilities must be integrated together into a new type of SOC operation. It's not enough to implement these strategies in isolation; they need to work in unison to provide a comprehensive security posture. We need to fundamentally rethink how SOC operations function and reconsider which functions should be insourced versus outsourced. Some roles, especially those requiring close collaboration with other IT teams or even business units (such as privacy or identity governance), may be difficult or less effective to outsource, requiring a thoughtful evaluation of which functions are better kept in-house.

Another shared characteristic of these modern tactics is the "shift left" mentality. This signifies a move away from traditional reactive security measures toward proactive strategies that embed security earlier in the development and deployment lifecycle. For instance, the focus is no longer solely on providing runtime security; we must now integrate security into the very fabric of the build and design process. This proactive stance allows organizations to identify and mitigate potential risks before they materialize into full-blown threats, thereby enhancing their overall security posture.

The "shift left" approach in security involves moving security considerations earlier in the development and deployment lifecycle. Traditionally, security was often an afterthought, addressed only after systems and applications were built and deployed. This reactive approach left organizations vulnerable to vulnerabilities and misconfigurations that could be exploited by attackers.

Shifting left, on the other hand, integrates security into the design and development phases, ensuring that security is built into systems and applications from the ground up. This proactive approach helps prevent vulnerabilities and misconfigurations, reducing the attack surface and minimizing the risk of security incidents. It also enables faster and more efficient remediation, as security issues are identified and addressed earlier in the development process.

By shifting left, security becomes an integral part of the development and deployment process, rather than a separate and often siloed function. This fosters collaboration between security teams and developers, ensuring that security considerations are incorporated into every stage of the lifecycle. The result is a more secure and resilient IT environment that can better withstand modern threats.

12.7.1 Integrating Capabilities into the SOC: The Fusion Center Approach

Integrating all these capabilities into your SOC requires a change in how we manage SOCs today. One way to tackle this challenge is to overhaul the capabilities of your SOC and create a "fusion center". A fusion center is a collaborative environment where information from various sources is gathered, analyzed, and shared to facilitate informed decision-making and coordinated action. The concept originated in law enforcement and intelligence agencies, where fusion centers were established to share information and intelligence across different agencies and jurisdictions to combat terrorism and other threats.

In the cybersecurity context, a fusion center brings together security analysts, threat intelligence experts, incident responders, and other stakeholders to collaborate and share information about potential threats and vulnerabilities. This centralized approach enables a more complete view of the security posture and facilitates a coordinated response to security incidents. Most fusion centers, as built today, don't include risk or privacy as a function. So even though a fusion center would be well suited for CTEM and CNAPP integration, if you were to build one to solve for the challenges of these modern tactics, you'd still have to figure out who would own the risk management that would be required for a proactive approach.

If you were to build out a fusion center, the process of integrating the various functions involves several key steps. Initially, you need to clearly define the scope and objectives of the fusion center. This includes determining the specific security functions that the center will encompass and outlining its primary goals and objectives. Next, it is imperative to identify the key stakeholders who will be actively involved in the fusion center and establish their respective roles and responsibilities. Once the stakeholders and their roles are defined, the next step is to establish robust processes and procedures for information sharing and collaboration within the fusion center. This encompasses determining how information will be collected, analyzed, and disseminated, as well as outlining communication protocols and escalation procedures.

Building a fusion center offers several compelling advantages. Centralizing information and intelligence provides a better view of the security estate, enabling better identification and assessment of potential threats. A fusion center can also foster a collaborative environment that encourages information sharing and coordination between different teams and stakeholders. This enhanced collaboration leads to a more coordinated and effective response to security incidents, minimizing their impact and improving overall organizational resilience. Through the utilization of advanced analytics and threat intelligence, a fusion center enables proactive threat detection and response, allowing organizations to identify and mitigate potential threats before they can cause any major harm. The real-time information and insights provided by a fusion center empower decision-makers to make more informed and timely choices about security investments and risk mitigation strategies.

However, establishing and maintaining a fusion center also present its share of challenges. The financial and logistical complexities associated with setting up and operating such a center can be substantial, requiring sizable investments in technology, personnel, and processes. Furthermore, the sheer volume of data generated from various sources within a fusion center necessitates robust data management capabilities to ensure data accuracy, integrity, and accessibility. Fusion centers aren't currently designed to take on risk management, which will be a key function for the advanced tactics discussed in this chapter. Finally, breaking down silos and fostering a culture of collaboration among diverse teams and stakeholders can be a hurdle, requiring a concerted effort to shift organizational culture and attitudes toward a more open and cooperative approach to security.

12.7.1.1 Examples of Fusion Centers

Several organizations have successfully implemented fusion centers, including the following:

- **Financial Services Information Sharing and Analysis Center (FS-ISAC):** A global nonprofit organization that facilitates information sharing and collaboration between financial institutions to combat cyber threats.

- **National Cybersecurity and Communications Integration Center (NCCIC):** A US government organization that serves as a central hub for sharing information and coordinating responses to cyber threats.

The construction of a fusion center requires extensive planning and execution. The initial step involves clearly defining the scope and objectives, encompassing the specific security functions it will handle and the intended outcomes. Next, it's imperative to identify the key stakeholders who will actively participate in the fusion center, including security analysts, incident responders, and threat and vulnerability, identity, cloud, and data experts. The establishment of a governance structure and leadership team is the next step, delineating roles and responsibilities while ensuring transparent communication and decision-making channels. The development of well-defined processes and procedures for information sharing, collaboration, and incident response within the fusion center is key. Fostering a culture of collaboration and information sharing within the fusion center is equally important, encouraging open communication and teamwork among the diverse stakeholders.

12.7.2 Alternative to Fusion Centers: The Risk Exposure Center

Creating a fusion center may be challenging for organizations with small SOC teams or those that outsource their SOC capabilities to a managed security service provider (MSSP). While some MSSPs offer fusion center capabilities, these services can be expensive, often starting in the range of seven figures per year. This cost, while potentially justifiable when compared to building an in-house fusion center, can still be a large portion of the budget for most security teams.

As an alternative, John Candillo, a Field CISO at CDW, suggests the concept of a "risk exposure center" (REC). While researching John's REC idea, I found that Brinqa published a blog in September 2023 with a similar idea and called it a risk operations center (ROC). I think I prefer ROC because it fits in nicely with the pre-existing SOC and network operations center (NOC) nomenclature.

The REC/ROC would function similarly to a SOC but focus on risk reduction instead of forensic investigation and would incorporate specialties like identity, threat and vulnerability management, data protection, and cloud security. This cross-functional team would focus on the risk reduction strategy, perform the engineering for advanced solutions, and assist the SOC team in tuning and understanding how to consume alerts and outputs from the advanced toolsets. They would strategically manage policies and vulnerabilities for the entire attack surface, managing threat exposures from the infrastructure, cloud, application security, and vulnerability management.

The ROC versus SOC: The REC/ROC is meant to be proactive and reduce risk "left of boom", where the SOC is reactive and detects and investigates incidents "right of boom". The REC would handle risk prioritization and mitigation, whereas the SOC handles threat detection and incident response. The ROC would look for indicators of risk exposure as opposed to indicators of compromise.

The REC/ROC would also handle the risk management functions related to CTEM, CNAPP, and modern Zero Trust-based access controls. This would include user risk scoring for contextualized just-in-time access. The REC/ROC is not only about moving from point-in-time risk assessment to a continuous model, but it's also about enhancing your vulnerability management program in a way that prevents the relentless pursuit of every single vulnerability or the constant firefighting of high-impact vulnerabilities. Instead, the REC/ROC champions a more strategic and measured approach, providing a panoramic view of risk across the entire business while maintaining the laser focus needed to pinpoint and neutralize the most critical threats before they can be exploited.

The REC/ROC would own the first half of the risk lifecycle outlined in Chapter 6, where a threat exploits a vulnerability, leading to risk that can damage an asset and cause an exposure that can be addressed by a control that protects from the threat. The REC would be responsible for mapping threats to vulnerabilities and assets to discover risk. This is similar to what we discussed in the explanation of the Risk Register 3.0, but the REC would manage the risk identification, and the CISO 3.0 would determine the quantification and treatment, which would be cataloged in the Risk Register 3.0. However, since the treatment for many of the identified risk is mitigation via the existing tools built into the same CTEM and CNAPP platforms used to identify the risk, the REC team could apply those control changes to mitigate the risk without waiting for the CISO. The CISO would only need to define treatments for identified risks that didn't already have control options in place.

A REC/ROC could be structured in various ways, depending on the organization's specific needs and resources. It could be a dedicated team within the security organization or a virtual team comprising individuals from different departments who collaborate on risk-related initiatives. The REC/ROC would typically be populated by individuals with expertise in various security domains, such as identity and access management, threat and vulnerability management, data protection, and cloud security. These individuals would work together to identify, assess, and prioritize risks; develop mitigation strategies; and implement security controls.

The specific roles and skills needed within a REC/ROC would depend on the organization's risk profile and the scope of its security program. However, some common roles might include the following:

- **Risk Analyst:** Responsible for conducting risk assessments, identifying threats and vulnerabilities, and quantifying potential impact.
- **Identity and Access Management Specialist:** Focuses on implementing and managing IAM controls, such as MFA, PAM, and IGA.
- **Threat and Vulnerability Management Specialist:** Responsible for identifying, prioritizing, and remediating vulnerabilities in systems and applications.
- **Data Protection Specialist:** Focuses on implementing and managing data protection measures, such as encryption, DLP, and data access governance.
- **Cloud Security Specialist:** Ensures the security of cloud environments and workloads, leveraging tools like CSPM and CNAPP.

Building a REC/ROC could be a cost-effective alternative to a full-fledged fusion center, especially for organizations with limited resources. Leveraging existing personnel and expertise from different departments, organizations can create a cross-functional team that is focused on risk reduction and security enablement.

The establishment of a REC/ROC could have some advantages. If you can assemble experts from diverse security domains, the REC/ROC could offer a more integrated perspective on the organization's risks, enabling a deeper understanding of potential threats and vulnerabilities. The collaborative nature of a REC/ROC fosters seamless information sharing and coordination between different teams, resulting in a more cohesive and effective security program. The REC's proactive focus on identifying and mitigating risks before they can be exploited empowers organizations to adopt a more strategic and forward-thinking approach to security. Finally, the REC/ROC presents a cost-effective alternative to a full-fledged fusion center, particularly for organizations with limited resources. Leveraging existing personnel and expertise from different departments, organizations establish a cross-functional team dedicated to risk reduction and security enablement, ultimately contributing to the organizations' overall success by aligning security initiatives with business objectives.

The CISO 3.0, as a strategic leader, is uniquely positioned to leverage these capabilities to enhance their organization's security posture and enable business success. Whether through the establishment of a fusion center or an REC/ROC, the integration of these modern security tactics into the SOC is essential for meeting modern security challenges.

12.7.3 Integrating Identity into SOC Operations

As we look to modernize our SOC, we have to figure out how to integrate identity-centric capabilities such as IAM and identity governance into daily operations. The shift toward identity-centric security is necessary for all the reasons we discussed earlier. Which is why SOCs must adapt by making identity a core pillar of their threat detection and response efforts.

Access management is all about controlling who has access to sensitive systems, data, and applications. SOCs can integrate IAM solutions by monitoring access controls and identity-based security events in real time. For example, unauthorized access attempts, privilege escalations, or violations of least-privilege policies should trigger alerts within the SOC, allowing security teams to take swift action. For example, it's currently uncommon for SOCs to ingest the logs and alerts from PIM/PAM tools like CyberArk, and it's even more rare to see management tools like Okta and governance tools like SailPoint integrated into SOC operations.

To implement IAM effectively, SOC teams should work closely with IAM administrators to ensure access policies are enforced consistently and that any anomalies are investigated promptly. Continuous monitoring of identity logs can help detect signs of compromised credentials or insider threats. Bringing the logs from your identity solutions into your SIEM can help the SOCs analyze and correlate suspicious identity-based activities across the IT environment. This is also one of the first steps toward insider threat detection capabilities.

Identity governance provides the framework for managing user identities and access rights within an organization. Governance platforms create a framework for and sometimes even automate user provisioning and deprovisioning, as well as access recertifications. Integrating identity governance into SOC operations allows for automated oversight of user access and entitlements. This involves implementing policies that define who should have access to what resources, enforcing role-based access control, and regularly auditing user permissions.

By embedding identity governance into SOC workflows, security teams can receive automated alerts when access policies are violated or when employees or attackers posing as employees gain unauthorized privileges. Regular access reviews can help ensure that permissions align with business needs and reduce the risk of excessive access rights being granted. SOCs can also play a role in ensuring that identity lifecycle management, from onboarding to offboarding, is properly enforced to minimize risks associated with former employees retaining access to systems.

Identity centricity goes a step further by making identity the focal point of security architecture. This approach involves treating user identities as the key control point for security. Integrating identity-centric security into SOC operations means moving beyond just monitoring access control violations to continuously verifying users' behaviors and access patterns.

A key aspect of identity-centric security is the use of a Zero Trust architecture that includes adaptive authentication and continuous verification of user actions. SOCs should collaborate with IAM and governance teams to ensure that security measures, such as MFA and risk-based access controls, are consistently applied. When anomalies are detected, such as a user accessing sensitive data from an unfamiliar location or device, the SOC can immediately investigate and respond.

In a Zero Trust model, the SOC is central to detecting anomalies that inform just-in-time access decisions to critical resources. To fully integrate Zero Trust into SOC operations, the SOC needs to add user risk-scoring capabilities. This real-time scoring allows Zero Trust decision points to leverage risk profiles when granting or denying access. For example, if a user exhibits high-risk behavior, such as attempting to access resources from an unverified device or location, the SOC can automatically restrict access or escalate the issue for further investigation.

User risk scoring is also foundational for building insider threat detection capabilities within the SOC. By continuously evaluating users' behavior and assigning risk levels, SOC teams can identify patterns that might indicate insider threats, such as data exfiltration or policy violations. By integrating IAM, identity governance, identity centricity, and risk scoring into SOC operations, organizations create a proactive, layered defense where identity-based risks are continuously monitored, and security decisions are informed by real-time data.

12.7.4 Integrating Data Privacy into SOC Operations

As we modernize SOCs, integrating data privacy, privacy operations, and privacy breach response is becoming increasingly essential. With the growing overlap between privacy and cybersecurity, SOCs can no longer focus solely on detecting and responding to cyber threats. Privacy regulations like GDPR and CCPA require organizations to meet stricter standards for handling personal data, meaning CISOs must ensure that privacy breach responses are built into the overall security strategy.

One approach to integrating privacy into SOC operations is to expand the SOC's role to include oversight of privacy breaches and compliance monitoring. Traditionally, SOCs have concentrated on cyber threats, but with the rise of breaches tied to privacy issues, their scope needs to broaden. Privacy breach response should become a standard part of incident response protocols. SOC teams need to be trained to recognize and respond to privacy-specific incidents, such as unauthorized access to personal data or violations of privacy laws. This also includes having workflows in place for timely notification to authorities and affected individuals, as required by regulations.

Another key component of integrating privacy into SOC operations is incorporating privacy risk assessments and data mapping into the existing threat management processes. Using tools like DSPMs, SOCs can automate the discovery, classification, and monitoring of sensitive data. This ensures that personal data is properly safeguarded and that any privacy-related vulnerabilities are detected early. Additionally, SOCs must implement privacy-focused logging and auditing tools to track how data is accessed and used, ensuring compliance with data retention policies and providing an audit trail in case of an investigation.

However, many of these functions, including breach incident response, privacy risk assessments, and legal compliance, are typically the responsibility of the DPO. To effectively integrate these capabilities into SOC operations, deep collaboration with the privacy team is required. This may involve embedding privacy team members within the SOC to ensure seamless coordination between the two teams. Additionally, formal updates to the incident response plan will be needed to clearly delineate how privacy breaches are handled compared to other cyber incidents, ensuring privacy issues receive the attention and care necessary under legal mandates.

We also need to ensure collaboration and build strong partnerships between the SOC and legal or compliance teams. Privacy operations, including responding to data subject requests and handling privacy breach notifications, often fall outside the traditional purview of the SOC. Close coordination is required to ensure compliance with privacy laws. Joint tabletop exercises involving the SOC and privacy teams can help foster a culture of collaboration and ensure that both teams are prepared to handle privacy-related incidents effectively.

12.8 CONCLUSION

The old ways of doing security just don't cut it anymore. It's time to ditch the outdated, reactive approach and move toward a proactive, multilayered strategy that tackles threats head-on. The CISO 3.0 gets this. They're not just about putting out fires or building a fortress that's tough to crack. Think of it like a well-trained army, using a combination of tactics to defend the kingdom. You've got your identity-centric security, making sure only the right people get access to the right things. You've got continuous monitoring, keeping a watchful eye on the battlefield 24/7. You've got data governance, ensuring your most valuable assets are protected. And you've got Zero Trust and cloud-native protection, building layers of defense that are adaptable and resilient.

These tactics are powerful on their own, but when you combine them and align them with your company's goals, that's when the magic happens. It's like a symphony orchestra, each instrument playing its part to create a harmonious and powerful defense.

One way to achieve this harmony is by creating a fusion center or a risk exposure center. These are collaborative hubs where your security experts can come together, share information, and proactively manage risks. It's about breaking down silos and working together as a unified force. Even if you choose not to take on one of those approaches, you will still have to rethink how these modern tactics fit into your SOC operations, how to expand your current SOC capabilities, and which portions of your SOC can and should be insourced or outsourced.

And let's not forget about the "shift left" mentality. It's about baking security into every step of the development and deployment process, not just bolting it on at the end. This proactive approach helps you catch vulnerabilities early, reducing your attack surface and making it harder for the bad guys to get in.

12.8.1 Key Takeaways

1. **Modern Security Tactics as Strategic Enablers**
    - Modern security tactics integrate advanced frameworks and technologies, transforming the role of the CISO into a strategic enabler aligned with business goals.
    - The focus shifts from reactive to proactive approaches, emphasizing capabilities like identity centricity, Zero Trust, CTEM, cloud-native protection, data governance, and privacy.
2. **Identity-Centric Security**
    - The transition from network-centric to identity-centric security ensures that access control focuses on individuals and devices rather than a static perimeter.
    - Identity governance supports Zero Trust by verifying and continuously monitoring users, minimizing risks of insider threats and compromised credentials.
    - Robust IAM practices, such as MFA and PAM, are essential pillars.
3. **Zero Trust and Granular Security**
    - Zero Trust operates on the principles of "never trust, always verify" and granular access control, minimizing the attack surface.
    - Implementation of Zero Trust requires careful planning, starting with identity management, device security, and micro-segmentation.
    - Frameworks like NIST and CISA guide organizations in assessing and improving Zero Trust maturity.
4. **Continuous Threat Exposure Management**
    - CTEM broadens traditional TVM to continuously assess the organization's attack surface, uncovering known and unknown risks.
    - Tools such as EASM and BAS enhance visibility, validation, and prioritization of risks.

5. **Cloud-Native Application Protection Platforms**
   - CNAPPs consolidate cloud security functions, including posture management, workload protection, and identity entitlement management, into unified solutions.
   - These platforms enable organizations to address cloud security challenges like misconfigurations, data breaches, and compliance gaps effectively.
6. **Data Governance and Privacy**
   - Effective data governance integrates lifecycle management, metadata controls, and privacy operations, ensuring compliance and strategic data utilization.
   - DSPs offer automated discovery, classification, and protection of sensitive data across environments.
7. **Fusion Centers and RECs**
   - Fusion centers and RECs integrate advanced security functions to enhance operational efficiency and proactive risk management.
   - These centers foster collaboration across specialized teams like TVM, identity governance, and cloud security, creating a unified security posture.
8. **The Role of the CISO 3.0**
   - CISO 3.0 transcends traditional roles by aligning cybersecurity with business strategies, driving organizational resilience, and fostering a culture of proactive security.
   - Emphasis on continuous learning, adaptability, and collaboration positions the CISO as a strategic leader and risk mitigator.

12.8.2 Reflection and Exploration Questions

Understanding the Concepts

1. How do the SOC and the REC/ROC complement each other in enhancing security?
2. How do identity centricity and Zero Trust complement each other in enhancing security?

Connecting the Dots

3. How do the proactive elements of CTEM align with earlier discussions on roadmap quantification and continuous improvement?
4. In what ways do CNAPPs support broader data governance and privacy initiatives?

Real-World Application

5. Can you identify examples of organizations successfully implementing Zero Trust or CTEM? What were the outcomes?
6. How might DSPs or similar tools transform the way your organization manages sensitive data?

Personal and Professional Growth

7. How can you apply the principles of a fusion center or REC to improve collaboration and efficiency within your security operations?
8. What steps can you take to enhance your understanding and implementation of cloud-native protection strategies?

Further Exploration

9. What additional frameworks or tools could be integrated into your organization's security strategy to address emerging threats?
10. What steps could you take to begin leveraging modern more modern tactics?

13 Modern Cyber Resilience

13.1 INTRODUCTION

Today, organizations face the constant challenge of maintaining uptime, ensuring that their critical systems and services remain available and operational. The consequences of downtime can be severe, ranging from financial losses to operational disruptions and even safety risks.

I've had countless clients experience outages, but I'll never forget one client whose customer loyalty system was locked out by ransomware and was down for more than 14 days. Customers were angry, employees were frustrated, and the business suffered financial losses. That's when I truly realized the importance of cyber resilience and started to work on the concepts of modern resilience because it's not about preventing breaches but about being able to bounce back quickly and minimize the impact of downtime.

For the CISO 3.0, uptime is not only an IT concern but also a critical business imperative. They understand that cybersecurity helps ensure resilience and minimizes the impact of cyber incidents on business continuity. As we discussed earlier, productivity uptime is one of the top business challenges for nearly every organization, whether it's a manufacturing line, shipping and logistics, or billing and invoicing. Business continuity is a top priority for most organizations. That's because, in this age of digital transformation, every business is a digital business, whether they realize it or not.

Take, for instance, the fictional example of Joe's Corner Tire Shop. Joe believes he is primarily in the tire business, but like all modern businesses, he runs an IT business that sells a product, and that product happens to be tires. Joe advertises online, has to manage his search results via search engine optimization (SEO), and monitor his online reviews. Customers book appointments online and through a mobile app; the point-of-sale machine, payment cards, and the loyalty program are all Internet-connected apps. He orders his tires and does inventory management via a web app and uses software-as-a-service (SaaS) applications for HR, payroll, and accounting. If any one of those things went down due to an outage or cyberattack, major portions of the business would grind to a halt.

So, as you can see, the bulk of Joe's business, like most businesses, is an IT service delivery business. Studies suggest that a single hour of downtime can cost most enterprises around $300,000, and 44% of large and midsize enterprises spend more than $1 million per hour. A conservative estimate per incident downtime costs is $43.2 million, which is much higher than the average ransom payment of $1.5 million.

Now, stop and ask yourself, "What percentage of your budget is spent on proactive versus reactive measures?"

If you're like most chief information security officers (CISOs) I talk to, it's probably close to 50/50. But is that really where you want to be? Wouldn't it be better to

DOI: 10.1201/9781003510789-18

shift more of your spending toward prevention? Ideally, yes. But in order to do that, we need to reduce both the need for and the cost of detection, response, and recovery.

Throughout this chapter, we'll explore ways to do exactly that. It's worth noting, though, that these strategies will require some upfront investment in response and recovery capabilities. While these are proactive steps, they may initially skew your spending toward reactive measures in the short term. However, once these systems are fully in place, you'll be able to refocus on strengthening your proactive controls, setting the foundation for a more resilient and balanced cybersecurity approach. This is the mindset of the CISO 3.0, looking beyond just buying tools and focusing on building capabilities that ensure the business can withstand disruptions, adapt, and thrive even through events and incidents.

13.2 BLACK SWAN RISKS

In Chapter 6, we defined black swan risks as described by Nassim Taleb, but we didn't spend much time learning to quantify unexpected and existential risks. That's because while it is possible to use statistics to calculate the long tails of a loss exceedance curve, sometimes referred to as "tail risk", I don't believe it's worth the effort.

It's difficult and confusing to calculate tail risk, and this doesn't provide much value. Remember, when you are calculating risk, you are predicting the future, and what is the point of spending a lot of effort trying to predict what are, by definition, unpredictable corner cases? It is, however, important to know that black swans are possible and to shield yourself against existential risk.

Modern cyber resilience is the best defense against such risks. Assessing risk isn't always necessary when its existence is already known. While many methods exist to assess security programs and create roadmaps, understanding your critical assets and their uptime risks can guide your actions without extensive penetration testing or compliance assessments.

Consider a critical process or application vital to your business. A risk appetite discussion can estimate its impact. Ask stakeholders how long the business could function without it and when losses would begin to drastically affect the bottom line. The answer is likely hours or days. You can even inquire about the time it would take for the business to fail entirely.

If the answer is less than 21 days, then action is needed because the average ransomware attack disrupts operations for approximately 21 days. If your organization cannot tolerate such an outage, improvement is necessary. To reduce downtime to 1 day would require a 2,100% improvement over the average, making your program world-class.

13.2.1 ANTIFRAGILITY

Nassim Taleb's book *Antifragile: Things That Gain from Disorder* offers a way to combat black swans. Antifragility is a property of systems that thrive and grow when exposed to volatility, randomness, disorder, and stressors. In contrast, fragile systems break down under stress, and resilient systems resist or recover from shocks but remain unchanged.

Examples of Antifragility

The Human Body: Regular exercise and exposure to minor illnesses strengthen the immune system and overall health, making the body more resistant to future challenges.

The Economy: Recessions and market downturns can force businesses to innovate and adapt, leading to a more robust and efficient economy in the long run.

The Internet: Attempts to censor or control the internet often led to developing new technologies and decentralized networks, making it even more resilient.

Antifragility concepts enhance modern cyber resilience by encouraging organizations to embrace change and uncertainty. Instead of predicting and preventing every threat, the focus shifts to building adaptable systems that improve when facing unexpected challenges. Organizations should learn from failures, viewing cyber incidents as opportunities for growth. Decentralization and distribution are key to avoiding single points of failure by spreading critical systems and data across multiple locations and platforms. Finally, fostering a culture of continuous improvement and experimentation is required for developing novel solutions and strategies to address emerging threats.

Resilience is the ability to withstand and recover from shocks, while antifragility goes further by thriving and growing stronger from adversity. In cybersecurity, resilience focuses on prevention and recovery, while antifragility aims to create systems that benefit from attacks through learning and adaptation.

Therefore, we should refer to it as "modern cyber antifragility" instead of just "modern cyber resilience".

13.3 CYBER RESILIENCE

Cyber resilience is the ability to anticipate, withstand, recover from, and adapt to adverse conditions, stresses, attacks, or compromises that affect systems relying on cyber resources. A truly resilient system is designed to minimize disruption, maintain operational effectiveness, and continue delivering business value despite incidents.

Cyber resilience is not merely about technology: It involves all types of resources, including processes, policies, and people working together to maintain business continuity. This means ensuring that security is embedded across operations, integrating recovery and continuity efforts with business objectives, and continuously improving defenses against evolving threats.

A key aspect of cyber resilience is *antifragility*; rather than just recovering from cyber incidents, antifragile security systems evolve by learning from attacks and adapting to future threats. This mindset shifts organizations from a purely defensive posture to one of continuous improvement, using every disruption as an opportunity to enhance security, streamline recovery processes, and reinforce business continuity mechanisms. Decentralization, redundancy, and iterative stress-testing are fundamental elements of this approach, ensuring that cybersecurity frameworks not only withstand threats but also emerge stronger after each challenge.

However, in our efforts to maximize efficiency within our processes, we inadvertently create very fragile systems and have very little resilience baked in. The pursuit of efficiency often leads to single points of failure, rigid dependencies, and over-optimized workflows that break down under stress. True cyber resilience requires balancing efficiency with redundancy, ensuring that systems can absorb disruptions and continue functioning under adverse conditions.

13.3.1 Building Cyber Resilience

Developing cyber resilience is not a one-time initiative; it is a continuous process that requires structured governance, thorough analysis, strategic design, and rigorous operational practices. Organizations often fall into the trap of thinking cyber resilience is about buying the latest technology, but the reality is that real resilience requires very little "stuff". Later we will talk about cyber recovery, much of which is driven by technology, but real resilience is generally more focused on people and process.

A strong cyber resiliency program starts with *cyber resiliency management*, which establishes governance structures, accountability, and long-term strategic oversight. One of the biggest pitfalls organizations encounter is failing to designate clear ownership of resilience initiatives. Without defined leadership and explicit executive sponsorship, these programs stagnate, leaving critical gaps in preparedness.

Cyber resiliency analysis is another cornerstone of the strategy, focusing on identifying critical business processes, dependencies, and vulnerabilities. Many organizations make the mistake of assuming they know what their critical assets are, only to discover during an incident that they lack visibility into key interdependencies. Resilience planning must be built on data-driven insights, mapping out business functions that must be preserved at all costs.

Cyber resiliency design ensures that architectural frameworks and alternative business processes support operational continuity. This step often gets overlooked in favor of efficiency, leading to brittle systems that cannot gracefully withstand failure. Designing for resilience means implementing redundancy, fault-tolerant applications, and failover mechanisms that reduce the impact of disruptions.

Cyber Resiliency Engineering and DevOps operationalizes resilience by developing and managing systems designed for rapid recovery. Organizations that embrace automation and continuous testing vastly improve their ability to withstand cyber incidents, yet many enterprises still rely on outdated manual recovery processes that introduce delays and inconsistencies.

Cyber resiliency planning and testing validate the effectiveness of the entire strategy. Without rigorous, real-world simulations – tabletop exercises, penetration testing, and full recovery drills – organizations operate under a false sense of security. Many only discover weaknesses during a real crisis when it's too late to correct them.

Finally, *cyber resiliency operations* ensures that resilience remains an ongoing priority. Many companies make the mistake of treating resilience as a one-time initiative rather than an embedded practice. Operational resilience must be continuously improved, with learnings from every incident used to refine and strengthen the program.

13.4 QUICK WIN STEPS TO BEGIN YOUR CYBER RESILIENCE JOURNEY

Understanding that your organization faces uptime risks is the first step in moving toward a more resilient, antifragile state. The journey begins with a structured approach designed to align your cybersecurity efforts with your business's unique needs and risk tolerance.

13.4.1 Step 1: Risk Appetite Assessment

Start by assessing your organization's risk appetite. This involves working with the executive team to define how much risk is acceptable, identifying tolerances and limits. It's important to revalidate this with leadership to ensure that everyone is on the same page when it comes to balancing uptime and risk. From here, you can reset your uptime expectations based on these risk limits.

13.4.2 Step 2: Define the Minimum Viable Company

Next, identify your organization's minimum viable company (MVC), a concept drawn from the Operational Resilience Framework. The MVC defines the essential services and processes that must remain operational to keep the business running in a functional state during a crisis. Think of it as the bare minimum your company needs to continue serving customers, complying with regulations, and handling essential financial operations like payroll and accounts payable.

To define your MVC, ask key questions:

- What business processes are so critical that losing them could threaten the company's survival?
- Which systems, services, or applications can tolerate minimal downtime, and which would impact the business first if they failed?
- What data must be protected at all costs, and which assets need to be restored first to maintain operations?

This exercise involves mapping the IT systems that support these critical processes, identifying key dependencies both upstream and downstream, and documenting the infrastructure necessary to keep the business operational.

13.4.3 Step 3: Analyze and Document Current Capabilities

Once you've defined your MVC, it's time to assess your current cyber resilience measures. Identify and document your existing backup, recovery, and continuity strategies, including

- business continuity plans (BCPs),
- disaster recovery (DR) measures,
- cyber recovery capabilities.

This includes everything from alternative processes that can kick in during an IT failure to the infrastructure in place for disaster recovery. The goal here is to see how well your current capabilities align with the critical needs of your MVC and where gaps might exist.

13.4.4 Step 4: Identify Opportunities for Improvement

With a clear picture of your current resilience measures and MVC in hand, it's time to spot areas for improvement. Large-scale attacks often expose weak points in existing tools, governance, and recovery plans. Many organizations overestimate their preparedness, relying on outdated recovery strategies that can't withstand the reality of modern cyberattacks, such as ransomware. Identifying these vulnerabilities early allows you to strengthen your defenses and build more effective recovery plans.

13.4.5 Step 5: Set Targets and Test Your Plan

Once the gaps are identified, you can set new goals for both your current and future states. Review your BCP/DR/cyber recovery plans and determine if your Recovery Time Objective (RTO) is realistic. If not, revise your policies and plans to meet your organization's risk tolerance and uptime goals.

After adjustments, conduct a tabletop exercise to test the feasibility of your RTO goals, such as hitting a 7-day recovery window. This practice run allows you to refine your processes and ensure the plan is actionable in the event of a real incident.

By following these steps, you'll begin building an antifragile cybersecurity strategy, one that not only recovers from disruptions but also grows stronger in the face of adversity.

13.5 CYBER RECOVERY

Cyber resilience and cyber recovery are often discussed together, but they serve distinct functions in an organization's cybersecurity strategy.

Cyber recovery is required when cyber resilience fails. Since achieving 100% resilience is impossible, recovery becomes essential when systems cannot simply shift or adapt to an incident and must be restored. However, recovery does not have to be a point-in-time activity that only occurs when an incident happens. Instead, recovery should be an automated and continuous process.

Cyber recovery is focused on restoring operations after a disruption, often relying on backup systems, redundancy, and restoration protocols. Unlike traditional IT disaster recovery, cyber recovery assumes that an adversary has targeted systems in a way that compromises trust in existing infrastructure, requiring more rigorous validation before restoration. Understanding the difference between IT disaster recovery, cyber recovery, and business continuity is essential for developing a robust resilience strategy.

IT Disaster Recovery is traditionally focused on responding to large-scale, natural, or man-made disasters such as hurricanes, floods, or terrorist attacks. These recovery plans are typically centered around relocating systems from an affected facility to a pre-designated location. The key assumption here is that the core IT infrastructure, like networks and critical services, remains intact and trustworthy. IT disaster recovery relies on near real-time data replication to achieve aggressive Recovery Point Objectives (RPO) and Recovery Time Objectives (RTO). Because these scenarios are predictable and well-understood, disaster recovery procedures are often standardized, regularly tested, and repeatable.

Cyber Recovery, on the other hand, deals with the more complex and unpredictable challenge of malicious attacks aimed at compromising data, deleting information, or disrupting essential systems. Cyberattacks can affect multiple applications and infrastructure across different locations, including foundational services like active directory (AD), domain name system (DNS), or core networks, which are often the first to be compromised. In such cases, the usual IT disaster recovery procedures may no longer apply, as the systems you rely on to restore operations may also be infected or unreliable.

Cyber recovery solutions are designed to address these high-stake scenarios where foundational services can't be trusted. They require more flexible and advanced recovery capabilities, such as handling partial recoveries or selectively restoring business-critical functions. These solutions may need to handle high-bar recovery requirements, including rapid RTOs and short-loss RPOs. Recovery might need to be faster than serialized pipelines allow, and it might involve recovering some, but not all, business functions or portions of business functions. These solutions must also be able to support various environments, whether on-premises, cloud-based, or mainframe systems, and must account for more complex attacks that traditional disaster recovery processes are not equipped to handle.

Business Continuity is the ability to ensure that critical business operations can continue even while IT systems are disrupted. It involves the development of processes, procedures, and contingency plans that enable organizations to operate manually or with alternative methods during recovery efforts. For instance, reverting to manual or paper-based processes ensures that essential functions, such as processing transactions or managing supply chains, can proceed while IT systems are being restored. Business continuity acts as a stopgap, preventing the business from hemorrhaging money during the recovery process, safeguarding operations, and protecting the organization's reputation.

It is necessary to have plans in place for all three pillars: *IT Disaster Recovery*, *Cyber Recovery*, and *Business Continuity*. Disaster recovery ensures that you can restore services in the event of a natural disaster like a fire or flood. Cyber recovery ensures that you can restore services in a hostile environment, such as a ransomware incident where systems are compromised. Business continuity, meanwhile, ensures that the business can continue to function, even if only at a reduced capacity, while recovery efforts are underway.

These three pillars are distinct yet complementary. Each addresses a different aspect of resilience, and having robust plans for all three is critical for an effective organizational strategy. However, it is rare for organizations to have fully developed plans across all three pillars, and even rarer for those plans to be at similar levels of maturity. To achieve this balance, organizations must develop, test, and refine their plans through regular tabletop exercises and simulations, ensuring that each pillar can effectively support the organization in its time of need.

This comprehensive approach to cyber recovery, combined with IT disaster recovery and business continuity plans, ensures that organizations can bounce back effectively from a range of disruptions, whether natural, malicious, or operational.

13.5.1 Recovering from Immutable Backup

If your organization isn't already using immutable backups, it's time to make it a priority. If you do have them in place, it's important to regularly check their health and ensure they meet best practices for immutability, indelibility, encryption, and redundancy. Additionally, confirm that your backup scope and retention policies align with your recovery expectations.

Many organizations mistakenly believe that implementing an immutable backup solution is enough to complete their recovery strategy. But immutability, while an important first step, only ensures that backups cannot be altered. It doesn't mean they can't be deleted or manipulated. This is where *indelibility* comes in. While attackers may not be able to change immutable backups, they could still access backup systems and adjust retention periods, leaving you with compromised data. Attackers can even snoop around in backups, extracting valuable information such as insurance policies or incident response plans. That's why layering indelibility and encryption into your backup solution is essential to prevent unauthorized access and manipulation.

Recovery timelines also need to be realistic. The industry average downtime for a cyberattack is 21 days. I know your backup vendor may have provided you with an optimistic RTO of 12 hours, but it is important to understand that the clock doesn't start ticking until the recovery process can actually begin. Before that, numerous steps occur. The first few hours or days after an incident are often spent confirming the attack, deciding whether external help is needed, and initiating a forensic investigation. Recovery can't start until that investigation is complete, which often takes several days. Then, there's the challenge of setting up a clean recovery environment. You don't want to restore systems into a potentially compromised environment, and you need to avoid disrupting any functional systems, assuming the FBI hasn't seized your hardware as evidence. Consequently, you will need to set up an alternate recovery environment. This environment will need an AD domain to join machines. However, attackers often compromise AD with invalid credentials, potentially including domain admin access. Without knowing the last clean AD backup, rebuilding AD from scratch might be necessary, which takes several more days. After that you can start your restore, but even then it won't be running in production. You still have to make sure it's clean functional and trustworthy in the restore environment before migrating it to the production environment. That migration can have its own challenges like shared IP space or slow data transfer speeds, which all take more time, and just like that, 21 days have passed.

Is 21 days a reasonable amount of downtime for your organization? Could your organization tolerate 21 days of outages? What would be a more reasonable RTO? 1 day, 8 hours, 4 hours? Let's say you wanted to get down to a 1-day RTO. In order to do that, you'd have to be 2100% better than the industry average. Is your program 2100% better than average at anything? You'd have to be a world-class program to be that much better at most security capabilities. The good news is that when it comes to resilience, it is totally possible.

To have a 2100% better than average RTO, you have to do as much of the work as possible in advance. You'll start by ensuring that your backups must be immutable, indelible, and encrypted. Once you've health-checked and secured your backups, you will need to design and implement a recovery architecture that includes vaults, clean rooms, and non-production restoration sites. Essentially, you will be creating a cyber recovery management framework.

13.5.2 On-Demand versus Continuous Recovery

On-demand recovery entails that services/applications and their supporting infrastructures are not recovered until a cyber incident occurs. The pre-incident workflow is limited to vaulting backups and using Vault-supported tools to evaluate those vaulted copies of backup data and metadata. Recovery testing is performed as needed or required, and recovery is queued and executed in sequence. This approach is suitable for profiled services/applications and their dependencies that do not require the fastest recovery (they can be queued and recovered in order).

In contrast, continuous recovery involves regularly recovering services/applications and their supporting infrastructures under normal operations. The pre-incident workflow includes partial or full recovery to a recovery room. Recovery testing is performed under normal operations, and partial or full recovery is pre-staged before any cyber incident. This approach is suited for shared services and profiled services/applications that require the most rapid recovery and cannot afford to be queued and recovered in order.

You can leverage your backup solution's continuous data protection (CDP) function, which constantly safeguards your data by capturing every change as it happens. Unlike traditional backups, CDP ensures that you can recover data to any moment, minimizing any potential losses. This real-time protection is achieved by instantly saving a copy of each file modification, creating an ongoing backup stream alongside your live data.

However, CDP alone doesn't fully realize continuous recovery. It's the bridge, not the destination. True continuous recovery requires a system where your critical systems and data aren't just backed up – they're *always* ready to take over in case of a disruption. Think of it as having a shadow IT infrastructure, constantly mirroring your production environment, waiting in the wings to step onto the stage at a moment's notice. For true continuous recovery, you will likely need a platform like Trillo, *or you can leverage the latest advancements in containerization.* While previously limited to containerizing individual services, it's now possible to containerize entire operating systems. This opens up the possibility of containerizing not only all of your critical apps but all of your critical infrastructure as well, including load balancers, firewalls, and databases, creating a fully redundant, resilient ecosystem. You can seamlessly stop one container and spin up another without any failover. This empowers you to constantly refresh an entire critical app and its runtime environment, ensuring maximum uptime and resilience.

Similar to the tactics discussed in the last chapter, a move toward continuous recovery helps shift left and make the tactical more strategic. Continuous recovery aligns with the CISO 3.0's focus on proactivity and agility. It enables organizations to minimize downtime and disruption by having critical systems and data pre-staged and ready for recovery, reducing the time it takes to resume operations after an incident. It also improves resilience and responsiveness, allowing organizations to quickly adapt to changing threats and vulnerabilities, ensuring they can maintain operational effectiveness even in unexpected challenges. It enhances confidence and trust by demonstrating a commitment to preparedness and rapid recovery, building trust with customers, partners, and stakeholders.

But containerization is only half the equation. To manage this dynamic environment, you need orchestration tools like Kubernetes. Kubernetes automates the deployment, scaling, and management of containerized applications, allowing you to switch between containers without any downtime seamlessly. While containerization and orchestration lay the groundwork, it's automation that truly drives continuous restore. Here's where tools like Ansible, Puppet, or Chef come in.

Ansible: Ansible uses playbooks, which are simple, human-readable YAML files, to automate tasks like configuration management, application deployment, and orchestration. In a continuous recovery scenario, Ansible playbooks can be used to define the desired state of your containerized infrastructure and automatically spin up new containers or reconfigure existing ones in response to a failure.

Puppet and Chef: Similar to Ansible, Puppet and Chef also automate configuration management. They use a declarative approach, where you define the desired state of your infrastructure, and the tool ensures it stays that way. In a continuous recovery scenario, these tools can continuously monitor your containerized environment and automatically remediate any deviations from the desired state, ensuring that your systems remain resilient.

By strategically combining CDP, containerization, orchestration, and automation, organizations can build a robust and resilient infrastructure that enables continuous recovery and restore. While it may be hard to accomplish, the rewards of minimized downtime and enhanced resilience are worthwhile.

13.5.3 Building Cyber Recovery

Unlike traditional disaster recovery, which assumes the infrastructure is intact and data is trustworthy, cyber recovery operates under the assumption that systems have been compromised, backups may be corrupted, and normal recovery procedures are unreliable. This means that cyber recovery must be purpose-built, integrating dedicated architectures, specialized processes, and rigorous validation mechanisms to ensure that restored systems are clean, secure, and operational.

Many organizations mistakenly believe that having a disaster recovery (DR) plan is sufficient, only to find out, often during a crisis, that their DR infrastructure is just as vulnerable as their primary systems. Cyber recovery planning must anticipate worst-case scenarios, such as ransomware locking out critical operations or attackers manipulating backups to embed long-term persistence. Without a tested and validated cyber recovery framework, organizations risk prolonged downtime, financial loss, and reputational damage.

Additionally, most DR plans are designed to restore entire data centers and assume that the new environment will become the permanent primary infrastructure, making it difficult to migrate data and applications from recovery environments back into production. Many organizations are surprised to discover that their recovery solutions lack the granularity to restore a single application or workload, preventing them from prioritizing the most critical things first. This oversight can result in significant delays, operational inefficiencies, and increased exposure to further risks. To address this, organizations must ensure that their cyber recovery plans include the flexibility to recover at multiple levels, ranging from full-site failovers to granular application-specific restores, allowing for a more agile and efficient recovery process.

The following sections outline the key components of an effective cyber recovery strategy, detailing how organizations can develop, implement, and maintain a recovery architecture that minimizes operational impact while ensuring the integrity of restored systems.

A comprehensive cyber recovery strategy follows a structured approach, mirroring the key steps of cyber resilience while adding layers of security and validation to address the unique challenges of cyber incidents.

- **Cyber Recovery Management:** Oversees the integration of recovery services into broader business operations, ensuring seamless governance of the recovery process.
- **Cyber Recovery Analysis:** Evaluates business functions and infrastructure to identify the specific requirements for recovery tailored to each organization's needs.
- **Cyber Recovery Design:** Translates the analysis into practical recovery plans, detailing how to restore affected systems and data.
- **Cyber Recovery Engineering/DevOps:** Builds and manages the technology platforms that will support the cyber recovery plan, using insights from the recovery profiles.
- **Cyber Recovery Planning:** Develops detailed recovery plans, ensuring they meet the organization's objectives and can be effectively executed in the event of an attack.
- **Cyber Recovery Testing:** Ensures that the recovery procedures are sound, with multiple levels of testing to guarantee systems can be restored quickly and effectively.
- **Cyber Recovery Operations:** Executes the recovery plan when needed, restoring critical business functions with minimal downtime.

My friend and colleague Gary McIntyre created a cyber resilience framework and a cyber recovery model, a portion of which you can see in Figure 13.1. That model underpins much of the content in this chapter.

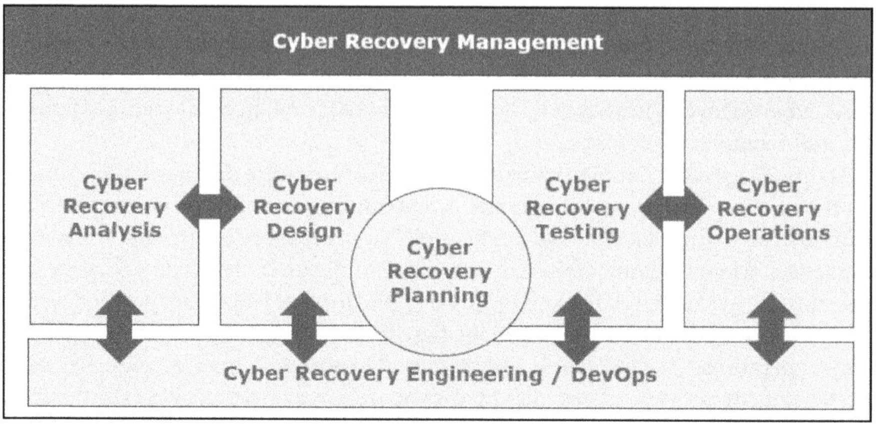

FIGURE 13.1 Cyber Recovery Model.

A well-designed cyber recovery architecture can mean the difference between a minor setback and a complete business disruption. I've seen firsthand organizations that have invested in cyber recovery and made the 2100% RTO improvement jump.

It is also necessary to have a cyber recovery-specific architecture in place in advance to allow for a quick recovery. This Cyber Recovery architecture is a blueprint designed to help ensure your organization can bounce back from a cyberattack, even when traditional disaster recovery measures fall short. A strong cyber recovery architecture should include four main components: vaults, clean rooms, recovery rooms, and well-documented processes.

13.5.3.1 Vault Solutions

Vaults are secure, isolated storage environments designed to safeguard critical data and system backups. They are typically air-gapped or logically separated from the production network to prevent unauthorized access or modification, especially during a cyberattack. Modern vaults often leverage immutable storage technologies, ensuring that backups cannot be altered or deleted, even by malicious insiders or sophisticated ransomware. Vaults provide the last line of defense against data loss or corruption. In the event of a successful cyberattack, having clean and uncompromised backups stored in a vault allows organizations to recover critical data and systems, minimizing downtime and ensuring business continuity.

13.5.3.2 Clean Rooms

Clean rooms are controlled environments where recovered data and systems are thoroughly analyzed and validated before being reintroduced into the production environment. They enable security teams to identify and remove any lingering malware, backdoors, or other compromises that might have evaded detection during the initial attack. Think of it as a quarantine zone for your data. Clean rooms are necessary for ensuring the integrity and trustworthiness of recovered data and systems. Without them, organizations risk reintroducing malicious codes or compromised data into their production environment, leading to reinfection or further damage.

13.5.3.3 Alternate Recovery Rooms

These are isolated, secure environments where critical business processes can be restored and operated independently of the primary production environment. They provide a safe haven to resume operations while the main environment is being investigated, cleaned, and rebuilt following a cyberattack. Alternate recovery rooms ensure business continuity by enabling organizations to continue serving customers and fulfilling critical functions, even if the primary infrastructure is compromised or unavailable. They minimize downtime and prevent material financial loss and brand damage.

13.5.3.4 Standardized Processes

This encompasses the documented procedures and workflows governing the cyber recovery process. This includes regular backup and recovery processes to ensure data is consistently protected. It also involves incident response and triage protocols to identify and contain cyberattacks quickly. It includes recovery testing and validation procedures to ensure that backups are usable and recovery plans are effective. Finally, it incorporates security hardening guidelines for recovery environments to prevent further compromise. Standardized processes bring order and predictability to the often chaotic aftermath of a cyberattack. They ensure that recovery efforts are efficient, consistent, and aligned with the organization's business objectives and risk appetite.

Each of these four components are involved in creating a robust cyber recovery architecture. Vaults provide a secure haven for critical data, clean rooms ensure the integrity of recovered systems, alternate recovery rooms enable business continuity, and standardized processes guide the entire recovery effort. Together, they form a recovery platform that enables organizations to effectively respond to and recover from cyberattacks.

13.5.4 Breach Simulation

Breach simulation, also known as attack simulation or red teaming, is a proactive cybersecurity strategy where real-world attacks are simulated to pinpoint vulnerabilities, test defenses, and enhance an organization's overall security stance. It's about learning from controlled failures to improve your defenses and response. In essence, breach simulation is a way to move from resilience toward true antifragility, where your systems not only withstand attacks but also actually *improve* because of them.

Netflix's Chaos Monkey project embodies the spirit of antifragility in the realm of IT infrastructure. Chaos Monkey is a tool that randomly terminates instances in their production environment to test the resilience of their systems. The idea is to proactively inject failure into the system to uncover weaknesses and force engineers to build more robust and fault-tolerant architectures. Applying this concept to security, we can use breach simulations as our "Chaos Monkey". By regularly simulating attacks, we can identify vulnerabilities, test our defenses, and force ourselves to adapt and improve. This proactive approach not only strengthens our security posture but also fosters a culture of continuous improvement and innovation.

Breach simulation platforms utilize various techniques to mimic the tactics of real-world attackers. These can involve automated attacks launched by the platform, such as phishing emails or network intrusions, to test the effectiveness of current security controls. Alternatively, experienced security professionals, or "red teamers", can conduct manual penetration testing to exploit vulnerabilities and assess the organization's detection and response capabilities. The platform can also simulate the behavior of specific threat actors to gauge defenses against their known tactics. Platforms like AttackIQ and SafeBreach offer a range of attack simulations, from ransomware and malware to data exfiltration scenarios.

By proactively simulating attacks, you can uncover weaknesses in your defenses before real attackers can exploit them. These simulations allow for the evaluation of security controls and incident response capabilities under realistic conditions, promoting continuous improvement and resilience. Regular simulations foster a culture of security by raising awareness and encouraging employees to be vigilant and proactive in identifying and reporting potential threats.

The insights gained from breach simulations can be leveraged to strengthen protective measures by addressing identified gaps in security controls. They also inform the design of more effective incident response plans, recovery procedures, and communication strategies.

Integrating breach simulation with automation takes this a step further, creating something like a "cyber immune system" that works like a vaccine to expose your system to a weakened threat so that you can proactively build a response. Security tools can be configured to respond to simulated attacks automatically, and orchestration tools like Kubernetes can automate recovery processes. Continuous improvement is achieved by regularly analyzing simulation results and using them to refine security controls, incident response plans, and automation workflows.

Breach simulation complements the other elements of a cyber recovery architecture, creating a well-documented and proactive approach to security. Immutable backups of critical data and systems are stored in secure vaults, isolated from the production environment. Clean rooms are utilized to analyze and validate recovered data and systems after a simulated breach, ensuring no lingering threats. Alternate recovery rooms provide a space to test recovery procedures and validate the restoration of critical business processes. Finally, the insights gained from simulations inform the documentation and refinement of incident response, recovery, and communication plans, with automation tools like Ansible, Puppet, or Chef codifying these processes for consistent execution.

13.5.4.1 Example

I once had a health insurance client who made attack simulation one of the cornerstones of their strategy. They adopted a methodical approach, focusing on one attack category at a time. By analyzing their gaps in protection and response across the entire kill chain for that specific threat, they gained invaluable insights. This information guided their prioritization of vulnerabilities and controls, ensuring that the most exploitable weaknesses tied to real threats were addressed first. Additionally, they used simulation results to uncover skill gaps within their team and tailor training programs to address those specific needs.

Once they were confident in their ability to handle one attack category, they moved on to the next, iteratively strengthening their defenses. This approach allowed them to focus on mitigations that truly mattered rather than blindly applying controls based on a generic checklist that treated every vulnerability equally.

By combining breach simulation with these components and leveraging automation and orchestration, CISO 3.0s can begin to create a truly antifragile security strategy. This approach goes beyond mere resilience, actively using simulated attacks to identify weaknesses, improve defenses, and ultimately build a stronger, more adaptable security posture. You've probably heard the saying, "We have to be right 100% of the time, and the bad guys only have to be right once". But that mindset shouldn't hold true. For years, we've had defense-in-depth strategies designed to layer defenses in a way that forces attackers to be right multiple times before they can inflict real damage. Breach simulation takes this concept even further, enabling us to map controls and remediations to each step of the kill chain. This makes it even more difficult for attackers to "be right" enough times and in the right sequence to achieve their malicious objectives. With this proactive approach, the odds shift more in favor of defenders, creating a security posture that is not only reactive but also anticipatory and dynamic.

13.6 CONCLUSION

In this chapter, we've taken a deep dive into modern cyber resilience, exploring how organizations can not only survive cyberattacks but also actually emerge stronger from them. We've looked at the constant struggle to keep systems up and running and how downtime can cripple a business. But we've also gone beyond mere survival, exploring the concept of "antifragility", which is the idea that we can build systems that actually *benefit* from challenges and disruptions.

The CISO 3.0 is at the forefront of this transformation. They're the ones with the vision and leadership to build a proactive, adaptable recovery strategy. It's about more than just bouncing back: It's about being ready to hit the ground running with critical systems and data at your fingertips.

We've also stressed the importance of a solid cyber recovery architecture; you can think of it as your emergency bunker, equipped with everything you need to weather the storm. We've talked about vaults to protect your data, clean rooms to scrub it clean, alternate recovery rooms to keep the business running, and standardized processes to guide you through the chaos.

Cyber recovery isn't just a technical challenge: It's a whole new way of thinking. It's about recognizing that traditional disaster recovery plans might not cut it in the face of a sophisticated cyberattack. It's about building dedicated teams and capabilities that can handle the unique challenges of cyber recovery.

Utilizing the principles of antifragility and building a solid cyber recovery program can turn adversity into an advantage. Organizations can even emerge from cyber incidents not only unscathed but also stronger and more resilient than before. And the CISO 3.0 is the key to guiding them toward a truly secure and antifragile future.

13.6.1 Key Takeaways

1. **Modern Cyber Resilience and Antifragility**
   - *Cyber resilience* is different from business continuity and disaster recovery.
   - Cyber resilience is not only about preventing breaches but also about recovering quickly and minimizing the impact of disruptions to maintain uptime and operational efficiency.
   - The concept of *antifragility*, drawn from Nassim Taleb's work, emphasizes building systems that grow stronger when exposed to challenges, rather than merely resisting or recovering from them.
   - Antifragility transforms traditional resilience strategies by promoting adaptability, learning, and growth through adversity.
2. **Uptime as a Business Imperative**
   - In today's digital-first environment, uptime is no longer just an IT concern but a critical business necessity, directly tied to financial performance and customer satisfaction.
   - Proactive strategies, such as recovery architecture and risk appetite assessments, help organizations align resilience with operational and strategic goals.
3. **Phases to Achieve Cyber Antifragility**
   - **Risk Appetite Assessment:** Define acceptable risk levels in collaboration with leadership to align with business objectives.
   - **Minimum Viable Company (MVC):** Identify critical processes, services, and infrastructure required for operational continuity during crises.
   - **Capability Analysis:** Assess and document current cyber resilience measures, including Business Continuity Plans (BCP), Disaster Recovery (DR), and Cyber Recovery capabilities.
   - **Improvement Opportunities:** Identify gaps in current capabilities and develop a roadmap to strengthen defenses and recovery strategies.
   - **Targets and Testing:** Set realistic recovery goals (e.g., RTOs) and validate them through tabletop exercises and simulations.
4. **Advanced Cyber Recovery Tactics**
   - **Immutable Backups:** Ensure that backups are immutable, encrypted, and indelible, safeguarding them against tampering or deletion.
   - **Cyber Recovery Architecture:** Build a robust framework including secure vaults, clean rooms for validation, alternate recovery rooms, and standardized processes.
   - **Continuous Recovery:** Enable pre-staged recovery processes and utilize automation tools for rapid response and minimal downtime.
5. **Breach Simulation for Proactive Resilience**
   - Simulating real-world cyberattacks (e.g., red teaming and penetration testing) helps organizations identify vulnerabilities and refine security controls.
   - Breach simulations enhance antifragility by transforming incidents into learning opportunities, improving defenses, and fostering continuous adaptation.

6. **The Role of the CISO 3.0**
    - The CISO 3.0 is pivotal in guiding organizations toward cyber resilience by integrating antifragility principles and aligning strategies with business objectives.
    - Their role extends to building trust with stakeholders, fostering a culture of preparedness, and leading innovative recovery practices.

13.6.2 Reflection and Exploration Questions

Understanding the Concepts

1. What are the core differences between traditional resilience and antifragility in cybersecurity?
2. How does the concept of uptime as a business imperative redefine the role of the CISO?

Connecting the Dots

3. How does the concept of MVC connect to earlier discussions about aligning cybersecurity with business priorities?
4. In what ways does breach simulation complement other resilience tactics like recovery architecture?

Real-World Application

5. Can you identify examples of organizations that demonstrated antifragility following a significant cyber event? What contributed to their success?
6. How might you apply immutable backup strategies to improve your organization's disaster recovery plan?

Personal and Professional Growth

7. How can you integrate continuous recovery principles into your own cybersecurity practices?
8. What steps can you take to strengthen collaboration between IT and leadership in defining risk appetite and resilience strategies?

Further Exploration

9. What emerging technologies or frameworks could further enhance the effectiveness of cyber resilience strategies?
10. How can you begin transforming cyber resilience into a competitive advantage?
11. What steps can you take to ensure that your DR plans include contingencies for cyber resilience?

14 AI and the Future of the CISO Role

14.1 INTRODUCTION

Artificial intelligence (AI) is no longer a futuristic concept; it's here, and it's shaking things up in a big way, especially in the world of cybersecurity. In fact, the rise of generative AI has been so mind-blowing that I had to add this whole chapter to the book just to keep up! We're talking about AI that can create stuff, text, images, and even code, and it's already changing the game in all sorts of fields.

Think about customer service, where AI chatbots are getting so good at handling inquiries that you might not even realize you're talking to a machine. Or look at software development, where tools like GitHub Copilot are helping developers write code faster than ever before. Heck, even in the creative world, AI is generating artwork, music, and even movie scripts. It's wild!

Honestly, I always thought AI would take over the boring jobs we didn't want, like taking out the trash. But here we are, with me still hauling my own garbage to the curb, while ChatGPT is out there writing chart-topping songs. It's enough to make you wonder if AI is going to take over *all* our jobs.

But the reality is, AI is more likely to change our jobs than replace them entirely. It's taking over the repetitive, mind-numbing tasks, freeing us up to focus on the strategic, creative stuff that humans are good at. And that's true even for CISOs. Imagine having an AI assistant that can whip up a risk assessment report in seconds, complete with fancy charts and graphs. Or think about using AI to sift through mountains of security logs, spotting those subtle signs of an attack that a human might miss. It's like having a superpower!

Of course, AI isn't all sunshine and rainbows. It comes with its own set of risks and challenges. In this chapter, we're going to explore the complex relationship between AI and security. We'll look at it from three angles: Security *for* AI, security *with* AI, and security *from* AI.

Security for AI: This includes both internal use cases, such as securing the use of generative AI for marketing or sales teams, and external use cases, like your development team adding large language models (LLMs) into processes and products for customers, like the customer service example mentioned earlier.

Security with AI: This encompasses the use of AI to enhance security within your organization. Laying AI and automation into your security operations center (SOC).

Security from AI: This addresses the threats posed by malicious actors using AI. AI has permeated nearly every part of the cyber kill chain, necessitating proactive defenses.

 DOI: 10.1201/9781003510789-19

We'll also discuss two main types of AI cyberattacks: *Target attacks* and *subject attacks*. As of the publishing of this book, these attack types haven't been formally named elsewhere, so I've taken the liberty of categorizing them as such.

Target attacks are instances where AI-generated content is weaponized against your network. Think of sophisticated phishing emails crafted with AI to sound incredibly convincing or automated tools relentlessly probing for vulnerabilities in your systems. The threat also extends to deepfake technology, where AI-generated voices or images can trick individuals or even automated systems, potentially leading to fraudulent payments or unauthorized access.

On the other hand, subject attacks occur when you or your organization become the subject, although not the direct target, of an attack. A prime example is the case of an F50 energy company whose CEO's voice was cloned to create a deepfake used in a fraudulent earnings call. The motive could range from manipulating the stock market to tarnishing the company's reputation. These attacks are particularly insidious because they don't directly hit your network, making them incredibly challenging to detect or prevent. Throughout this chapter, we'll explore how organizations are addressing these threats and examine what they mean for today's chief information security officer (CISO).

It's a brave new world out there, and the CISO 3.0 needs to be ready for it. So, let's dive in and explore how AI is changing the game of cybersecurity.

14.2 AI OVERVIEW

The history of AI is a captivating story of humanity's relentless quest to create machines that mirror our own intelligence. From the early days of expert systems, where knowledge was meticulously encoded into rules, to today's sophisticated machine learning (ML) and deep learning algorithms, AI has experienced remarkable progress, especially over the last couple of years.

Back in the 1960s, the seeds of AI were planted with the development of expert systems. These systems functioned as digital libraries, housing extensive knowledge on specific topics. Developers could then craft rules to query these knowledge bases, leading to the creation of decision trees and diagnostic workflows. Imagine teaching a computer to identify a cat by asking a series of questions: "Is it an animal? Does it have fur? A tail? Whiskers?" Early expert systems like Dendral (used to identify organic molecules) and Mycin (a diagnostic tool for diseases) showcased the potential of AI to assist in complex tasks.

ML propelled AI to new heights. Instead of relying on predefined rules, ML algorithms learned to identify patterns and classify data based on examples. Show an ML model countless pictures of cats, and it will learn to recognize what makes a cat a cat. Introduce pictures of dogs, and it learns to distinguish between the two. These systems are deterministic: Give them the same input, and they'll produce the same output every time.

The latest breakthrough, exemplified by generative pre-trained transformer (GPT) style systems, is generative AI. The big difference with the new GPT-style systems we've seen recently is that they are generative or nondeterministic. You'll get a different newly generated output every time you enter the same input. It's primarily based on layers of ML, where the output of one ML layer becomes the input for the next layer so that it can determine the context. It's not really "creating" something new; it's guessing at the most likely next word. Like the prompt on your texts, you can create a sentence by hitting the middle button. "We are headed to a variety of meetings that may be too much for the next steps". It's a sentence, but it doesn't make sense because the middle button doesn't know the context of all the previous words. GenAI takes the previous outputs as inputs when guessing the next best word to build real sentences. GenAI takes that concept and then trains it with feedback to give responses that are deemed the best and most appropriate, but GenAI can't REALLY write a novel; it can just infer what order to put words in based on the way Stephen King, Shakespeare, and Jackie Collins did it. This reliance on patterns necessitates training on massive amounts of data, which is, essentially, all the data in the world.

As AI has grown in complexity, so too has the need for massive amounts of data to fuel its learning. The field has shifted from curated knowledge bases to the uncharted territory of "all the data everywhere". The era of open, public, all-encompassing AI models might be nearing its end. Major tech companies are advocating for stricter controls, intellectual property concerns are escalating, and commercial viability remains uncertain. The future seems to lie in private AI models, trained on proprietary and customer data, tailored for specific business needs. Data itself is becoming a commodity, licensed and monetized. This places a premium on data governance, ensuring data is used responsibly and securely.

The specter of AI as an existential threat looms large in public discourse, but it's important to distinguish between general AI (a theoretical, all-knowing superintelligence) and the narrow AI we have today. While narrow AI carries risks like data breaches, bias, and potential job displacement, it's not about to initiate a robot uprising.

AI includes a diverse set of technologies that allow machines to carry out tasks that traditionally rely on human intelligence.

Computer Vision: This involves identifying and classifying objects in images and videos, with applications in self-driving cars and facial recognition.

Cognitive Computing: This aims to mimic the human brain's reasoning and judgment capabilities, exemplified by IBM Watson.

Natural Language Processing (NLP): This is the ability to understand and process human language, enabling chatbots and virtual assistants like Siri.

Machine Learning: A subset of AI that uses task-specific algorithms to mimic human learning through categorization and classification. It can be descriptive (explaining what happened, like in fraud detection), predictive (forecasting what will happen, like in next-gen antivirus), or prescriptive (making suggestions, like in recommendation systems).

Deep Learning: Similar to ML but layered to be more like the human brain, where the output of one layer is the input of the next. This allows it to perform complex tasks like predicting if a kicked ball will go into a net. There's no consensus on exactly how deep learning works.

Neural Networks: These are the architecture at the heart of deep learning, inspired by the human brain's nodes and neurons. Nodes have input and output layers and learn how to connect via weights (like a route table). They learn from training data and improve over time. Google's search algorithm is an example of a neural network.

Generative AI (GenAI): This is the current hype, but it's still early days. We're just at the beginning of how organizations will monetize all the AI touchpoints in our lives.

AI is no longer just a buzzword; it's transforming the way we do business, and the changes are happening fast. In the world of e-commerce, AI-powered chatbots are like your personal shopping assistants, offering tailored product recommendations that make you feel like you've got your own stylist. Content creation is getting a makeover, too, with AI automating the production of news articles, blog posts, and marketing materials, freeing up human writers to focus on the big ideas. Even HR is getting in on the action, using AI to streamline the hiring process and find the perfect candidates for the job.

And it's not just about business. AI is popping up everywhere, from customer service chatbots that are available 24/7 to gaming and health apps that offer personalized experiences and insights. It's even making its way into fields like medicine and software development, helping doctors diagnose diseases and programmers write better code. The possibilities seem endless.

But of course, with great power comes great responsibility. AI's access to sensitive data like personal information and health records raises some serious ethical questions. We need to make sure it's used fairly and doesn't lead to discrimination or bias. And let's not forget about the technical challenges because developing, deploying, and securing AI systems is no walk in the park.

Despite these hurdles, GenAI has the potential to revolutionize the way we work and live. It can create content, automate tasks, and provide insights that we never thought possible. It's a tool that can boost productivity, streamline operations, and deliver amazing customer experiences. But it's up to us to use it wisely and responsibly.

14.3 SECURITY FOR AI

AI is a double-edged sword. It has the potential to revolutionize cybersecurity, but it also introduces new risks and vulnerabilities. As CISOs, we need to be aware of these risks and take steps to mitigate them. That means securing AI systems themselves, protecting the data they use, and ensuring that they're not being used for malicious purposes.

Security for AI is about enabling your organization's rapid adoption of these new productivity tools with confidence. It requires balancing rapid adoption with risk management. Unlike many security projects where tactical actions can be taken first and governance can follow later, AI demands a governance-first approach. Every organization needs an AI governance board or steering committee to determine how the organization wants to approach AI. Do you want to be an early adopter, or would you rather wait and see? Will you focus on productivity gains for employees or experience improvements for customers? The governance body acts as the legislative branch, guiding the CIO and CISO (the executive branch) on what to build.

14.3.1 AI GOVERNANCE

Establishing an effective AI governance board starts by bringing together key stakeholders from various departments such as IT, security, legal, compliance, HR, and business units that are deeply involved with AI. The board's purpose should be clearly defined, covering responsibilities like shaping AI strategy, setting relevant policies, and managing associated risks. Create a governance framework that guides AI adoption and use across the organization, ensuring that every aspect of AI implementation is carefully considered and monitored. Regular board meetings should be scheduled to provide ongoing oversight and review of AI initiatives. Additionally, it's important to promote awareness and understanding of AI governance throughout the organization, which can be achieved through clear communication and regular training programs.

The governance framework, illustrated in Figure 14.1, addresses key elements of effective AI governance. It starts with outlining the objectives and scope, focusing on the specific AI applications, systems, or processes under governance. Establish the structure, roles, and responsibilities of those involved in AI oversight; this may include a dedicated AI governance board or committee. It's crucial to define AI principles that reflect the organization's ethical standards, legal obligations, and compliance requirements.

Risk assessment and mitigation strategies must be built into the framework to address potential risks associated with AI technologies. Equally important is setting up data governance practices to ensure the quality, security, and ethical handling of data used in AI systems. The framework also emphasizes model oversight, including accuracy, traceability, interpretability, and explainability of AI/ML models.

To put this governance into practice, clear policies, procedures, and standards must be developed. Aligning these with established frameworks like the National

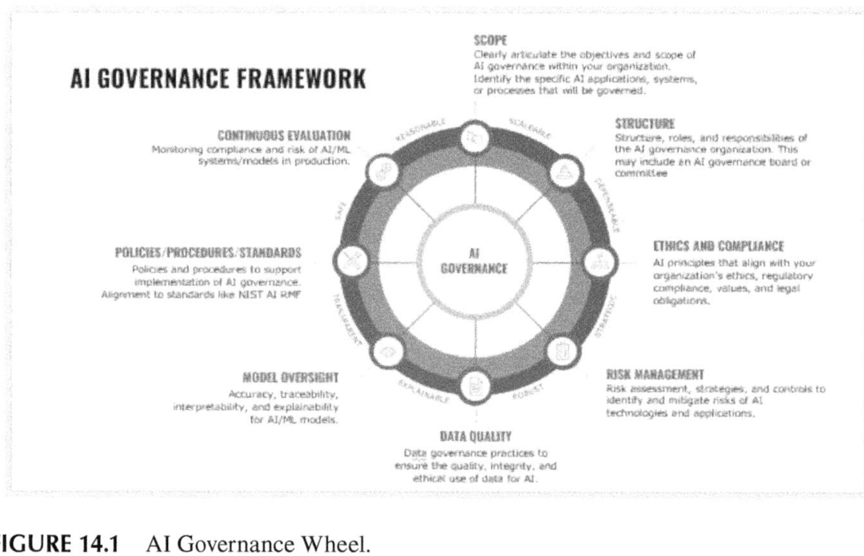

FIGURE 14.1 AI Governance Wheel.

Institute of Standards and Technology (NIST) AI risk management framework is a good starting point. Finally, ongoing evaluation and monitoring of AI systems in production will be essential to maintaining compliance and managing risks as the AI landscape evolves.

14.3.2 Policy

One of the first tasks for the AI governance committee is to create and review policies related to AI, particularly GenAI. Establishing clear policies and guidelines for the use of GenAI is a prerequisite for effectively managing the associated risks. A policy should be created that outlines who can access GenAI and under what circumstances. It's important to identify and categorize the types of data that GenAI can access and ensure that access is limited to those who need it to perform their job duties.

Developing effective GenAI policies requires addressing several key factors, such as data privacy, security, and regulatory compliance. Any use of GenAI must align with relevant laws and regulations, including General Data Protection Regulation (GDPR), HIPAA, and PCI-DSS. Organizations must also account for potential risks, such as the inadvertent exposure of sensitive information or the threat of malicious actors exploiting the technology. Engaging stakeholders from across the organization, including legal, compliance, and IT departments, is essential to ensure these policies reflect both security needs and broader business concerns. In addition, tools like web filtering can be implemented to control access to GenAI and prevent unauthorized use.

When crafting policies for AI and ML technologies, it's important to consider the full scope of usage, including internal services, broader GenAI tools, and third-party products. Selecting GenAI providers with strong security features and flexible policy options is crucial for protecting sensitive data. In some cases, exploring on-premises GenAI solutions may be a better fit for organizations handling highly sensitive information.

It's also necessary to revisit and update existing policies to incorporate the risks introduced by AI/ML technologies. This may mean revising policies on email writing, data sharing, and third-party code usage, ensuring they align with the landscape of AI-driven risks and responsibilities.

14.3.3 Regulations

Many international jurisdictions are developing AI legislation and assessment methodologies. Let's take a closer look at a few key examples and their impact on the use and development of AI within corporations.

In the United States, the "National AI Initiative Act of 2020" establishes a national AI initiative to accelerate AI research and development while addressing ethical, legal, and societal implications.

The guidance on the "Responsible use of artificial intelligence (AI)" in Canada outlines key considerations for the responsible development and use of AI in the Canadian government context. The "Algorithmic Impact Assessment Tool" in

Canada aids in assessing the potential impact of AI systems on individuals and society, promoting transparency and accountability. "Bill C-27, The Artificial Intelligence and Data Act (AIDA)" in Canada is a proposed legislation that aims to regulate the development and deployment of high-impact AI systems, focusing on mitigating risks and harms.

The European Union (EU) AI Act takes a risk-based approach, classifying AI systems into different risk categories. High-risk AI systems, such as those used for critical infrastructure, law enforcement, or employment decisions, face stricter requirements, including conformity assessments, quality management systems, and human oversight.

Certain AI practices deemed unacceptable due to their high risk of harm are prohibited under the act. These include social scoring by governments, real-time biometric identification systems in public spaces for law enforcement (with limited exceptions), and the use of subliminal techniques or manipulative techniques exploiting vulnerabilities to materially distort behavior and cause harm. The act would also prohibit certain discriminatory practices, such as using AI to target individuals based on their race, gender, sexual orientation, or other protected characteristics.

As you can imagine, some of these prohibited activities are precisely what governments and corporations might want to use AI for – things like classifying people for targeted marketing or assigning them social scores. It's conceivable that we will see a divergence in AI regulation between the United States and the European Union, similar to the differences in data privacy regulations between the GDPR and the current US model, which has been largely left to individual states.

Even so, it's imperative for your governance committee to consider where the activities you want to leverage AI for will fall on the EU AI Act's risk rating scale. Understanding this will help you determine whether you may want or need to comply with future EU regulations, even if your organization isn't based in the European Union. This forward-thinking approach can help avoid potential compliance issues down the line. While still in draft form, the EU AI Act signals a shift toward regulating AI, with a focus on mitigating risks and ensuring ethical and responsible AI development and use.

Take a moment to evaluate your organization's AI governance framework. Do you have clear policies and procedures in place to manage the risks associated with AI adoption? How can you strengthen your AI governance to ensure responsible and ethical AI use?

14.3.4 Enablement Risks and Technical Challenges

Introducing GenAI into your organization is like bringing a powerful but unpredictable tool into your operations. While it offers significant potential benefits, it also brings a host of risks that can impact every part of the business. From security challenges to legal complexities and operational concerns, there's much to consider.

On the security front, GenAI opens the door to new vulnerabilities. Think about unauthorized access to sensitive data, accidental leaks, or even malicious users tricking the AI into doing something harmful. As CISOs, we need to be proactive, putting strong controls in place, such as strict access permissions, prompt security, and even policies for those fancy AI-enabled gadgets that are popping up everywhere.

Then there's the legal minefield. We have to grapple with the ethics of AI, potential penalties for misuse, liability concerns, and a whole slew of regulations, both existing ones like GDPR and the new AI laws that are popping up faster than you can say "chatbot". And let's not forget about intellectual property rights and who owns what when AI is creating content.

It's not just about security and legal, either. GenAI has implications for every team in the company. HR is worried about job losses, marketing is excited about new possibilities but also nervous about brand risks, finance is trying to figure out the cost–benefit equation, and IT is scrambling to build the infrastructure to support it all. It's a complex web, and the CISO 3.0 needs to be at the center of it, making sure AI is integrated smoothly and securely across the entire organization (Figure 14.2).

Some of the biggest challenges we're facing include the following:

- Keeping user data private and secure.
- Navigating the ever-changing AI regulations.
- Protecting sensitive information and intellectual property.
- Dealing with the limitations of AI technology.
- Figuring out copyright and ownership issues.
- Preventing bias and discrimination in AI systems.
- Making sure AI integrates smoothly with our existing systems.
- Maintaining our cybersecurity resilience in the face of AI-powered attacks.
- Securing AI models themselves from being hacked.
- Building security into the development process from the start.

It's a lot to handle, but the potential rewards of GenAI are huge. It can boost productivity, streamline operations, and create amazing new products and services. The key is to approach it with a balanced mindset and take advantage of the opportunities while proactively managing the risks.

14.3.5 Enablement Risk Examples

14.3.5.1 Sensitive Data and Intellectual Property

- **The Issue:** The use of ChatGPT may result in the disclosure of sensitive data and intellectual property, potentially leading to brand damage and loss of competitive advantage.
- **The Risk:** Unauthorized access to or disclosure of sensitive data and intellectual property can lead to financial loss, legal and regulatory issues, and loss of competitive advantage.
- **The Guidance:** CISOs should implement access controls to limit access to ChatGPT and ensure that only authorized users can generate content. In addition, encryption tools can be used to protect any data transmitted or stored by ChatGPT. Organizations should also consider extending their data security capabilities, especially with cloud security access broker (CASB) and data loss prevention (DLP) tools to inspect submissions

14.3.5.2 Regulatory Compliance

- **The Issue:** The use of chatbots like ChatGPT in organizations can pose regulatory compliance risks. Chatbots that process personal data or other sensitive information must comply with various data protection regulations, such as the GDPR in the European Union or the California Consumer Privacy Act (CCPA) in the United States.

- **The Risk:** Failure to comply with these regulations can result in substantial fines and brand damage. For example, if a chatbot is used to process personal health information and is not HIPAA compliant, the organization could face penalties and lawsuits.
- **The Guidance:** To address these compliance risks, organizations should implement various controls to ensure that chatbots are designed, developed, and used in compliance with applicable regulations. Some of these controls include conducting a privacy impact assessment (PIA) before implementing a chatbot, ensuring that the chatbot's data processing activities are transparent to users, and conducting a thorough risk assessment to identify potential regulatory compliance risks associated with chatbot use. This should include an analysis of any applicable regulations or standards such as the GDPR, HIPAA, or PCI-DSS.

14.3.5.3 Privacy Concerns

- **The Issue:** ChatGPT may capture and store user data, including sensitive information, raising concerns about privacy violations.
- **The Risk:** Unauthorized access to or disclosure of sensitive user data can lead to brand damage, legal and regulatory issues, and loss of customer trust and confidence.
- **The Guidance:** CISOs should implement strict data privacy controls such as pseudonymization, data minimization, or anonymization to reduce the amount of personal data captured by ChatGPT. CISOs should use tools such as DLP solutions to monitor and prevent the unauthorized capture or transfer of sensitive data. Organizations should also consider using frameworks, such as the GDPR or CCPA, to ensure compliance with data privacy regulations.

14.3.5.4 Bias and Discrimination

- **The Issue:** ChatGPT may generate content that is biased or discriminatory, potentially leading to legal and reputational issues.
- **The Risk:** The use of biased or discriminatory content can lead to loss of customer trust and confidence, reputational damage, and legal and regulatory issues.
- **The Guidance:** CISOs should establish guidelines for the types of content that ChatGPT can generate and regularly test ChatGPT for any bias or

discriminatory language. In addition, regular training should be provided to employees to ensure that they are aware of the risks associated with using biased or discriminatory language. Organizations can also use tools such as sentiment analysis to detect any language that may be discriminatory.

14.3.5.5 Malicious Use

- **The Issue:** ChatGPT may be used by threat actors to generate malicious content or phishing messages, leading to financial loss, data breaches, and reputational damage.
- **The Risk:** The use of ChatGPT to generate malicious content can lead to financial loss, reputational damage, and legal and regulatory issues.
- **The Guidance:** CISOs should establish strict access controls and regularly monitor ChatGPT's usage logs to identify any suspicious or unauthorized activity. Regular security testing can be conducted to identify any vulnerabilities in the ChatGPT software. Organizations can also use frameworks such as MITRE ATT&CK to identify potential attack vectors and implement controls to prevent them. Email gateway policies may need to be tightened as end-user phishing training becomes less effective.

14.3.5.6 Technical Limitations

- **The Issue:** ChatGPT may not be able to generate accurate or relevant content in some situations, potentially leading to errors or misunderstandings.
- **The Risk:** Inaccurate or irrelevant content generated by ChatGPT can lead to misunderstandings, errors, and reputational damage.
- **The Guidance:** CISOs should establish best practices for training and testing ChatGPT, including training it on the organization's specific language and terminology and regularly testing the accuracy of its responses. Employees should also be trained to verify any content produced by ChatGPT before it is shared externally. Organizations can also use tools such as NLP and ML algorithms to improve the accuracy of ChatGPT's responses. If you find that chatbot software frequently misuses your organization's specific language and terminology, Digital Rights Management software like Vera or Microsoft AIP can be trained to identify and either correct or block common errors in outgoing emails.

14.3.5.7 Integration Challenges

- **The Issue:** Integrating ChatGPT into existing systems or workflows may pose technical challenges, such as incompatibility with legacy systems or difficulties in data integration.
- **The Risk:** Integration challenges can lead to delays, increased costs, and decreased efficiency in the organization's workflows.
- **The Guidance:** CISOs should work with IT teams to ensure that ChatGPT is compatible with existing systems and workflows. In addition, the

organization should establish clear communication channels between IT and business teams to identify and address any integration challenges early on. Organizations can also use tools such as APIs and webhooks to facilitate data integration between ChatGPT and other systems. To achieve these controls, CISOs can leverage a range of tools and technologies, including access controls, encryption tools, DLP solutions, sentiment analysis, NLP algorithms, and ML algorithms.

14.3.6 Implementing Controls

When it comes to implementing controls to mitigate these risks, while there are some new cutting-edge technologies to provide agent-based DLP specifically for public LLMs and APIs to git repos, it's mostly the same old controls and mitigations we should have had in place already. These include data security controls like data access governance, DLP, and Privacy Ops; application security controls like API gateways and web application firewalls (WAFs); resilience controls like immutable backup, business impact analysis (BIA)/MVC reporting, and business continuity planning/disaster recovery (DR) planning; identity and access control solutions like identity governance and access (IGA) controls; and cloud security controls like cloud-native application protection platforms (CNAPPs) and CASBs.

Controlling access to chatbots in a corporate environment is critical to mitigating potential risks. Access control measures can include implementing authentication and authorization mechanisms to restrict access to chatbots to authorized personnel only. Role-based access control (RBAC) can be used to assign specific permissions to users based on their job responsibilities. Implementing web filtering measures can also help prevent unauthorized access to chatbots by blocking access to chatbot applications from non-approved devices or networks.

Only after determining the risk versus the business value should a CISO begin to establish controls that include access controls, data privacy controls, content generation guidelines, usage monitoring, employee training, third-party risk management, and business continuity plan (BCP)/DR modifications. It's important to note that controls should be implemented in a risk-based approach and in alignment with the organization's overall security strategy. This means that controls should be proportional to the risks associated with the use of chatbots and should be integrated into the organization's overall security program. In addition to the controls mentioned, organizations should also consider implementing chatbot policies into their firewalls, intrusion detection and prevention systems (IDPS), and security information and event management (SIEM) to monitor and control access to chatbots. These tools can help detect and prevent unauthorized access attempts as well as provide alerts and reports on potential security incidents. It's also important to regularly review access control policies and monitor access logs to identify any unauthorized access attempts or suspicious behavior. Regular security audits and penetration testing can help identify potential vulnerabilities and ensure that access controls are functioning correctly.

Many organizations have what I call "Strategic Debt". It's basically the opposite of technical debt and refers to the things we should have done 5 or 10 years ago but were too hard or costly. Technical debt happens when we wait too long to remove existing technologies, and strategic debt happens when we wait too long to implement missing technology. Things like data governance and identity governance are prime examples of strategic debt within most organizations. Consequently, those organizations now owe this strategic debt as a prerequisite to all the cool new modern things they want to do now, like Zero Trust and AI security.

However, blocking access to chatbot applications may not be the best way to enable technological innovation for your organization and often leads employees to seek out alternatives, which causes shadow AI. That is why starting with the "legislative branch" policy instead of the "executive branch" enforcement and controls is so important. CISOs should work with business leaders to establish policies and guidelines for chatbot use within the organization. This should include setting clear expectations for acceptable use, defining roles and responsibilities for chatbot administration, and establishing procedures for reporting and responding to potential security incidents. An acceptable use policy may be all that is required. It's also important to regularly review and update policies to ensure that they remain effective in mitigating risks.

CASB is a critical tool for managing third-party GenAI. It helps manage shadow AI, provides DLP, and enforces access control. If you don't have a CASB, consider this your sign to go get one, and if you do have a CASB, you need to tune it for GenAI and chatbot use cases.

While controls to manage access to third-party GenAI like CASB are important to have in place, blocking access will most likely stifle innovation and lead to shadow AI. Providing access to sanctioned AI programs, like Microsoft Copilot, is a great way to reduce shadow AI usage and boost productivity. I can tell you firsthand that it greatly reduced my personal reliance on unsanctioned AI tools. I was an early adopter and integrated Zapier automation with ChatGPT for various use cases, from generating email responses to crafting after-action reports. However, once I had access to a sanctioned Copilot, I dropped all those other programs. MS Copilot, for example, can join Teams meetings, summarize discussions in real-time, capture action items, and identify unresolved questions. In Outlook, it can help triage your inbox, prioritize emails, summarize threads, and even generate replies. Copilot in Excel can analyze raw data, providing insights, trends, and suggestions, all of which are fantastic features that I use regularly.

The challenge with both MS Copilot and Google's Duet lies in access control. While they offer tenant isolation and don't use your business data to train their foundational LLMs, they do inherit the user's permissions within your tenant. This means Copilot can access any data the user has access to, including sensitive information like payroll data or the CEO's personal contact details. This creates a potential "live off the land" tool for attackers and elevates insider threat risks.

To combat this, you can opt for a less integrated sanctioned AI tool like Bing Chat Enterprise, which operates as a standalone GenAI portal without direct access to your private company data. Alternatively, you can strengthen your access governance controls to ensure least privileged access and proper data segregation. Microsoft offers a tool called Purview for this purpose, but it's currently bundled with the E5 enterprise license, whereas Copilot is often included in the E3 license. You might need to upgrade your licensing to get the necessary security functions for

Copilot. There are also robust third-party data access governance tools like Varonis that are well-suited for this purpose.

14.3.7 Third-Party Risk

One of the major risks associated with using AI, particularly chatbots, is the potential for third-party breaches. Third-party vendors often develop and maintain AI technologies, and their security practices might not align with your organization's standards. If a vendor suffers a breach, your sensitive data could be compromised, leading to reputational damage and potential legal liabilities. Recent examples include OpenAI's breach where hackers accessed payment records and the 2017 breach of [24]7.ai, a customer service chatbot provider, which exposed customer payment card information. To mitigate third-party risk, robust vendor management practices are fundamental. Conduct thorough due diligence on potential vendors, ensure they adhere to your security requirements, and include contractual provisions that hold them accountable for maintaining the security of their AI technology and data. Regular security audits and penetration testing are also essential components of a good third-party risk management program.

Contract management is a primary control for third-party AI tools. While cutting-edge solutions can scan LLMs for vulnerabilities, companies like Salesforce won't allow you to scan their software. Your best defense is strong third-party risk management practices that vet vendors' software development life cycles (SDLCs) and controls, coupled with robust contract management to ensure service-level agreements (SLAs) and remedies for any harms.

The CrowdStrike incident in mid-2024, where a software update caused a mass outage affecting machines running Microsoft operating systems, is a perfect example. The boilerplate contract only offered a refund as a remedy, even for major multimillion dollar outages. Companies that negotiated their liability terms fared better. This underscores the importance of contract negotiation and management within your third-party risk program.

You should already have third-party risk measures in place, including questionnaires to vet vendors' security practices. You should add AI and GenAI-specific language to these questionnaires.

14.3.7.1 Example Questions for Third-Party Risk Questionnaires

Data Security and Privacy

- How do you ensure the confidentiality, integrity, and availability of data used by your AI systems?
- What specific data privacy regulations do you comply with (e.g., GDPR, CCPA, or HIPAA)?
- How do you handle data anonymization and de-identification?
- What are your data retention and deletion policies for data processed by your AI models?

AI Model Security

- What measures are in place to prevent and detect adversarial attacks or data poisoning against your AI models?
- How do you ensure the explainability and interpretability of your AI models' decisions?
- Do you conduct regular audits and assessments of your AI models for bias and fairness?
- What is your process for updating and maintaining your AI models to address threats and vulnerabilities?

Incident Response and Liability

- What is your incident response plan in case of a security breach or malfunction involving your AI technology?
- What are your liability limitations and indemnification provisions in case of damages caused by your AI technology?
- How do you communicate security incidents or vulnerabilities to your customers?
- Do you carry cyber insurance to cover potential liabilities related to your AI technology?

AI Development and Deployment

- Do you follow a secure SDLC for your AI models and applications?
- What security testing and code review practices do you have in place for your AI development?
- How do you ensure the security of the infrastructure and platforms on which your AI systems are deployed?
- Do you have any certifications or attestations related to the security of your AI technology?

General AI Governance

- Do you have an established AI governance framework in place?
- How do you address ethical considerations and potential societal impacts of your AI technology?
- Do you have a process for ongoing monitoring and evaluation of your AI systems for potential risks and unintended consequences?
- How do you ensure transparency and accountability in the use of your AI technology?

By incorporating these and other relevant questions into your third-party risk assessments, you can gain a better understanding of vendors' AI security practices so you can select which vendors to partner with.

Remember, the goal is to enable the use of AI while mitigating its risks. Implementing a combination of technical controls, policies, and robust vendor management practices allows CISOs to help their organizations safely navigate the exciting but challenging world of AI.

14.3.8 AI Development

The use of chatbots for code development has gained traction in many organizations, offering a platform for rapid prototyping and automating repetitive tasks, which frees up developers to focus on more complex projects. However, this shift introduces potential security risks, such as the unintentional introduction of vulnerabilities or even malicious code. To address these concerns, it's critical to integrate security considerations into every phase of the SDLC.

CISOs should collaborate closely with development teams to establish clear security protocols when using chatbots for development. This includes secure coding practices like input validation and output sanitization, along with thorough testing and code reviews. Security testing tools, such as vulnerability scanners and encryption techniques, should be used to protect sensitive data. Additionally, static code analysis and penetration testing are essential to identify any weaknesses early on. Implementing robust access controls and authentication mechanisms is key to restricting chatbot use to authorized personnel, while audit logs can help track activity and detect potential security incidents. Following established security frameworks such as Open Web Application Security Project (OWASP) Application Security Verification Standard (ASVS), OWASP Mobile Application Security Verification Standard (MASVS), and NIST SP 800–53 can guide organizations in developing secure chatbot solutions while ensuring compliance with data privacy regulations.

When integrating GenAI technology into organizational products, there are clear benefits, including enhanced customer engagement and improved user experiences. However, chatbots, whether through an API, third-party module, or package, can present vulnerabilities that hackers may exploit to gain unauthorized access or execute malicious actions. A common example is the risk of malicious code injection through chatbot APIs, which could result in data breaches or system compromises. Similarly, chatbot APIs could expose sensitive information due to misconfigurations or exploited vulnerabilities.

To mitigate these risks, organizations need to take a security-first approach during the integration process. This starts with rigorous vendor evaluations and secure coding practices. Regular vulnerability scanning and penetration testing should be conducted to identify and remediate potential security flaws. Strong access controls and authentication mechanisms can limit access to chatbot APIs, and audit logs should be implemented to monitor activity and detect security incidents promptly.

With the rise of off-the-shelf LLMs like those from Hugging Face, development teams now have the ability to build AI-powered chatbots for a wide range of business applications. These tools, when used correctly, can drive efficiency and improve user experiences. However, they also carry risks if not developed and implemented securely. Hackers could exploit vulnerabilities in chatbot design or rely on compromised data sources to gain access to sensitive information.

To safeguard these chatbots, organizations must thoroughly test both third-party LLMs and any internally developed chatbots. This should be done following the same strict third-party risk management protocols applied to external vendors. Comprehensive security assessments, vulnerability scanning, and regular penetration testing should be conducted. Additionally, monitoring chatbot behavior for suspicious activity and ensuring the software is updated and patched regularly can help mitigate the risk of exploitation. By following secure development and deployment practices, organizations can leverage the power of LLMs while minimizing potential security threats.

14.3.9 GenAI in Software Development

The integration of GenAI into software development has revolutionized how organizations approach automation, efficiency, and innovation. With off-the-shelf models like those from Hugging Face, organizations can leverage pre-trained LLMs for a wide array of applications without incurring the significant costs and resource demands of developing models from scratch. This approach allows businesses to tap into cutting-edge AI capabilities while maintaining flexibility in customizing solutions to fit their specific needs.

14.3.9.1 Off-the-Shelf Models versus Custom Models

Off-the-shelf models are pre-trained LLMs provided by platforms like Hugging Face. These models are trained on diverse datasets and can perform a wide range of tasks, from text generation to sentiment analysis. The primary advantage of these models is that they eliminate the need for resource-intensive training cycles, making advanced AI accessible even to organizations without substantial computational resources or in-house AI expertise.

Custom models, on the other hand, are trained from the ground up or fine-tuned extensively to meet highly specific requirements. While they offer unparalleled customization, the cost of training such models – both in terms of computing power and data preparation – can be prohibitive for most organizations. As a result, many organizations opt for methods that allow for customization without full-scale retraining, such as injecting off-the-shelf models with custom data.

14.3.9.2 Injecting Licensed LLMs with Custom Data

Injecting LLMs with custom data enables organizations to tailor models to their unique contexts without retraining. Techniques like retrieval-augmented generation (RAG); modular reasoning, knowledge, and language (MRKL); and agent-based workflows provide powerful mechanisms for enhancing LLM functionality.

1. **Retrieval-Augmented Generation:**
   RAG integrates external knowledge bases with LLMs to provide contextually accurate and relevant outputs. When a user queries the model, the system retrieves relevant data from a vectorized database (often built using embeddings of custom documents) and includes it in the model's prompt. This method effectively enables real-time knowledge injection, allowing the model to deliver up-to-date, domain-specific responses.

2. **Modular Reasoning, Knowledge, and Language:**
   MRKL frameworks (pronounced miracle) break down complex problems into smaller modules. Each module handles a specific task, such as reasoning, accessing external APIs, or retrieving knowledge, and collaborates with the LLM to provide a cohesive response. This approach enhances the model's ability to handle multistep processes and complex queries.
3. **Agents:**
   Agents are autonomous components that use LLMs to make decisions and execute tasks. These agents can interact with external systems, call APIs, or perform calculations to achieve specific objectives. For example, an agent might fetch financial data from an API, analyze it, and generate a report.
4. **Multi-Agentic Workflows:**
   Multi-agentic workflows involve multiple agents working together to accomplish larger, more complex tasks. These workflows leverage the strengths of individual agents to handle specialized components of a process, ensuring efficiency and precision. For instance, one agent might retrieve data, another might validate it, and a third could generate a user-friendly summary.

14.3.9.3 Using LangChain to Build Workflows

LangChain is a robust framework designed to streamline the development of applications that leverage LLMs. It facilitates the integration of LLMs with external data sources, tools, and workflows, enabling developers to build powerful, multifunctional AI systems.

With LangChain, developers can do the following:

- Create pipelines that combine data retrieval, processing, and inferencing.
- Design workflows that incorporate RAG, MRKL, and multi-agentic architectures.
- Simplify the integration of APIs, databases, and external tools into AI-driven applications.

LangChain provides modular components for tasks such as prompt engineering, memory management, and interaction with vector databases. This modularity makes it a go-to tool for constructing sophisticated AI applications tailored to specific business needs.

14.3.9.4 Security Implications and Controls

While these techniques offer immense potential, they also introduce unique security challenges that must be addressed to ensure safe implementation:

- **Data Security:** Protecting the data injected into LLMs is critical. Sensitive information should be encrypted and de-identified, and access should be controlled through robust authentication and authorization mechanisms.
- **Input Validation:** Preventing malicious prompt injections requires strict validation of user inputs and API calls.

- **Output Filtering:** Ensuring that model outputs do not disclose sensitive information or contain harmful content is essential for maintaining trust and compliance.
- **Audit Trails:** Implementing logging and monitoring tools to track interactions with the model helps detect and respond to anomalies or security incidents.
- **RAG and Vector Database Security:** Securing the vectorized knowledge base involves encrypting stored embeddings, implementing access controls, and regularly updating the database to address evolving threats.
- **API Security:** APIs connecting LLMs to external systems should be protected using gateways, rate limiting, and strong authentication protocols.

By implementing these controls and fostering collaboration between developers and security teams, organizations can unlock the transformative power of GenAI while minimizing associated risks. As AI adoption continues to grow, a proactive and comprehensive security strategy will be critical for leveraging its full potential responsibly.

14.3.10 AI Architecture

Understanding the architecture of AI systems, particularly those utilizing LLMs, is important for implementing effective security measures. The OWASP has recognized this and created a Top 10 list of LLM vulnerabilities, similar to their renowned OWASP Top 10 for web vulnerabilities.

The OWASP Top 10 for LLM applications encompasses the following:

- **LLM01: Prompt Injection:** This involves manipulating an LLM through crafty inputs, causing it to perform unintended actions. Direct injections overwrite system prompts, while indirect ones manipulate inputs from external sources.
- **LLM02: Insecure Output Handling:** This vulnerability arises when an LLM's output is accepted without scrutiny, potentially exposing backend systems to exploits like cross-site scripting (XSS), cross-site request forgery (CSRF), server-side request forgery (SSRF), privilege escalation, or even remote code execution.
- **LLM03: Training Data Poisoning:** This occurs when the data used to train an LLM is tampered with, introducing vulnerabilities or biases that compromise the model's security, effectiveness, or ethical behavior. Common sources of training data include Common Crawl, WebText, OpenWebText, and books.
- **LLM04: Model Denial of Service:** Attackers can overload LLMs with resource-heavy operations, leading to service degradation or high costs. This vulnerability is amplified due to the resource-intensive nature of LLMs and the unpredictability of user inputs.
- **LLM05: Supply Chain Vulnerabilities:** The LLM application lifecycle can be compromised by vulnerable components or services, leading to

security attacks. Using third-party datasets, pre-trained models, and plugins can introduce additional vulnerabilities.

- **LLM06: Sensitive Information Disclosure:** LLMs might inadvertently reveal confidential data in their responses, leading to unauthorized access, privacy violations, and security breaches. Data sanitization and strict user policies are needed to mitigate this risk.
- **LLM07: Insecure Plugin Design:** LLM plugins can have insecure inputs and insufficient access control, making them susceptible to exploitation, potentially resulting in severe consequences like remote code execution.
- **LLM08: Excessive Agency:** LLM-based systems could unintentionally cause harm if they are given too much functionality, permissions, or autonomy, leading to unforeseen outcomes.
- **LLM09: Overreliance:** When systems or individuals rely too heavily on LLMs without adequate supervision, they risk spreading misinformation, causing miscommunication, encountering legal challenges, and exposing security gaps due to inaccurate or inappropriate content produced by these models.
- **LLM10: Model Theft:** This refers to the unauthorized access, replication, or extraction of proprietary LLMs.

Some of these LLM vulnerabilities pertain to the model itself, while others relate to its implementation and use. Since most organizations will likely never develop their own LLMs, vulnerabilities like "LLM03: Training Data Poisoning" won't be a direct concern. However, vulnerabilities like "LLM01: Prompt Injection" and "LLM02: Insecure Output Handling" will be highly relevant, as they directly impact how users interact with and utilize these models.

We will likely see a shared responsibility model emerge, much like in cloud computing, where the model developer is responsible for vulnerabilities inherent to the model itself, and the user is responsible for securing its implementation and use.

1. **Data Preprocessing/Embedding:** This initial phase involves preparing and organizing private data for future use. Documents are split into smaller sections, run through an embedding model, and stored in a specialized vector database for efficient retrieval.
2. **Prompt Construction/Retrieval:** When a user submits a query, the application assembles a prompt for the language model. This prompt usually combines elements like a structured prompt template, sample outputs (few-shot examples), data from external APIs, and relevant information retrieved from the vector database.
3. **Prompt Execution/Inference:** After constructing the prompt, it's sent to a pre-trained language model for processing. This step can utilize proprietary models, open-source solutions, or self-trained models. Operational components such as logging, caching, and validation can be incorporated at this stage to ensure smooth execution.

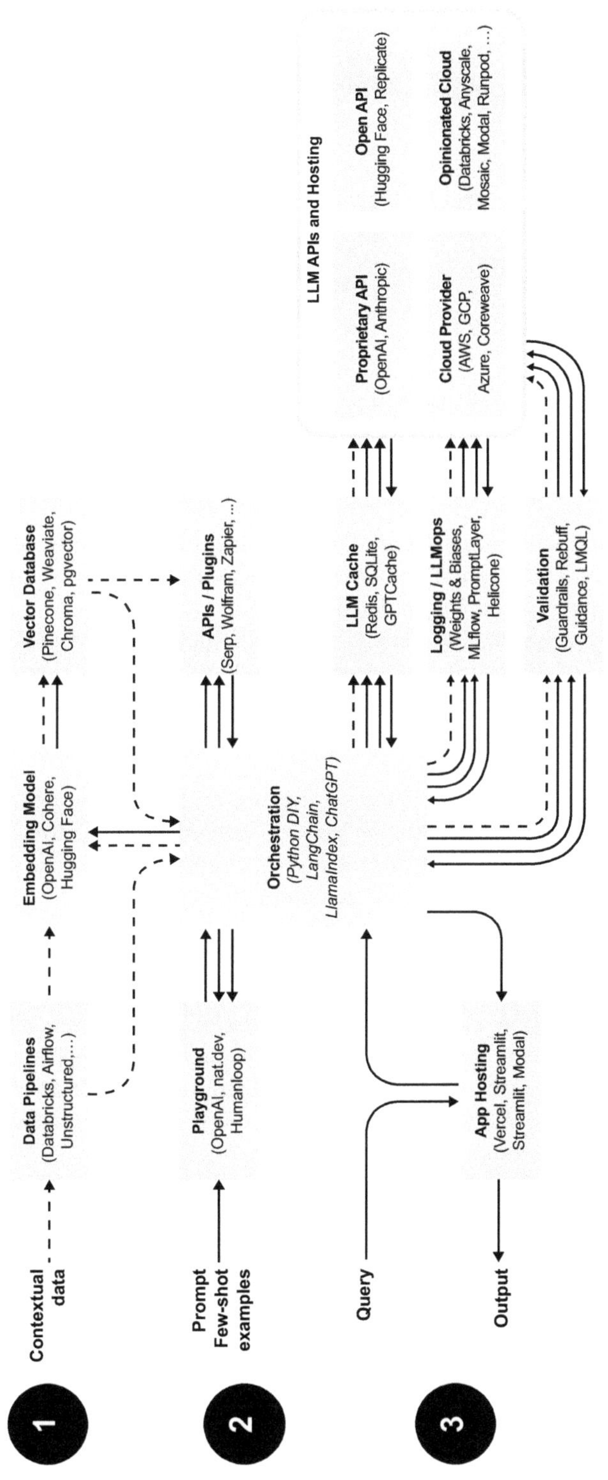

FIGURE 14.2 AI Architecture.

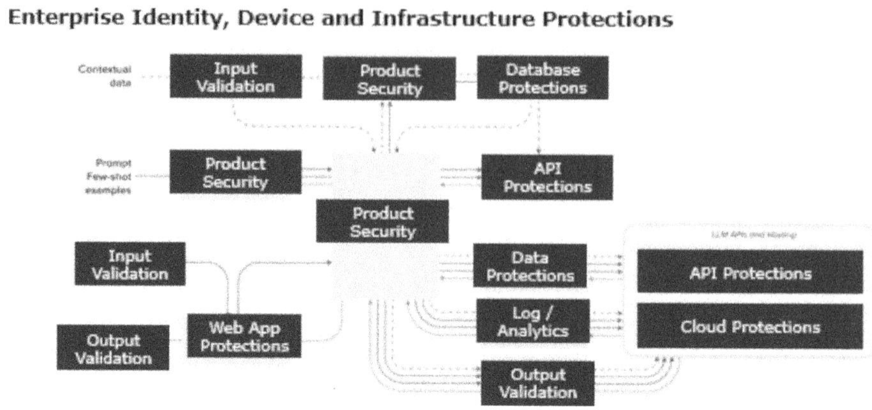

FIGURE 14.3 AI Security Architecture.

You'll also notice that within this AI architecture, there are API calls, redirects to the cloud, a web front end, prompt inputs, and model outputs. Each of these represents a potential security vulnerability, as illustrated in Figure 14.3.

To ensure reliable protection, we need to apply data security and data governance measures at Layer 1, where private data is stored and processed. At Layer 2, where prompt construction and retrieval occur, it's essential to implement input/prompt validation and API security to prevent malicious prompt injections and safeguard communication with external services.

In Layer 3, where prompt execution and inference take place, we need to apply OWASP protections to the web front end using a web application firewall. Cloud vulnerability and runtime environment protections, coupled with output validation, are vital to ensure the secure and reliable execution of AI models.

A range of security solutions can be employed to address these risks:

API Security: API gateways, content delivery networks (CDNs), DDoS protection, application delivery controllers (ADCs), and authentication mechanisms are essential for securing API calls and data transfers.

Web Application Firewall (WAF)/Web Application and API Protection (WAAP): WAFs and WAAP defend against common web app vulnerabilities like XSS, SQL injection, and forceful browsing.

CNAPP: CNAPPs offer vulnerability and runtime protection for cloud infrastructure, containers, and serverless functions.

SDLC Security: Throughout the SDLC, secure coding practices should be reinforced with static application security testing (SAST), dynamic application security testing (DAST), and software composition analysis (SCA) to safeguard AI models and applications.

Data Controls: Implement robust data security measures, including data lake/warehouse design, de-identification, DLP, access control (RBAC and MFA), data lake firewalls, data encryption/masking, and backup/recovery.

New cutting-edge solutions known as LLM firewalls are emerging to tackle specific security issues related to AI. These firewalls can validate inputs to prevent prompt injections and check outputs to identify sensitive data leaks, AI-generated errors (often referred to as hallucinations), and ethical violations. Some even have the ability to scan models for potential vulnerabilities.

Cloud providers offer built-in LLM firewall features when using their models, providing an additional layer of security:

- *AWS Bedrock* includes *guardrails*, which allow organizations to define security and compliance constraints for AI models. Guardrails can filter inputs and outputs to prevent model misuse, such as blocking harmful, biased, or inappropriate content.
- *Azure AI Foundry* offers *guardrails* that enable developers to enforce content filtering, ensuring AI-generated responses align with business policies.
- *Google Cloud's Vertex AI* includes *Safety Settings*, which perform input and output validation to prevent prompt injections and unwanted responses.

For example, if a company develops a financial chatbot to assist banking customers, it must ensure the chatbot stays within its intended function. Without guardrails, the AI might respond to off-topic inquiries, such as providing romantic advice, which could dilute brand credibility and introduce unintended risks. Leveraging AI guardrails allows businesses to block such outputs, ensuring AI products maintain strong ethical and functional boundaries.

14.3.11 Securing AI Applications through Threat Modeling

As AI-powered applications become integral to business operations, securing them requires a structured approach. One of the most effective methodologies for assessing security risks is **threat modeling**. Threat modeling is a proactive process that helps organizations identify, analyze, and mitigate security risks in AI-enabled applications. Given the complexity and evolving nature of AI, traditional security methods may not fully address the unique threats posed by ML models, data pipelines, and AI-driven decision-making processes. Threat modeling provides a systematic way to understand how attackers might exploit AI systems, enabling organizations to implement preemptive security measures. By incorporating established frameworks and AI-specific methodologies, security teams can better anticipate risks, develop countermeasures, and improve the overall resilience of AI applications.

14.3.11.1 The Four Key Questions of Threat Modeling

Threat modeling frameworks are built around four core questions:

1. **What are we working on?**
   - This step defines the AI system's architecture, data flows, dependencies, and trust boundaries. For AI applications, it includes understanding model inputs, training datasets, API integrations, and storage mechanisms.

2. **What can go wrong?**
   - Identifies potential threats to the AI system, such as adversarial attacks, data poisoning, model evasion, and API abuse. It also considers risks like unauthorized data access, biased decision-making, and model theft.
3. **What are we going to do about it?**
   - Establishes security controls to mitigate identified threats. This includes implementing data validation, access control mechanisms, monitoring for adversarial attacks, and employing explainability techniques.
4. **Did we do a good enough job?**
   - Evaluates the effectiveness of implemented security measures. Regular audits, penetration testing, and red-teaming exercises ensure AI applications remain secure against evolving threats.

14.3.11.2 Threat Modeling Frameworks for AI Security

Several threat modeling frameworks provide structured methodologies for securing AI systems:

- **Spoofing, Tampering, Repudiation, Information Disclosure, Denial of Service, Elevation of Privilege (STRIDE)**
  - STRIDE helps organizations identify security threats based on the potential impact on AI application components, including APIs, training datasets, and ML models.
- **Process for Attack Simulation and Threat Analysis (PASTA)**
  - PASTA follows a risk-based approach, integrating business context with technical risks. For AI applications, this ensures alignment between AI security measures and organizational goals.
- **Damage, Reproducibility, Exploitability, Affected Users, Discoverability (DREAD)**
  - DREAD scores threats based on severity and prioritizes mitigation efforts. It is particularly useful for evaluating AI risks like adversarial ML attacks.

Among these frameworks, *MITRE ATLAS* (*Adversarial Threat Landscape for Artificial-Intelligence Systems*) is specifically designed for AI security.

14.3.11.3 MITRE ATLAS: AI-Specific Threat Modeling

MITRE ATLAS is an adversary-focused knowledge base that maps AI security risks and attacks. It provides a structured way to understand how AI systems can be exploited and offers guidance for defensive measures. Unlike generic cybersecurity frameworks, MITRE ATLAS specifically categorizes AI-related attack techniques, enabling organizations to apply targeted defensive strategies to protect AI models and infrastructures from known adversarial threats.

MITRE ATLAS helps organizations by categorizing AI threats into distinct tactics and techniques similar to the MITRE ATT&CK framework. These tactics range from data poisoning and model evasion to adversarial perturbations that can manipulate AI decisions. By understanding these tactics, organizations can develop more comprehensive security strategies that address both direct and indirect AI threats.

One of the primary uses of MITRE ATLAS is mapping attacks to AI workflows. AI systems often involve complex workflows, including data ingestion, model training, inference, and deployment. Each stage of this workflow presents unique vulnerabilities. ATLAS helps organizations align known adversarial tactics with these stages, ensuring that security teams can proactively defend against potential threats. For example, model poisoning attacks occur during training, while adversarial perturbations target the inference stage. Mapping these threats helps organizations deploy security controls at the right points in the AI lifecycle.

Since AI security is a continuously evolving field, threat models must be updated regularly. MITRE ATLAS provides a dynamic resource that keeps pace with new AI vulnerabilities and attack techniques. Organizations leveraging ATLAS should integrate it into their existing security frameworks and regularly review its updates to ensure they remain protected against the latest adversarial threats.

14.3.11.4 Threat Syntax: Writing Effective Threat Statements

Threat statements help communicate risks clearly. A well-defined threat statement follows this syntax:

> *A [threat source] with [prerequisites] can [threat action] which leads to [threat impact], negatively impacting [impacted assets].*

For example, a malicious actor with access to training data can manipulate data inputs, leading to model poisoning. This type of attack degrades AI accuracy and impacts decision-making reliability, potentially resulting in financial or reputational losses for an organization. Proper controls, such as data validation and monitoring mechanisms, can mitigate these risks.

Similarly, an unauthorized user with API access can perform adversarial queries, causing model evasion. In this scenario, the AI model produces incorrect predictions and security misclassifications, creating vulnerabilities that attackers can exploit. Implementing strict authentication and access controls can help prevent such attacks.

By structuring threat statements using this syntax, security teams can systematically describe risks and ensure clarity in their mitigation strategies. Threat modeling frameworks like MITRE ATLAS further enhance this process by mapping potential attack vectors and aligning them with structured defense mechanisms.

14.3.11.5 Using Threat Models to Secure AI Applications

The first step in using threat models to secure AI applications is to answer the "*what are we working on?*" question by defining all AI system components and data flows. This involves mapping out the architecture; identifying how data is collected, processed, and stored; and determining how different components interact. Understanding these details is essential to recognizing potential vulnerabilities and attack vectors within the AI ecosystem.

Next, we have to answer the "*what can go wrong?*" question. We do that by identifying potential threats using established frameworks like STRIDE, PASTA, DREAD, and MITRE ATLAS. These frameworks provide structured methodologies for assessing security risks, enabling teams to systematically analyze how adversaries may exploit weaknesses in AI models, APIs, or data pipelines.

Once threats are identified, security teams must evaluate them using structured threat syntax. This helps in articulating each risk clearly, specifying threat sources, potential actions, and impacted assets. By formalizing threat descriptions, organizations can prioritize risks and develop targeted mitigation strategies.

After that we have to answer the "*what are we going to do about it?*" question by implementing mitigation strategies. These can include adversarial detection mechanisms, enhanced access control policies, explainability techniques to improve model transparency, and anomaly detection systems that flag unusual behavior. Deploying multiple layers of defense ensures that AI applications remain resilient against diverse attack vectors.

In order to answer the "*did we do a good enough job?*" question, after mitigating the threats discovered in the threat model, we must verify that those mitigations function as expected. This requires penetration testing on both a recurring time-based schedule and targeted penetration testing per release. Additionally, automated testing using SAST and DAST tools should be integrated into the security lifecycle.

Beyond verifying mitigations, we must also validate our processes by comparing our threat findings against established frameworks and reviewing them for accuracy, clarity, and completeness. Continuous testing, monitoring, and refinement of security measures are necessary for adapting to emerging threats. AI security is an evolving field, and regular security assessments, penetration testing, and incident response exercises help maintain a strong security posture. By iterating on threat models and integrating lessons from security events, organizations can proactively enhance the protection of their AI applications.

As businesses increasingly integrate AI, especially GenAI, into their operations and products, it continues to present both opportunities and challenges. It is the CISO's job to secure the responsible adoption of AI technologies. The key lies in understanding the unique security risks AI presents and applying both traditional and cutting-edge security controls.

Securing AI involves a complex set of challenges that require a proactive and comprehensive approach. CISOs must not only protect AI systems themselves but also mitigate risks related to AI development and third-party integrations. Balancing these technical and governance considerations involves building a strong AI governance framework, establishing clear policies, and utilizing both established and emerging security measures to ensure the safe adoption of AI technologies.

By addressing the risks of AI enablement, technical hurdles, and concerns around third-party use, CISOs can empower their organizations to fully harness AI's transformative power while reducing potential downsides. This proactive stance not only strengthens the company's overall security but also fosters trust and confidence among stakeholders, laying the groundwork for a secure, AI-driven future.

14.4 SECURITY WITH AI

AI offers a powerful set of tools that can significantly enhance cybersecurity efforts. ML algorithms can sift through massive volumes of data to detect subtle patterns and anomalies that may indicate a cyberattack, far exceeding human capabilities. AI can also automate routine security tasks, freeing up security professionals to focus on strategic initiatives. CISOs should actively explore and adopt AI-powered security solutions to proactively address emerging threats.

Adding AI technology to our security tool stacks is all about leveraging AI to improve our toolsets and capabilities. This can be done in a variety of ways, from using AI-powered threat intelligence to automate routine tasks to using AI to detect and respond to threats faster.

14.4.1 Security Use Cases for GenAI

GenAI holds the promise of transforming SOCs by enhancing the work of analysts and allowing them to focus on more strategic tasks. One area where AI can make a significant impact is vulnerability discovery. AI can assist in identifying weaknesses in code, systems, and applications, potentially uncovering hidden risks before attackers have a chance to exploit them. Additionally, AI can automate the collection and analysis of threat intelligence from various sources, providing real-time insights into emerging threats and enabling faster, more proactive responses.

For threat detection, AI can analyze network traffic, system logs, and user behavior to spot anomalies and potential threats that traditional security tools might miss. It can also streamline patch management, automating the prioritization and application of security patches, which helps reduce the window of vulnerability for organizations.

AI's role in identity and behavior analysis is another critical area of enhancement. By monitoring user and entity behavior, AI can detect deviations from normal patterns that could indicate insider threats or compromised accounts. It builds behavioral profiles for users and systems, allowing for early detection of unusual activities that may signal malicious intent.

One of AI's strongest capabilities is its ability to manage and analyze vast amounts of security data. AI can quickly sift through massive datasets to identify trends, patterns, and correlations that human analysts might overlook, helping detect potential breaches faster and more accurately.

Within the SOC itself, AI-powered chatbots can assist analysts in real time by automating routine tasks, answering questions, and freeing up human resources for more critical incidents. When it comes to threat hunting, AI can automate the search for indicators of compromise (IoCs) and other signs of malicious activity, significantly speeding up the process.

Other applications of AI in SOCs include anomaly detection, where AI rapidly identifies unusual patterns in large datasets, and predictive analysis, which allows AI to forecast potential threats and vulnerabilities based on historical data and current trends. AI can also enhance incident triage by automating the prioritization and categorization of security incidents, ensuring that the most pressing issues are addressed first. Finally, AI can automate parts of the incident response process, such as blocking malicious IP addresses or isolating infected systems, reducing response times and limiting the impact of security breaches.

GenAI is not just a tool for automating tasks; it has the potential to reshape how SOCs operate, making security efforts more efficient and allowing analysts to focus on higher-level strategic issues that ultimately enhance an organization's overall security posture.

14.4.2 AI versus Automation

While AI holds immense potential in security, it's important to distinguish between AI and automation. Many security tasks, such as anomaly detection and incident response automation, have already been achieved using ML and security orchestration, automation, and response (SOAR) platforms. For instance, ML models have been employed for years to identify patterns and flag unusual activity, while SOAR solutions automate various incident response processes.

What sets GenAI apart is its ability to go beyond routine automation, offering contextual insights and solutions that were previously challenging to automate. A GenAI tool can analyze a complex security incident, understand the broader context, and suggest a customized response plan, which is a capability that extends far beyond what traditional automation or basic ML could accomplish.

The key difference between ML and GenAI in security tools comes down to their functions. Traditional ML focuses on recognizing patterns in data, making predictions, and identifying anomalies. It is great for classification tasks and improving the efficiency of threat detection. GenAI, however, is designed to create new content based on patterns it has learned, which opens up opportunities in tasks like generating reports, summarizing threat intelligence, or even drafting phishing awareness content for employees.

A practical example of GenAI in security is CrowdStrike's Charlotte AI. This tool uses GenAI to provide natural language explanations of threat detections, suggest remediation steps, and produce custom reports, making it a valuable asset for security teams aiming to streamline processes while enhancing the quality of their responses.

14.4.3 Real-World AI Security Uses

14.4.3.1 AI-Powered Threat Intelligence

AI can significantly enhance the way organizations manage threat intelligence by automating the collection, analysis, and application of data. It can sift through vast amounts of information from various sources, such as the dark web, social media, and threat databases, to identify patterns, predict potential risks, and suggest preventive actions in real time. This capability allows security teams to respond faster and more effectively than would be possible with manual efforts alone.

AI's ability to analyze and correlate data across different sources helps identify trends and classify threats more accurately. ML algorithms, in particular, can detect subtle patterns that might go unnoticed by human analysts, enhancing both the precision and speed of threat detection.

Moreover, AI can streamline the distribution of threat intelligence to key stakeholders, including security analysts, incident response teams, and even other AI-driven systems. By automating this process, organizations can ensure that critical information reaches the right people quickly, fostering a more coordinated and timely response to emerging threats.

A key advantage of AI is its ability to learn continuously. As new threats emerge, AI systems can adapt, refine their models, and improve the relevance and accuracy of the intelligence they generate. This constant learning is essential in an ever-evolving threat landscape, where attackers frequently change their tactics and strategies.

In this way, AI not only boosts the effectiveness of threat intelligence but also helps organizations stay one step ahead of potential attackers.

14.4.3.2 AI-Driven Security Analytics

AI-driven security analytics involves using AI to analyze large volumes of security data to identify threats, vulnerabilities, and anomalies. AI can efficiently process and analyze vast volumes of security data that would be overwhelming for human analysts, providing real-time insights into the security posture, and enabling quick detection and response to threats. AI can leverage pattern recognition and ML to predict potential threats before they materialize, further enhancing proactive security measures.

Use cases for AI-driven security analytics include the following:

Anomaly Detection: AI systems can quickly detect anomalies or deviations from normal patterns in large datasets, flagging potential threats for further investigation.

Threat Hunting: AI can identify patterns and correlations in data to proactively hunt for threats that may have evaded traditional security tools.

Risk Assessment: AI algorithms can analyze data to assess risk levels and prioritize responses, enabling more effective risk management.

14.4.4 SOAR and Hyperautomation

Often, when discussing AI in security with CISOs and their teams, I find that the desired outcomes are actually automation use cases. Teams want to automate away slow, repetitive Tier 1 and 2 tasks within the SOC. The good news is that we don't need to wait for GenAI tools to achieve this; automation can be done today, and GenAI is making it even more powerful.

SOAR automation has evolved through several generations. The first generation offered open integration with various tools but required manual creation of playbooks and workflows. The second generation provided pre-built integrations and playbooks, sacrificing flexibility for ease of use.

The latest generation, hyperautomation, combines the best of both worlds. It leverages GenAI to create code, build integrations, and craft playbooks on the fly. You can simply describe what you want to accomplish in plain English, and the GenAI will help you build the automation.

14.4.5 Leveraging GenAI to Augment SOC Analysts

GenAI can be a powerful tool for augmenting the capabilities of SOC analysts. Tools like Microsoft Security Copilot, an AI-powered chatbot that leverages GPT-4 and Microsoft's threat intelligence, can assist analysts in detecting, investigating, and responding to cyberattacks more efficiently. It can provide context-aware insights, suggest remediation actions, and even generate custom reports.

Even without specialized security tools, SOC analysts can leverage general-purpose GenAI models like ChatGPT or Gemini to perform various tasks. Here are some prompt ideas:

Threat Triage and Prioritization: "Help me analyze and prioritize a list of detected security alerts based on their potential impact and relevance to our organization".

Incident Investigation: "Guide me through the process of investigating a suspicious network traffic pattern identified in our logs and help me determine its root cause".

Malware Analysis: "Assist me in analyzing a potentially malicious file or script to identify its behavior, potential risks, and recommended remediation steps".

Phishing Email Analysis: "Review a suspicious email and provide insights on its phishing indicators, including sender analysis, link analysis, and attachment scrutiny".

Vulnerability Assessment: "Help me evaluate the severity of a recently discovered software vulnerability and suggest strategies for mitigating the risk".

Security Policy Guidance: "Provide guidance on creating effective firewall rules to enhance network security and compliance with industry standards".

SOC Playbook Creation: "Assist in building a playbook for common security incidents, outlining step-by-step response procedures and recommended tools".

Threat Intelligence Gathering: "Retrieve the latest threat intelligence related to a specific malware family or attacker group, including known IoCs".

Log Analysis and Correlation: "Help me correlate logs from multiple sources to identify patterns and anomalies that may indicate a security breach".

User Behavior Analysis: "Analyze user behavior data to detect unusual activity or insider threats and suggest actions for further investigation".

Patch Management: "Provide recommendations for prioritizing and scheduling software patch updates based on current vulnerabilities and potential impact".

Endpoint Security Optimization: "Suggest strategies and configurations to enhance endpoint security and reduce the attack surface".

SOC Training and Education: "Offer resources and materials for ongoing SOC team training in the latest cybersecurity trends, threats, and best practices".

Security Awareness Campaigns: "Help me create an engaging security awareness campaign to educate employees about phishing risks and safe online behavior".

Regulatory Compliance Assistance: "Guide me through the process of ensuring compliance with specific data protection regulations, such as GDPR or HIPAA".

14.4.5.1 Example

Recently, I received a call from the CISO of a rapidly growing e-commerce company. Their SOC was drowning in a deluge of alerts and busy work that had pushed their team to the brink of exhaustion. The CISO, a visionary leader with a keen eye on the future, recognized the unsustainable nature of their current approach. He knew they needed a radical transformation, not just a band-aid solution. That's when he reached out to me, seeking guidance on how to modernize and automate their SOC to regain control and proactively defend against emerging threats.

Together, we tackled the implementation of hyperautomation. The initial phase was not without its hurdles. The system, while brimming with potential, required training on the company's unique processes and integrations. The team had to adapt their workflows to seamlessly integrate the AI's insights. However, the CISO's unwavering commitment to the project and the team's eagerness to tackle this new technology proved to be the driving force behind their success.

Within weeks, the AI system began to prove its worth. It effortlessly sifted through mountains of security data; its algorithms were able to perform tasks in moments that would have taken even the most seasoned human analyst hours. The initial headaches subsided, allowing the team to laser-focus their attention on genuine threats. The AI even started to predict potential attacks based on historical data and emerging threat intelligence, empowering the team to take proactive measures to thwart breaches before they could materialize.

The impact on incident response times was nothing short of transformative. What once took hours or even days to investigate and contain could now be addressed in a matter of minutes. The team's morale experienced a resurgence as they shifted from a reactive, firefighting mode to a proactive, threat-hunting stance. The CISO's decision to invest in AI-driven security analytics had not only fortified their security posture but also revitalized their team, fostering a sense of empowerment and purpose.

14.4.6 Chatbots in Security Automation

Chatbots are becoming an integral part of security automation, helping organizations streamline processes like incident response, threat hunting, and vulnerability management. For example, chatbots can quickly triage and categorize security alerts, allowing security analysts to focus on more critical issues. However, introducing chatbots into security workflows also introduces new risks, such as the potential exposure of sensitive data or unauthorized access to vital systems, which CISOs must address.

To ensure the secure use of chatbots in security operations, it's important to define clear security protocols and guidelines. Implementing access controls and maintaining audit logs will help restrict access to chatbots and enable monitoring of their usage, making it easier to detect any suspicious activity. Additionally, chatbots should be integrated into existing security systems, like SIEM, so their actions can be tracked alongside other security events, ensuring a holistic approach to threat detection.

Regular security assessments, including vulnerability scans, are necessary to ensure that chatbots remain secure and adapt to evolving threats. It's also crucial to verify that chatbots comply with relevant regulations, such as GDPR or PCI-DSS. This can be achieved through appropriate security measures like data encryption and secure storage practices, safeguarding the organization against both security risks and regulatory compliance issues.

14.4.7 Specialized AI Security Tools

For more advanced use cases, specialized AI security tools like shellGPT, TerminalGPT, and ChatGPT-Bypass for Kali Linux can be leveraged. These tools can automate various security tasks, such as network enumeration, vulnerability fuzzing, exploit crafting, and even creating phishing payloads. Red teams can utilize these tools to automate reconnaissance, vulnerability scanning, and exploit development, enabling them to identify and exploit weaknesses more efficiently. This, in turn, helps organizations stay one step ahead of potential attackers by proactively addressing vulnerabilities and strengthening their defenses.

Another promising tool is WhiteRabbitNeo, a security-focused LLM designed to assist security professionals in various tasks. It can generate security reports, analyze code for vulnerabilities, and even provide guidance on incident response. WhiteRabbitNeo can help security teams streamline their workflows, improve their efficiency, and enhance their overall security posture.

The integration of these specialized AI security tools into the security stack empowers CISOs and their teams to tackle the challenges of an increasingly complex and dynamic threats. However, you should remember that AI is a tool, not a replacement for human expertise. The ultimate responsibility for security still lies with the CISO and their team, who must exercise sound judgment and critical thinking when interpreting and acting upon AI-generated insights.

14.4.8 AI and Automation for the CISO Role

The CISO 3.0 can benefit from the capabilities of GenAI and automation. A considerable portion of the CISO's responsibilities, particularly those involving repetitive tasks, data analysis, and communication, can be augmented or streamlined through AI.

GenAI can serve as a powerful assistant for CISOs, helping them perform various tasks more efficiently and effectively. For instance, in risk assessment and reporting, AI can analyze vast amounts of security data, identify patterns, and generate risk assessment reports replete with actionable insights, allowing CISOs to focus on strategic decision-making rather than data compilation.

When it comes to policy and procedure creation, AI can assist in drafting clear and concise documents that are aligned with industry best practices and regulatory requirements. This not only saves time but also ensures consistency and accuracy in policy documentation. Communication and reporting can also be enhanced through GenAI. It can help CISOs create compelling narratives and presentations for board members and executives, effectively translating complex technical concepts into easily understandable language, thereby fostering better understanding and support for security initiatives. Threat intelligence analysis is another area where AI can provide immense value. It can efficiently sift through large volumes of threat intelligence data, identify relevant information, and generate concise summaries for CISOs to review, enabling them to stay abreast of the latest threats and vulnerabilities without being overwhelmed by information overload. In incident response, AI can assist in incident triage and response, providing recommendations based on historical data and best practices. This can help security teams respond to incidents more quickly and effectively, minimizing their impact on the organization.

14.4.8.1 Example Prompts for CISOs

Here are some specific prompts a CISO could use to leverage GenAI for various tasks, tailored to elicit more relevant and actionable information, potentially incorporating attachments or additional context where necessary:

Risk Assessment

- "Analyze the attached vulnerability scan report and our current asset inventory to generate a risk assessment, prioritizing the top five risks based on potential business impact and likelihood of occurrence".
- "Considering our industry and recent cyber threats, identify and assess emerging risks that might not be fully captured in our existing risk register".

Policy Creation

- "Draft a revised acceptable use policy that addresses the use of generative AI tools within our organization, ensuring compliance with GDPR and our internal data handling standards".
- "Review our current incident response plan and suggest updates to incorporate best practices for handling AI-related security incidents".

Board Presentation

- "Using the attached financial data and security metrics, create a presentation for the board highlighting the ROI of our cybersecurity investments and the key risk areas we are addressing".
- "Develop talking points for a board presentation on the potential impact of AI on our industry and our cybersecurity strategy to address these emerging threats".

Threat Intelligence Summary

- "Analyze the attached threat intelligence reports and provide a concise summary of the latest TTPs (tactics, techniques, and procedures) used by ransomware groups targeting organizations in our sector".
- "Identify any new IoCs associated with these ransomware attacks and suggest how we can proactively detect and mitigate these threats".

Incident Response

- "Given the attached incident report on the recent phishing attack, outline a detailed response plan, including containment, eradication, and recovery steps, as well as recommendations for improving our defenses against similar attacks".
- "Analyze the incident data and identify potential gaps in our current security controls that allowed this attack to succeed. Suggest improvements to prevent future incidents".

GenAI has the potential to be a game-changer for CISO 3.0s. For instance, AI can assist greatly with quantitative risk assessments and the creation of registers as described in Chapter 6. It analyzes data related to vulnerabilities, threats, and potential impacts, giving CISOs a clearer, data-driven understanding of risk.

In addition, AI can help CISOs craft compelling narratives and presentations that resonate with board members and executives, which we will discuss further in Chapter 16. AI can help bridge the communication gap between the CISO and other stakeholders by translating technical jargon into clear and concise language, facilitating informed decision-making and garnering support for security initiatives.

AI can also assist in developing meaningful key risk indicators (KRIs), which is covered in Chapter 15. AI can analyze security data and then identify and track those indicators, providing actionable insights into the organization's security posture and enabling CISOs to proactively address potential risks.

AI can also assist in the creation and maintenance of IT balance sheets or balanced scorecards, which we will explore in Chapter 16, automating data collection and analysis, and providing visualizations to track performance and identify areas for improvement. This allows CISOs to effectively communicate the value and impact of security investments to the board and other stakeholders.

Using AI technologies and integrating them into their workflows, CISOs can free up time for strategic thinking, enhance their decision-making capabilities, and ultimately strengthen their organizations' security posture. While AI will undoubtedly automate many aspects of the CISO role, it's important to remember that it's a tool, not a replacement for human judgment and leadership. The CISO 3.0 will need to strike a balance between leveraging AI for efficiency and maintaining a human touch in critical decision-making and stakeholder engagement.

14.5 SECURITY FROM AI

The rise of AI-powered cyberattacks is a real concern. A strong security strategy must consider the cyber and business risk of both adopting and defending against AI technologies. As AI becomes increasingly accessible, threat actors are leveraging it to launch novel and potent attacks. These attacks are becoming more sophisticated and harder to detect, and they have the potential to cause real damage. As CISOs, we need to stay ahead of the curve and develop new strategies to defend against these threats. It's a constant arms race, but we can't afford to lose.

As discussed earlier, we can categorize AI-powered cyberattacks into two primary types:

Target AI Attacks: These attacks directly target your organization. AI is used to generate or accelerate attacks against your people, network, or assets. Examples include intelligent malware that can adapt and evade detection, AI-augmented phishing attempts that are highly personalized and convincing, and automated vulnerability exploitation tools that can rapidly identify and exploit weaknesses in your systems.

Subject AI Attacks: In these attacks, your organization becomes the subject, although not necessarily the direct target. The AI-generated attack or content is disseminated outside your network or assets. This can include deepfakes used for extortion, where a CFO's voice is convincingly replicated to authorize fraudulent transactions at partner organizations, or disinformation campaigns aimed at damaging your brand value through fabricated news articles or social media posts.

14.5.1 Target Attacks

GenAI, while a powerful tool for productivity and innovation, can also be exploited for malicious purposes. For instance, ChatGPT, with its ability to write code, can be manipulated to generate malicious scripts once its ethical filters are bypassed. This includes creating sophisticated SQL injection attacks that were once difficult to automate and integrating directly into penetration testing tools like Kali Linux.

AI has infiltrated almost every stage of the cyber kill chain, as illustrated in Figure 14.4. In the social engineering phase, GenAI can be used to craft sophisticated phishing and smishing attacks, making them more difficult to detect and resist. Fraudulent activities can also be enhanced through intelligent captcha attacks and voice cloning, enabling attackers to bypass security measures and impersonate individuals.

GenAI can craft highly convincing phishing emails and smishing texts, lowering the barrier of entry for non-native English speakers to launch social engineering attacks. The same automated vulnerability detection tools we use to improve our security posture can be leveraged by threat actors to identify and target weaknesses in our systems.

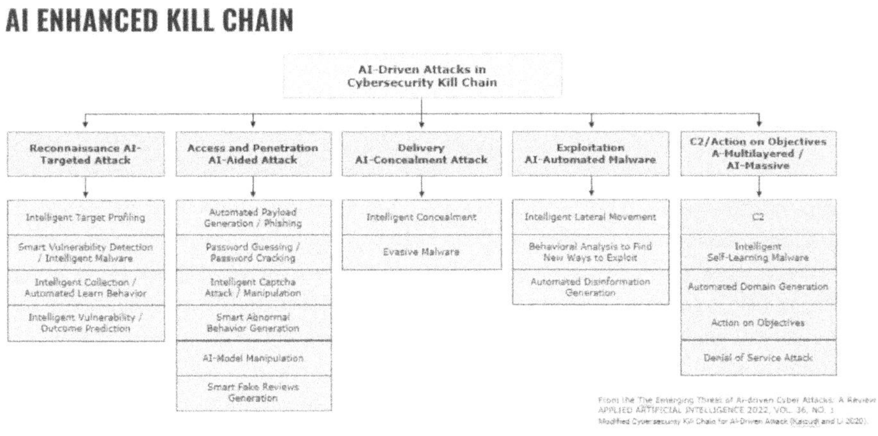

FIGURE 14.4 AI Cyber Kill Chain.

Spoofing attacks targeting individuals or organizations can leverage deepfakes, encompassing text, social media posts, audio, video, and even live streams, to deceive and manipulate. Password attacks can be automated and accelerated through AI, enabling more efficient credential stuffing, password guessing, and password cracking attempts.

In the malware domain, AI can create polymorphic and metamorphic malware capable of altering their code or even rewriting themselves to evade detection and adapt to changing environments. Lastly, in the vulnerability discovery and exploitation phase, AI-powered fuzzing can accelerate the identification of vulnerabilities, and AI can even be used to write exploits for these vulnerabilities, making attacks more potent and widespread.

Subject attacks, as mentioned earlier, are a particularly challenging type of AI-powered attack. They involve the use of AI-generated content to target individuals or organizations indirectly, often through disinformation and reputational damage.

Deepfakes are a prime example of this. Deepfakes leverage AI and ML algorithms to create hyper-realistic but fake images, videos, or audio, often to the point where they're nearly indistinguishable from authentic media. These can be used in social engineering attacks to impersonate trusted individuals, or to spread disinformation and damage reputations by creating fake evidence or spreading false narratives.

Other forms of subject attacks include disinformation campaigns that utilize real but edited or out-of-context media, fake accounts spreading misinformation on social media, and review bombing to manipulate public perception.

The various forms of media that can be manipulated or generated by AI include the following:

- *Text*, such as fake documents, phishing emails, smishing texts, fake reviews, and fake social media posts.
- *Images*, including fake IDs and compromising or misleading photos.
- *Audio*, such as vishing attacks (voice phishing) for transfer fraud, extortion, password resets, and the creation of fake statements.

- *Video*, ranging from real-time video manipulation in collaboration tools to fake ads, fake news, and fake statements.

While consumers express concerns about deepfakes and voice clones, there's uncertainty about the readiness of institutions to combat them. Recent studies indicate that a significant portion of Americans feel uncertain about how well businesses are prepared to handle deepfake technology. While there's more trust in sectors like banking and healthcare, confidence drops when it comes to news and social media, even though these platforms are often where people encounter deepfakes. It appears that the public's expectations for deepfake protection might be higher than what companies are currently delivering. Tools for detecting deepfakes are emerging, but for target attacks, the best controls often remain low-tech. This includes ensuring that help desks verify identities and call back on known phone numbers before taking any action.

Subject attacks pose a more complex challenge. While potential controls like watermarking known good communications exist, their effectiveness depends on consumer awareness and vigilance. Metadata solutions are in development, but currently, one of the best defenses is promoting media literacy and critical thinking. Incorporating deepfake and AI attack strategies into security awareness training for all employees is necessary.

In summary, the rise of AI, particularly GenAI, presents a double-edged sword for cybersecurity. While it offers powerful tools for defense, it also empowers malicious actors with new capabilities. CISOs must proactively address both target and subject attacks, leveraging a combination of technical controls, policies, and user education to safeguard their organizations.

14.6 CONCLUSION

The AI revolution is in full swing, and it's shaking up every industry, including ours. In this chapter, we've explored the complex dance between AI and cybersecurity, and how it's transforming the role of the CISO. We've seen not only the incredible potential AI offers but also the new risks and challenges it brings. It's clear that CISOs need to adapt and evolve to stay ahead of the curve.

We've broken down AI security into three key areas: securing AI systems themselves, using AI to boost our defenses, and protecting ourselves from AI-powered attacks. We've also talked about the two main types of AI cyberattacks: Those that directly target your organization, and those that use your company or its people as a weapon, even if you're not the primary target. We've seen how GenAI can be misused, how AI is changing the cyberattack playbook, and the dangers of deepfakes and disinformation.

For the CISO 3.0, this chapter is a call to action. It's time to fully accept AI, not only as another tool in the tech stack but also as a strategic asset that shapes the future of security leadership. This means going beyond technical expertise and stepping into the role of a visionary leader, capable of navigating both the opportunities and challenges AI brings. You need to understand not only the technical risks but also the ethical implications, building a governance framework that ensures AI is used responsibly. At the same time, it's about using AI to your advantage and leveraging its capabilities to strengthen your defenses and stay ahead of evolving threats. The future of cybersecurity is inseparable from AI, and as a CISO 3.0, you are at the forefront of this shift.

14.6.1 Key Takeaways

1. **AI as a Transformative Force in Cybersecurity**
   - AI has revolutionized cybersecurity, with applications ranging from predictive threat detection to automating repetitive tasks.
   - AI not only presents opportunities for innovation and efficiency but also introduces new risks, including malicious use by threat actors.
2. **Three Dimensions of AI Security**
   - **Security for AI:** Safeguarding AI systems and ensuring their secure and ethical deployment.
   - **Security with AI:** Using AI technologies to enhance cybersecurity measures, such as threat detection and incident response.
   - **Security from AI:** Defending against AI-powered cyberattacks, including those leveraging GenAI for malicious purposes.
3. **AI Governance as a Foundational Element**
   - Establishing robust AI governance frameworks is critical for balancing innovation with ethical and secure AI adoption.
   - AI governance boards guide organizational strategy, risk assessment, and compliance, ensuring AI initiatives align with business goals and regulatory requirements.
4. **Risks and Challenges of AI Implementation**
   - Risks include data breaches, regulatory noncompliance, bias in AI outputs, and the misuse of GenAI by malicious actors.
   - Addressing these challenges requires technical controls, such as LLM firewalls, and robust risk management practices.
5. **The Rise of AI-Powered Cyberattacks**
   - Threats include target attacks (e.g., AI-generated phishing) and subject attacks (e.g., deepfakes for disinformation).
   - The cyber kill chain now includes sophisticated AI-driven tools for social engineering, vulnerability discovery, and automated malware development.
6. **Leveraging AI as a Strategic Tool**
   - AI enhances SOC capabilities through anomaly detection, threat intelligence, and predictive analysis.
   - Tools like GenAI assist CISOs in tasks such as creating policies, assessing risks, and communicating with stakeholders.
7. **Ethical Considerations and Responsible AI Use**
   - CISOs must navigate ethical dilemmas related to AI, including its impact on jobs, privacy concerns, and the potential for discrimination.
   - Proactive measures, including regular audits and stakeholder engagement, ensure AI use remains transparent and aligned with organizational values.

8. **The Evolving Role of the CISO 3.0**
   - The CISO of the future is a strategic leader who integrates AI into cybersecurity while addressing its risks and leveraging its potential for competitive advantage.
   - Leadership extends to fostering an organizational culture that embraces innovation and resilience in the face of AI-driven challenges.

14.6.2 Reflection and Exploration Questions

Understanding the Concepts

1. What are the three dimensions of AI security, and how do they address the challenges posed by AI in cybersecurity?
2. How does AI governance contribute to responsible and secure AI adoption?

Connecting the Dots

3. How do AI-powered attacks align with earlier discussions on threat landscapes and risk management strategies?
4. In what ways can AI complement traditional cybersecurity tools to enhance an organization's security posture?

Real-World Application

5. Can you identify examples where organizations have successfully implemented AI to improve their cybersecurity measures?
6. How might you apply AI-powered SOC tools in your organization to address current security challenges?

Personal and Professional Growth

7. How can you use GenAI to streamline your cybersecurity responsibilities and improve decision-making?
8. What steps can you take to strengthen your organization's AI governance framework?

Further Exploration

9. What additional skills or technologies would help you integrate AI into your cybersecurity strategy effectively?
10. How might future regulatory developments shape the use of AI in cybersecurity?
11. What ways can you begin to utilize AI in your role as a cybersecurity leader?

Part 4

Bringing It All Together

15 Developing Modern Metrics

15.1 INTRODUCTION

To truly get your message across at the executive level, you need a solid system of metrics that speaks the language of business. Metrics aren't just about taking a snapshot of your security posture: They're about tracking your progress, spotting trends, and making smart decisions about where to invest your resources and how to tackle risks head-on.

In this chapter, we're going to explore cybersecurity metrics, with a special focus on two key players: key performance indicators (KPIs) and key risk indicators (KRIs). We'll break down the difference between these two, look at how to develop them, and give you some tips on how to communicate them effectively to the folks upstairs. We'll also point out some common pitfalls to avoid, so you can build a metrics system that truly delivers.

Developing and communicating KRIs is where the CISO 3.0 really shines. While KPIs are great for tracking your team's performance and making internal improvements, KRIs are your ticket to the boardroom. They help you translate those technical security details into something the executives can understand: The potential impact on the business. It's about taking those raw numbers, like the number of vulnerabilities, and turning them into a story about potential financial loss. That's what gets their attention and helps them see the big picture.

Defining and monitoring KRIs is also the seventh step in the risk strategy framework outlined in Chapter 6. Strong KRIs allow you to establish a baseline tolerance for risk and measure against it, providing clear visibility into how your risk buy-down activities are progressing. Well-defined KRIs aren't just useful for internal tracking; they are fundamental to effective, risk-based communication at the board level. By framing cybersecurity in terms of measurable risk reduction, you can make a stronger case for security investments and ensure alignment with business objectives.

By the end of this chapter, you'll have the tools you need to build a metrics system that tells a compelling story. You'll be able to show the value of your security program, justify your budget requests, and foster a culture of risk awareness throughout the organization. It's about taking your cybersecurity expertise and turning it into a powerful tool for driving business value and securing your place as a strategic leader.

Take a moment to reflect on your current approach to measuring and communicating cybersecurity performance and risk. Are you effectively using metrics to inform decision-making and drive improvements? How well are you communicating these metrics to different stakeholders, including executive leadership and the board?

DOI: 10.1201/9781003510789-21

15.2 KEY PERFORMANCE INDICATORS

KPIs are quantifiable measurements used to evaluate how well an organization or specific function is performing. They function as benchmarks to assess progress, identify trends, and empower informed decision-making. In cybersecurity, KPIs are critical for gauging the effectiveness of your security program and tracking your progress toward established goals.

KPIs are like the dashboard of your security program. They tell you how you're doing and where you need to improve. But it's important to choose the right KPIs and track them consistently. I've seen organizations get bogged down in a sea of metrics, losing sight of the big picture. The CISO 3.0 focuses on a few key indicators that truly reflect the health of the security program. A good KPI should provide a focused lens into a specific facet of your security posture. It helps you answer targeted questions about your program's effectiveness, such as the speed of incident detection and response, the efficacy of vulnerability management and patching processes, and the level of security awareness among your employees.

Measuring cybersecurity performance using KPIs is essential for several reasons. KPIs ensure that cybersecurity efforts align with the organization's tactical goals, demonstrating how security practices support the overall program. They provide a clear view of the cybersecurity program's performance, including the effectiveness of security measures and incident response times.

By monitoring KPIs, CISOs can identify and mitigate threats and vulnerabilities more effectively, make informed decisions about resource allocation, and adapt tactics to evolving threats. KPIs also offer a quantifiable way to communicate the performance of the cybersecurity program to stakeholders, gaining support for security initiatives. Regular tracking and analysis fosters continuous improvement by allowing you to find trends and make course corrections. Moreover, KPIs help demonstrate regulatory and policy compliance, providing evidence that controls are in place, monitored and functioning as expected.

The CISO 3.0 must regularly evaluate the effectiveness of KPIs, using the data to refine security strategies and communicate results to stakeholders. KPIs should be adapted to reflect evolving threats and benchmarked against industry standards to measure performance effectively. Through proper documentation, reporting, and resource allocation based on KPI insights, organizations can enhance their cybersecurity posture and ensure that their strategies remain aligned with business goals.

When implemented correctly, KPIs serve as powerful tools to drive continuous improvement, enhance accountability, and ensure that the security program remains on course to achieve its operational objectives. By focusing on a curated set of meaningful KPIs, CISOs can not only monitor their programs effectively but also clearly communicate progress and needs to executive leadership.

15.3 KEY RISK INDICATORS

KRIs, on the other hand, are proactive, forward-looking metrics that identify and track potential risks before they materialize into actual incidents. They serve as early warning signals, allowing you to anticipate and address emerging threats and vulnerabilities before they impact your organization.

A KRI is a predictive metric that provides insights into the likelihood and potential impact of a specific risk. It helps you answer questions such as: Is our organization's financial exposure to cyber risk increasing due to changes in our business operations or the threat landscape? Are we observing a rise in employee negligence or risky behaviors that could compromise sensitive data or lead to a security incident? Are our key third-party vendors demonstrating adequate cybersecurity practices, or are they potentially introducing vulnerabilities into our supply chain?

KRIs should be used continuously to monitor risks and identify emerging threats. KRIs are indispensable in risk management, helping organizations anticipate high-risk areas and take timely, proactive steps to address them. Through the use of KRIs, firms can gain a clear understanding of their current risk exposure and spot new trends, which allows them to stay ahead of potential threats. KRIs can also highlight weaknesses in the existing risk treatment strategy or even specific controls, providing opportunities to strengthen them and improve the overall risk posture. They also help in facilitating risk reporting and escalation processes, ensuring that relevant information reaches the appropriate levels of management for timely decision-making. Ultimately, the effective use of KRIs contributes to operational risk management, adding value to the firm by enabling proactive risk mitigation and informed decision-making. KRIs should be utilized at all levels of the organization, from the security team to executive leadership and the board of directors.

KRIs are like the early warning system for your security program. They help you identify emerging threats and vulnerabilities before they can be exploited. I've seen KRIs help organizations prevent breaches and mitigate the impact of attacks. But it's important to choose the right KRIs and monitor them closely. A KRI that's not actionable is just noise.

Developing good KRIs can be challenging. One key challenge is identifying the most relevant risks that could impact the organization's objectives. Another hurdle lies in data collection and analysis; you need access to reliable and relevant data sources to track KRIs effectively. Developing KRIs that accurately predict future risks requires a combination of historical data, threat intelligence, and expert judgment, adding another layer of complexity.

To overcome these challenges, you need to conduct thorough risk assessments and utilize the quantification techniques laid out in Chapter 6 to identify and prioritize the highest risks. Collaboration between enterprise and IT risk teams is key to developing and validating KRIs, ensuring their accuracy and relevance.

While KRIs are valuable for proactive risk management, they are not without their limitations. Predicting future risks inherently involves a degree of uncertainty; KRIs can provide valuable insights, but they are not foolproof. There's also the possibility of false positives or negatives, where KRIs might trigger unnecessary alarms or fail to detect certain risks. Therefore, careful monitoring and adjustment of KRIs are necessary. Remember what we discussed in Chapter 6: Risk quantification is attempting to predict the future and should only go as far as necessary to obtain sufficient data for informed decision-making.

15.3.1 Leading and Lagging Indicators

Leading indicators are forward-looking metrics that help predict future events or behaviors. They are derived from data that can offer insights into potential risks and opportunities. For example, tracking the frequency of employees clicking on simulated phishing links can be a leading indicator of potential susceptibility to actual phishing attacks. On the other hand, lagging indicators are based on historical data and help identify trends within the firm. An example of a lagging KPI would be the number of backup failures in the previous month, which can reveal patterns and potential weaknesses in the backup process. Current indicators provide a snapshot of the present state, such as the number of employees who have completed security awareness training, offering insights into the current level of security preparedness.

15.4 KPIS VERSUS KRIS

KPIs and KRIs, while both are essential for cybersecurity measurement, serve distinct purposes and cater to different audiences within the organization. However, these metrics are often confused, and it's common for CISOs to inadvertently develop KPIs when their intention is to create KRIs. This confusion can lead to misaligned reporting and ineffective communication of security performance and risk exposure.

TABLE 15.1

Indicator Metric	What Does It Measure?	What's the Purpose?	Who's the Audience?
Key performance indicator (KPI)	KPIs measure how effectively the organization is achieving its business objectives	They provide directional insight into how you're processing toward strategic objectives or the effectiveness of specific business processes or control objectives	**Strategic KPIs** Most often executive management and the board **Operational KPIs** Most often managers, operational process owners, and department heads
Key risk indicator (KRI)	KRIs measure how risky certain activities are in relation to business objectives	They provide early warning signals when risks (both strategic and operational) move in a direction that may prevent the achievement of KPIs	**Strategic KPIs** Most often executive management and the board **Operational KPIs** Most often managers, operational process owners, and department heads
Key control effectiveness indicator (KCI)	KCIs measure how well controls are working	They provide direct insight into a specific control activity, procedure, or process that wasn't implemented or followed correctly	Most often front-line control activity owners

KPIs are generally backward-looking lagging indicators, focusing on past performance and measuring progress toward established goals. They provide insights into the effectiveness of the security program and its processes. In contrast, KRIs tend to be forward-looking leading indicators, identifying potential risks and predicting future outcomes. They offer early warning signals of emerging threats and vulnerabilities. That said, it is possible to create KRIs that are lagging and/or backward-looking, and vice versa.

In terms of target audience, KPIs are typically used internally by the security team and reported to executive leadership to demonstrate the operational effectiveness of the security program. KRIs, with their focus on potential business impact, are more suitable for communication with the board of directors and senior executives who are responsible for strategic decision-making and risk oversight.

When it comes to what they measure, KPIs focus on specific aspects of security performance, such as incident detection and response times, vulnerability management metrics, and security awareness training completion rates. KRIs, conversely, measure the likelihood and potential impact of identified risks, such as the financial loss associated with a data breach or the probability of a successful ransomware attack.

By understanding the distinct roles of KPIs and KRIs and using them appropriately, you can create a balanced and effective approach to cybersecurity measurement that supports both operational efficiency and strategic risk management.

15.4.1 Real-World Example: From KPIs to KRIs

Here's a real-world example from a client I worked with. The names and specifics have been changed to protect the innocent, but this scenario is a common challenge I see within security organizations.

This organization set out to develop a set of KRIs to communicate with the board of directors. They started with metrics showing outgoing emails containing unprotected sensitive data (account numbers and credit card numbers) and the number of incoming malicious emails (malware or phishing attempts) blocked by their email controls.

These aren't KRIs; they're KPIs, and worse, they're operational metrics. You should never share raw operational metrics with your board. When presenting metrics, interpret their meaning and explain the story they tell.

Recognizing the need for more meaningful KRIs, the organization, with the help of consultants, refined their metrics to include the following:

- The number of outgoing automated emails containing unprotected sensitive data.
- The number of outgoing invoices sent via email containing unprotected sensitive data.
- The number of outgoing customer communications sent via email containing unmasked or unprotected credit card numbers.
- The number of incoming malware attempts not detected by Microsoft Office 365.
- The number of incoming phishing attempts not blocked by Microsoft Office 365.

Original KRI Description	Updated KRI Description
# of outgoing emails that contain unprotected highly sensitive data (i.e., Account numbers, credit card numbers)	# of outgoing ***automated*** emails that contain unprotected highly sensitive data (i.e., Account numbers, DOB)
	# of outgoing ***invoices*** sent via email containing unprotected highly sensitive data (i.e., Credit Card numbers, DOB)
	# of outgoing ***customer communications*** sent via email containing unmasked or unprotected credit card number
# of incoming malicious emails (i.e., malware or phishing attempts) that were detected and blocked by Microsoft Office 365	# of incoming malware attempts that ***were not detected*** by Microsoft Office 365
	# of incoming phishing attempts that ***were not blocked*** by Microsoft Office 365

FIGURE 15.1 KPI Example.

While these metrics were now "measurable and reportable", they remained operational KPIs, albeit more specific. In their next board meeting, they shared that they had 150,000 noncompliant emails. The board, lacking context, agreed the number sounded high and instructed the security team to fix it ASAP. This left the CISO in a difficult position, responsible for reducing noncompliant emails without the budget or buy-in to achieve it.

This is where I stepped in. I explained that they had created KPIs, not KRIs, and had made the mistake of sharing operational metrics with the board without providing the necessary context (Figure 15.1).

We started by converting those KPIs into KRIs. One simple way to achieve this is by setting a baseline target. This transforms a performance story into a compliance story. The new question is: How compliant are we against our target?

We developed two new KRIs:

- % Variance against a target for outgoing emails containing unprotected sensitive data per business unit.
- % Variance against target compliance for outgoing emails containing unprotected sensitive data.

Over the next quarter, noncompliant emails dropped from 150,000 to 90,000. While still a seemingly large number, we could now show a 40% reduction against the target of zero.

These were better metrics, but they still didn't tell the full story. What is the risk exposure from these activities and the impact of our efforts? To achieve this, we performed simple quantification by multiplying the number of noncompliant emails by the average cost per breach record (approximately $160, according to the Ponemon Institute).

With a 40% reduction of 150,000 emails, that's

*60,000 emails * $160/email = $9.6 Million Exposure Reduction,*

with $14.4 million in residual risk.

Updated KPI Description	Business KRI Translation
# of outgoing ***automated*** emails that contain unprotected highly sensitive data (i.e., passport numbers, DOB)	**Reduce risk exposure by measuring % variance against a target** for outgoing emails that contain unprotected highly sensitive data (i.e., account numbers, credit card numbers)
# of outgoing ***manifests*** sent via email containing unprotected highly sensitive data (i.e., passport numbers, DOB)	Per Business Unit % Variance against target compliance for outgoing emails that contain unprotected highly sensitive data (i.e., passport numbers, credit card numbers)
# of outgoing ***guest communications*** sent via email containing unmasked or unprotected credit card number	Target was set to 0, however current is 150K, so reduction target for Q1 was set to -40%
	Quantified Exposure Reduction # of emails that contain unprotected highly sensitive data (i.e., passport numbers, credit card numbers) multiplied by $160 (the average per record cost of a data breach per the Ponemon Intitute)
	Example: at target a 40% reduction of 150k emails. 60,000 emails x $160 = $9.6M Exposure Reduction

FIGURE 15.2 KRI Example.

Now, that was a compelling story. Communicating a $9.6 million risk reduction was impactful, but even more powerful was highlighting the remaining $14.4 million risk and showing which business units were the least compliant. This shifted the conversation, ultimately garnering budget for the program and buy-in for an organization-wide culture shift in email handling (Figure 15.2).

15.5 DEVELOPING MEANINGFUL METRICS

15.5.1 Developing Meaningful KPIs

Developing good KPIs can be challenging; a lack of clarity in defining KPIs and aligning them with your specific security goals and objectives can lead to confusion and misinterpretation. Data availability is another hurdle; you need access to reliable and relevant data to measure KPIs accurately. Additionally, quantifying certain security aspects can be inherently difficult, making it a challenge to develop truly meaningful KPIs.

To overcome these obstacles, collaborate with key stakeholders throughout the KPI development process. Involving them ensures the metrics align with business priorities and resonate with decision-makers. Leveraging security tools and platforms that offer built-in reporting and analytics can simplify data collection and analysis. Above all, focus on KPIs that provide actionable insights that not only track performance but also guide improvements to your security strategy. Choose KPIs that not only measure performance but also provide clear guidance for improvement and optimization of your security program.

While KPIs are useful for measuring progress, they have their limitations. They often focus on past performance, which may not be a reliable predictor of future outcomes. KPIs also tend to measure specific areas in isolation, which can overlook broader risks or vulnerabilities. Also note that without clear definition and communication, there's a risk that KPIs could be misinterpreted or manipulated.

To develop impactful KPIs, follow a structured process:

1. Define your security program's goals and objectives with precision.
2. Identify key metrics that directly measure progress toward these goals.

3. Establish baseline values for each KPI to facilitate tracking progress over time.
4. Leverage your security tools and platforms to collect and analyze the relevant data.
5. Communicate the findings regularly to stakeholders, highlighting successes and areas for improvement.

It's important to be selective in your KPI development. Focus on tracking only the most relevant and impactful indicators that align directly with your organization's long-term strategic objectives and overall direction.

Several data sources can inform KPI development. Security tools and platforms like SIEM, endpoint detection and responses (EDRs), and vulnerability scanners often offer built-in reporting and analytics features. Incident response data can track metrics related to incident detection, response, and recovery times. Security awareness training platforms provide insights into employee engagement and knowledge retention through training completion rates and assessment scores. Third-party risk assessments can gather data on vendor performance and compliance, informing risk-related KPIs.

15.5.2 Common Metric Data Sources

- Networks (intrusion detection/protection systems (IDS/IPS), network access control (NAC), malware detection)
- Data (DLP, mail filtering, gateway firewalls, threat intel, and content security)
- People (IDS/IPS, NAC, lightweight directory access protocol (LDAP) or other identity access management, user provisioning, ticketing, and cybersecurity awareness)

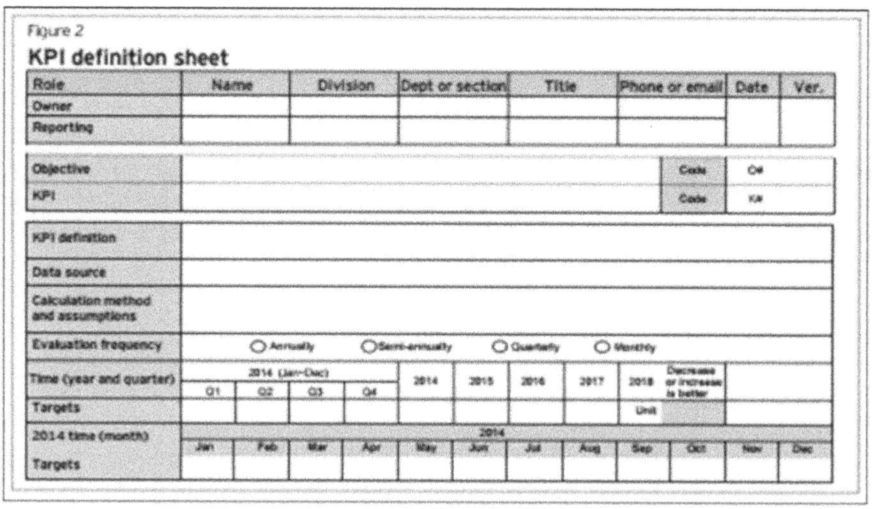

FIGURE 15.3 KPI Definition Sheet.

- Applications cloud/app security (Azure, AWS, and Blue Coat)
- Hosts (asset and vulnerability management, endpoint security, and patch management)
- Other (manual survey and custom applications)

Here is a list of example KPIs and their formulas (Figure 15.3):

1. **Mean Time between Incidents (MTBI):** This represents the average time, measured in days, that elapses between verified security incidents. It is calculated from the moment a formal incident case is created until the next incident occurs, encompassing the time needed for detection and triage.
2. **Percent Difference in MTBI (Monthly):** This KPI tracks the month-to-month change in the MTBI for a specific group of incidents, expressed as a percentage.
   *((Current Month MTBI – Previous Month MTBI)/Previous Month MTBI) * 100*
3. **Mean Time to Detect (MTTD)/Dwell Time:** This measures the average time, in hours, days, weeks, or months, it takes to detect malicious activity. It's calculated from the timestamp of the first evidence of the activity to the date and time it's detected, and a formal incident case is created.
4. **Percent Difference in MTTD (Monthly):** Similar to MTBI, this KPI tracks the month-to-month change in MTTD, measured as a percentage.
   *((Current Month MTTD – Previous Month MTTD)/Previous Month MTTD) * 100*
5. **Alarm Time to Triage (TTT):** This KPI measures the latency in your team's ability to inspect an alarm immediately, providing insight into real-time threat responsiveness. It's measured within alarm priority bands (e.g., high/medium/low or risk score bands) and can indicate whether your team has the capacity to handle additional monitoring or needs to narrow its focus.
   TTT = Date/Time Alarm Inspection – Date/Time of Alarm Creation
6. **Alarm Time to Qualify (TTQ):** This metric tracks how long it takes to thoroughly review and classify an alarm, offering insight into your team's ability to process and assess potential threats. By breaking it down by alarm priority and outcome (such as false positive, benign issue, or actual incident), you can pinpoint inefficiencies or capacity issues. It also highlights areas where your security tools may fall short in providing effective search, data analysis, or contextual information needed for quick, accurate threat assessment.
   TTQ = Date/Time of Alarm Closure or Addition to Case – Date/Time of Alarm Creation
7. **Threat Time to Investigate (TTI):** This key performance indicator tracks the time it takes your team to thoroughly investigate a verified threat, providing insight into your team's capacity to manage different types of incidents. By categorizing investigations based on threat types, such as those outlined in MITRE ATT&CK, you can uncover areas where your process or tools may be slowing down the investigation. This metric also helps pinpoint inefficiencies in your security operations platform, such as issues

with search functionality, data analysis, or collaboration tools that could be hindering faster response times.

TTI = Date/Time of Case Closed or Elevated to Incident – Date/Time of Case Creation

8. **Time to Mitigate (TTM):** This metric tracks how long it takes to mitigate an incident and neutralize the immediate risk to the organization, reflecting your team's ability to contain or slow down an active threat quickly. TTM should be measured based on different threat and incident types, and it can reveal delays within your technology stack, such as challenges with evidence capture, playbook execution, automation, or team collaboration.

   TTM = Date/Time Incident Mitigated – Date/Time Incident Determination
9. **Time to Recover (TTR):** This measures how long it takes for your team and other departments to fully recover from an incident, providing insight into the efficiency of your overall incident recovery process. It can highlight operational bottlenecks or collaboration issues across different teams. Like TTM, this metric should be analyzed by threat type to pinpoint slowdowns in areas such as evidence handling, playbook execution, or team coordination.

   TTR = Date/Time of Recovery from Incident – Date/Time of Incident Mitigation
10. **Incident Time to Detect (TTD):** This key performance indicator measures how long it takes to detect and confirm an incident after it initially occurs, offering insight into the effectiveness of your security operations. TTD should be broken down by threat types and detection methods (e.g., threat hunting and behavioral analytics), and it can help uncover weaknesses in your detection and qualification capabilities.

    TTD = Date/Time Threat Qualified for Investigation/Case Creation – Date/Time of Initial Indicator of Threat
11. **Incident Time to Response (TTIR):** This measures the time taken to investigate and mitigate a confirmed incident, another key indicator of security operations effectiveness. It should be measurable based on threat/incident types and might highlight slowness or weakness in your technology solution's threat investigation and mitigation capabilities.

    TTIR = Date/Time of Incident Mitigation – Date/Time Initiated of Investigation
12. **Security Ticket – Mean Time to Resolution:** This metric tracks the average time, in minutes, it takes for the security team to fully resolve or close a security ticket, starting from the moment it's submitted until the ticket is officially closed. It provides insight into the team's efficiency in addressing security concerns.
13. **Security Ticket – Total Number of Requests Opened:** This is the total number of security requests or tickets submitted to the security team within a defined timeframe. The count begins as soon as a request is received, offering a clear view of workload volume.
14. **Security Ticket – Percentage of Requests Not Resolved within SLA:** This metric reflects the proportion of security tickets that remain unresolved beyond the agreed service-level agreement (SLA) timeframe,

shown as a percentage of the total number of issues resolved during that period. It highlights potential challenges in meeting response deadlines and helps identify areas for process improvement.

15. **Percent Difference in Critical Vulnerabilities (Monthly):** The month-to-month change in the total number of common vulnerabilities and exposures (CVEs) discovered on systems rated critical was measured as a percentage for the group of systems being examined.
16. **Percentage of Critical and Sensitive Systems without Up-to-Date Patches/ Patch Coverage Rate:** The total number of systems flagged as critical or sensitive (all deployed instances) without up-to-date patches, expressed as a percentage of total critical system end-user devices/workstations.
17. **Total Number of Critical System Backup Failures:** The total number of critical system backup processes that failed (didn't run, weren't captured in full, had errors, etc.) during the measurement period.

Remember, these are just examples, and you should tailor your KPIs to your organization's specific needs and goals.

15.5.3 Developing Meaningful KRIs

Developing effective KRIs is a more complex task than developing KPIs, as it necessitates a deep understanding of your organization's risks and its most critical assets. The first step involves identifying these critical assets and processes that require protection. Next, you need to assess the potential threats and vulnerabilities that could impact them. Based on this assessment, define specific metrics or risk indicators that can provide early warning signals of potential risks. For each KRI, establish thresholds that, when exceeded, trigger alerts or actions. Then, continuously monitor KRIs and report findings to relevant stakeholders, enabling proactive risk mitigation.

Various data sources can be leveraged to inform the development of KRIs. Threat intelligence feeds can provide valuable information about emerging threats and vulnerabilities, while security incident data analysis helps identify patterns and predict future risks. Vulnerability scan results enable you to track the number and severity of vulnerabilities in your systems and applications. Third-party risk assessments evaluate vendor performance and compliance to identify potential supply chain risks. Also, industry benchmarks and reports offer a valuable context for comparing your organization's risk profile to industry standards and best practices. By themselves, most of the metrics directly from these data sources will be KPIs but can be the foundation for crafting meaningful KRIs by establishing thresholds and tying them to quantified risks.

For KRIs to be truly effective, they should possess several key characteristics. They need to be measurable, using quantifiable metrics such as numbers, counts, percentages, or dollar amounts. They should be predictable, providing early warning signals of potential risks. They should be comparable, allowing you to track trends over time. Lastly, KRIs should be informational, offering insights into the current status of risks and controls.

15.5.4 KRI Roadmap

Below is a high-level roadmap for establishing a KRI framework (Figure 15.4):

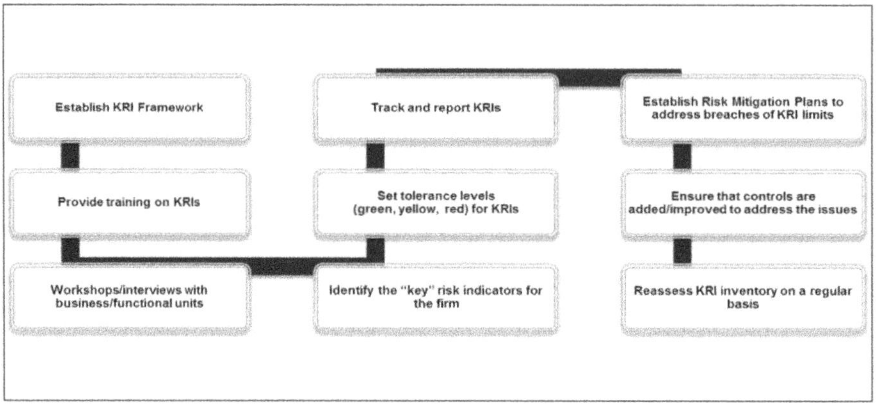

FIGURE 15.4 KRI Process.

15.5.5 KRI Identification

The first step in identifying KRIs is to review the metrics your organization is already tracking. This helps establish a baseline and spot gaps in your current risk monitoring efforts. From there, it's important to assess how relevant and effective these metrics are at predicting and measuring risks. If gaps are found, new metrics should be created or existing ones refined to better align with the organization's risk landscape. Setting baselines and targets for each KRI is key to making comparisons over time and tracking progress.

Another key aspect of KRI identification is conducting risk and control self-assessments (RCSAs) and interviewing business units, as described in Chapter 6. These activities help uncover potential risks and identify key indicators that can signal changes in the risk profile or the effectiveness of controls. However, it's important not to rely solely on RCSAs and interviews. Instead, focus on indicators that provide genuine insights into changes in risk and control effectiveness rather than tracking a large number of metrics that may be less useful.

When identifying KRIs, it's essential to concentrate on the highest risks and their root causes. Consider both forward-looking (leading) and historical (lagging) indicators to gain a thorough understanding of the risks. The types of metrics used can vary, including absolute values, numbers, ratios, percentages, and aging, depending on the nature of the risk being measured. Collecting data on KRIs consistently is essential. While monthly reporting is typical, the frequency should be determined by the nature of the risk. Quantifying the data – especially in terms of monetary impact – can improve communication and decision-making across the organization.

When selecting KRIs, ensure they are measurable, meaningful, and predictive. Tracking too many indicators can lead to inefficiency and make it hard to manage the process effectively. Focus on those that provide the most valuable insights into your organization's risk profile.

Once KRIs have been selected, you need to determine and validate trigger levels or thresholds for each. These thresholds should be based on industry tolerance levels or the organization's internal risk appetite, and they should ideally be approved by the board of directors. The thresholds should also align with the organization's risk appetite statement, ensuring consistency in risk management practices.

Once KRIs are selected, set thresholds or trigger levels for each one. These should align with the organization's risk appetite and be approved by the senior management or the board. Thresholds must be carefully calibrated to reflect both internal risk tolerances and industry standards.

KRIs should be tracked regularly, with the frequency depending on the specific risks involved. Some might need daily or weekly attention, while others may only require monthly monitoring. Reporting KRI outcomes to management should be streamlined to avoid information overload, focusing on the most critical risks and their potential impacts. Escalation processes should be in place to address any breaches of thresholds, with different KRIs having varying levels of escalation based on the severity of the risk.

As described in Chapter 6, risk treatment plans (RTPs) should be developed for high-risk items identified through KRI monitoring. For risks where their RTP includes some form of mitigation, you should develop a risk mitigation plan (RMP). These plans should outline specific actions to be taken to mitigate the identified risks and enhance the effectiveness of existing controls. It's important to track the progress of RMPs to ensure that controls are enhanced and risks are mitigated effectively. Progress on these plans should be tracked and reported to the management and the board to ensure accountability and transparency in risk management.

Clear roles and responsibilities are required for successful KRI management. Risk management teams should lead the development of the KRI framework, provide guidance, and oversee reporting and escalation when thresholds are breached. Business units play a role in selecting relevant KRIs, setting appropriate thresholds, and monitoring the data. The internal audit function provides assurance and validation, incorporating KRIs into their audit plans to assess the effectiveness of controls when risks breach or approach thresholds.

Establishing and maintaining an effective KRI framework can be challenging. Some common obstacles include gaining buy-in from business units, demonstrating the positive impact of KRIs on the organization, potential challenges in identifying relevant KRIs, and the need for adequate resources to track and manage them effectively. Overcoming these challenges requires strong leadership, collaboration across different functions, and a commitment to continuous improvement in risk management practices (Figures 15.5–15.7).

15.5.6 Example metrics

Patch policy compliance	Cyber budget as % of IT
% Systems w/severe vulns	Budget allocation
% Critical applications	% of incidents detected internally
Risk assessment coverage	% of security exceptions
Security testing coverage	% Changes with security review

ASSOCIATED RISK	MEASURABLE KRI	NATURE	WHY YOU SHOULD TRACK THIS
Vendor service interruption	Number of applications currently running in the organization without a service level agreement (SLA).	All	Without an SLA, your organization may be engaging with a high-risk vendor. The vendor may not adhere to your regulations or they could end service at any moment, causing a disruption in the business.
ISP failure	Number of ISP outages.	Leading	High numbers of outages can be an indicator that it's time to change providers. Especially if you provide online services, outages can mean business comes to a full stop.
Loss of data	Number of system backup failures.	Lagging	New or upgraded software can cause backup failure, or there could be misconfigurations due to overly customizable software that result in backup failures.
Lack or misappropriation of IT budget	Total discrepancy (dollars) of IT budget versus actual.	Lagging	Over-spending in IT can mean critical or new tools go unfunded. Under-spending can mean IT is overlooking important investments, or isn't budgeting accurately.
Lack or misappropriation of IT personnel	Average amount of time to resolve IT support requests.	All	Higher time to close tickets can indicate a lack of resources, which may lend itself to larger, undiscovered issues which could cause business interruptions.

FIGURE 15.5 KRI Examples 1.

ASSOCIATED RISK	MEASURABLE KRI	NATURE	WHY YOU SHOULD TRACK THIS
Terminated employees accessing systems	Average time between employee termination and disabling of accounts/ termination of access to all systems.	Lagging	Allowing terminated employees to continue to access data and systems could lead to serious data breaches.
Unaddressed critical incidents	Time to resolve a critical incident and the number of critical incidents.	Lagging	Extended time to resolve a critical incident may infer that the organization's critical incident procedure requires an overhaul.
Loss of hardware/physical assets	Number of company-issued phones without monitoring software installed.	Lagging	Monitoring software can locate a lost or stolen phone, and wipe the data before it gets into the wrong hands. All company-issued phones should have this software installed.
Anonymous data leak	Number of active default database administrator accounts.	Leading	Pre-configured default database administrator accounts means if an event were to happen, you can't tie it back to an individual and resolve the issue.
Breach of GDPR compliance	Time to respond to requests for personal data.	Lagging	Massive fines can be issued for organizations who breach GDPR. This could cause serious financial and reputational damage.

FIGURE 15.6 KRI Examples 2.

ASSOCIATED RISK	MEASURABLE KRI	NATURE	WHY YOU SHOULD TRACK THIS
Shared login credentials	Number of concurrent system logins using the same ID.	Lagging	Could indicate an employee has shared their login credentials with an unauthorized individual who shouldn't have access to confidential information.
Improper security assignments	Total number of users with similar roles but dissimilar security assignments.	All	This could indicate that one employee may be accessing customer data files that they shouldn't.
Malware	Number of employees who click on IT-sent phishing emails.	Leading	By setting up and testing employees with fake phishing emails, you can identify those employees that require additional security training.
Employees unaware of what defines confidential information	Pass/fail results for employee information security training.	Leading	Employees who fail or don't complete security training regularly increase the risk of customer information being shared.
Non-compliance and data breaches	Frequency of review of high, elevated (privileged) permissions on IT systems.	Lagging	These accounts are more likely to be targeted by cyber attackers to gain access to confidential or customer data.

FIGURE 15.7 KRI Examples 3.

15.5.7 Project Execution KRIs

1. **Planned Value**
   Also known as budgeted cost of work scheduled (BCWS), this is the estimated cost of project activities that are planned or scheduled by a certain reporting date. You can compare planned value (PV) with other key metrics to see if your project is on track.
   PV = (The Hours Left Scheduled on the Project) x (Project Worker's Hourly Rate)
2. **Actual Cost**
   Also called actual cost of work performed (ACWP), this metric shows how much money has been spent on the project so far, including any time and materials that have been used or billed.
   Actual Cost (AC) = Sum of All the Project-Related Expenses to Date
3. **Earned Value**
   This is the approved budget for all completed project work and activities as of a specific date. It helps you see what work has been completed and the budget used to finish that work.
   Earned Value (EV) = Cost of Completed Work
4. **Cost Variance**
   This metric compares the planned budget to the actual budget at a given point in time. It shows whether the project is over or under budget compared to the original forecast.
   Cost Variance (CV) = Planned Budget versus Actual Budget
5. **Scheduled Variance**
   Scheduled variance (SV) measures the difference between EV and PV. It helps determine if the project is running ahead or behind schedule by comparing the budget spent to the estimated budget at the start.
   $SV = EV - PV$

15.5.8 Security Program KRIs

1. **Control Effectiveness**
   Leverages key control indicators (KCIs) to measure the percent reduction to the likelihood or impact of a stated risk by applying a mitigating control.
   Control Effectiveness (CE) = % Reduction of Specific Risk Likelihood and/or Impact
2. **Residual Risk**
   Measures the amount of inherent risk that remains after controls are applied.
   *Residual Risk = Inherent Risk * CE*
3. **Loss Exceedance Curve**
   This graphical measurement compares risk tolerance, inherent risk, and residual risk to determine if the current risk exposure falls within the acceptable risk limits defined in the formal risk appetite statement.
   Loss Exceedance Curve (LEC) = (Risk Tolerance) + (Inherent Risk) + (Residual Risk) Line plotted on a graph where the likelihood is the x-axis and financial impact is the y-axis.
4. **Return on Security Investment**
   This metric compares the total benefits of a security project to its costs. A positive return on security investment (ROSI) means the benefits outweigh the expenses, indicating a worthwhile investment.
   Return on Investment (ROI) = AC and EV Metrics versus Risk Reduction Measures
5. **Risk Buy-Down**
   Compares the cost of control to the amount of risk reduction provided by that control.
   *Risk Buy-Down = AC * CE*
6. **Unallocated License Ratio (or Unused Software Ratio)**
   This metric can be used to determine tool utilization. It highlights software purchases that aren't being used and, therefore, aren't adding value to the organization. Ideally, this ratio should be as close to zero as possible, indicating that most, if not all, purchased licenses are being fully utilized.
   Unallocated License Ratio = (Unused License Quantity)/(Total License Quantity)
7. **Post-Implementation Risk Assessments**
   Continuous risk audit – a targeted NIST cybersecurity framework (CSF) risk audit will be internally conducted after each improvement phase.
8. **Risk Reassessment**
   A NIST CSF risk assessment will be conducted by a third-party security firm after the final phase of the improvement plan.
9. **Post-Implementation Security Posture Assessments**
   Continuous attack, surface management penetration tests will be conducted against a static subset of systems and network devices. These assessments will expose security posture changes as controls are deployed.

Targeted penetration assessments will be conducted by a third-party security firm against any modified control that contains access rules or access control lists.

A third-party security firm will conduct penetration assessments annually.

15.5.9 Additional KRI Ideas

Insider Threat Contributors (Metrics Most Used in Insider Scenarios)

- # or % database administrators
- Security awareness training levels
- DLP metrics

Environmental/Hygiene Exposure Factors (Overall and by Platform)

- Aggregate platform hygiene performance (vulns, footprint, etc.)
- Platform-specific hygiene (specific to vendor or technology types)

Third-Party Exposures (Metrics Most Used in Third-Party Risk Scenarios)

- Partner security updates (recommended, dependent on available data)
- # of missing or inadequate third-party responses
- % of vendors' assessments not updated annually
- Business continuity plan vendor survey status

15.5.10 Metric Development Frameworks: DIKW and GQIM

Several frameworks can guide the development and communication of meaningful metrics. Two helpful frameworks that can aid in the development are the DIKW Pyramid and the GQIM (goals, questions, indicators, and metrics) Framework.

15.5.11 The DIKW Pyramid

The *DIKW Pyramid*, a concept often credited to Russell Ackoff, represents the progression from *Data* to *Information*, *Knowledge*, and, finally, *Wisdom*. It shows how raw data can be transformed into meaningful insights that support better decision-making.

Data: These are raw facts and figures collected from various sources but not yet organized.

Information: When data is structured and given context, it becomes useful and relevant.

Knowledge: This is the result of analyzing and interpreting information, helping you recognize patterns and relationships.

Wisdom: The ability to apply knowledge to make informed, data-driven decisions and take effective action.

The DIKW Pyramid is especially useful when developing metrics. It reminds us that raw data alone doesn't lead to good decisions. To be useful, data must be organized, analyzed, and interpreted in a way that aligns with your security goals and objectives, turning it into actionable insights.

15.5.12 Applying DIKW to Create KPIs and KRIs

Data: Collect raw data from various sources, such as security tools, incident reports, and vulnerability scans.

Information: Organize and structure the data, categorizing it by type, source, and relevance to security objectives.

Knowledge: Analyze the information to identify patterns, trends, and potential risks.

Wisdom: Use the knowledge gained to develop meaningful KPIs and KRIs that inform decision-making and drive improvements in the security program.

15.5.13 The GQIM Framework

The GQIM framework, originally developed by Gartner, stands for goals, questions, indicators, and metrics. It provides a structured approach to aligning metrics with strategic goals, ensuring that the metrics you develop are relevant and actionable and contribute to the organization's overall success. The GQIM framework is particularly useful in the early stages of metric development, helping ensure that your metrics are directly linked to your strategic goals.

The first step in the GQIM framework involves establishing clear and well-defined cybersecurity goals. These goals could range from reducing the risk of data breaches to improving the efficiency of incident response times. Once these goals are established, the next step is to formulate key questions that can help assess the progress made toward achieving them. These questions should be specific and measurable, enabling you to track your performance and identify areas that need attention. For instance, if your goal is to reduce data breaches, you might ask, "How often are we experiencing data breaches?" or "What are the primary causes of these breaches?"

Once you have these questions, the next step is to identify the specific indicators that can provide answers. These indicators should be measurable and directly related to your goals, offering insights into both your performance and potential risks. For example, the number of data breaches or the mean time to respond to incidents could serve as indicators for the questions mentioned earlier. The final step is to develop quantifiable metrics to measure these indicators and track progress over time. These metrics become your KPIs, providing a tangible way to assess your performance and identify areas for improvement. Finally, you can translate these KPIs into forward-looking KRIs by considering the potential impact of negative trends or deviations from your established targets. This allows you to anticipate and proactively address emerging risks before they escalate into material incidents (Figures 15.8–15.9).

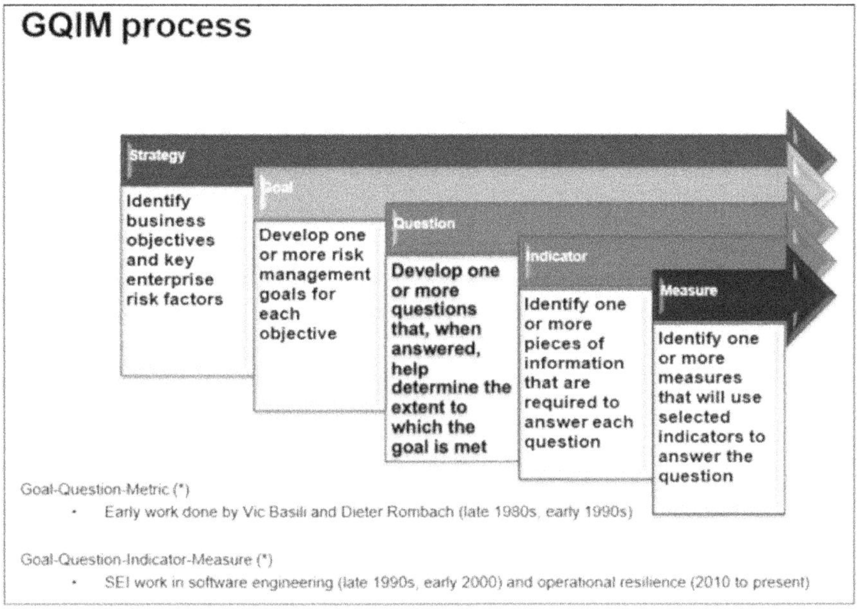

FIGURE 15.8 GQIM Process.

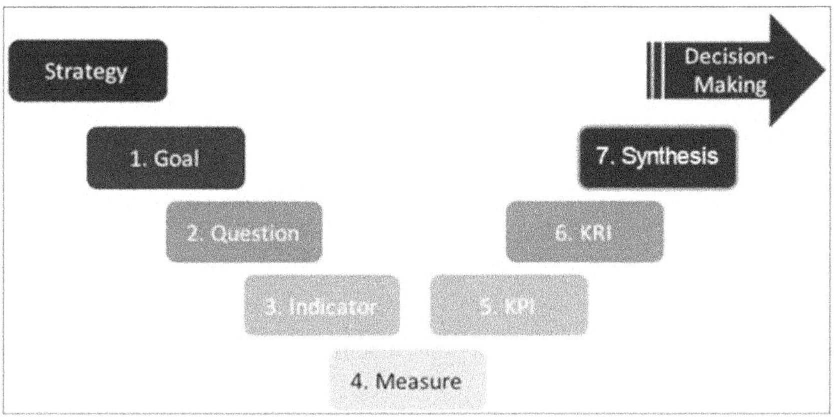

FIGURE 15.9 GQIM Flow.

15.6 HOW TO COMMUNICATE KRIS

The effectiveness of KRIs depends heavily on clear and impactful communication. The CISO 3.0 recognizes the importance of tailoring messages for different audiences to ensure the insights from KRIs are both understood and actionable. Effective KRI communication requires not only clarity but also alignment with the priorities of diverse stakeholders.

15.6.1 Tailoring KRI Communication to Stakeholder Audiences

To maximize the impact of KRI reporting, it is essential to align the communication approach with the needs and priorities of different audiences:

1. Executive Communications (for the Board of Directors and Executive Leadership)

What to Share

- KRIs that highlight critical risks and their potential business impacts.
- Summaries of significant incidents and strategic recommendations for risk mitigation.
- Trends in risk exposure and their alignment with business objectives.

Why It Matters

The board and executives are focused on high-level oversight and strategic decision-making. KRIs provide a concise view of the organization's risk posture and how it aligns with business goals, financial performance, and regulatory expectations.

How to Present

- Use plain language and focus on strategic impacts.
- Provide visual aids such as heatmaps, dashboards, and summary charts for clarity.
- Emphasize alignment with business objectives and financial implications.

When to Share

- Quarterly to coincide with board meetings and strategic planning cycles.
- Additional updates following major incidents or regulatory changes.

2. Cyber Operations Reports (for Cybersecurity Teams and IT Management)

What to Share

- KPIs such as incident response times, patch management effectiveness, and threat detection rates.
- Detailed analyses of incidents, including root causes and resolutions.
- Updates on ongoing security initiatives and technical performance metrics.

Why It Matters

Operational teams require detailed metrics and analyses to assess the effectiveness of their efforts, prioritize tasks, and ensure alignment with organizational strategies.

How to Present

- Include technical details in structured formats such as tables and graphs.
- Provide actionable insights to guide team improvements and decision-making.
- Highlight performance trends and areas needing attention.

When to Share

- Weekly or monthly, depending on the scope of activities and operational goals.
- Immediately after significant incidents to facilitate post-incident reviews.

3. Compliance and Audit Briefs (for Legal Teams, Auditors, and Regulators)

What to Share

- Documentation demonstrating adherence to relevant laws, regulations, and standards (e.g., GDPR, HIPAA, and NIST).
- Evidence of control effectiveness, audit results, and remediation actions.
- Updates on regulatory changes and their implications for the organization.

Why It Matters

Compliance and legal stakeholders need comprehensive, evidence-based reporting to demonstrate regulatory adherence, prepare for audits, and mitigate legal risks.

How to Present

- Provide detailed documentation, compliance matrices, and audit findings in organized formats.
- Summarize complex regulatory changes into actionable updates.
- Highlight remediation plans for any identified gaps.

When to Share

- Annually or biannually, aligning with audit cycles and regulatory deadlines.
- Additional updates shared as needed after regulatory changes or audit findings.

4. Incident Summaries (for Incident Stakeholders across Departments)

What to Share

- Detailed accounts of incidents, including their scope, causes, and affected systems.
- Actions taken during the response, timelines, and resolutions.
- Impact analyses, including financial, reputational, and operational consequences.
- Lessons learned and recommendations for preventing future incidents.

Why It Matters

Cross-departmental stakeholders need transparency about incidents to understand their implications, foster collaboration, and improve future response efforts.

How to Present

- Use structured incident timelines and root cause analyses.
- Summarize technical details for nontechnical audiences when applicable.
- Clearly outline lessons learned and next steps in a dedicated section.

When to Share

- Immediately following an incident and as part of a post-incident review process.

5. Security Awareness Updates (for All Employees)

What to Share

- Updates on phishing attempts, security training participation rates, and emerging threats.
- Success stories of employees mitigating risks, such as reporting phishing attempts.
- Guidelines for best practices and reminders of security policies.

Why It Matters

Engaging employees builds a security-aware culture, reinforcing their role in protecting the organization from threats and reducing human error.

How to Present

- Use engaging formats such as newsletters, infographics, and short videos.
- Tailor content to specific roles and responsibilities within the organization.
- Highlight positive examples to encourage proactive behaviors.

When to Share

- Monthly or quarterly, based on the frequency of updates and organizational needs.

6. Project Progress Reports (for Senior Management and Project Stakeholders)

What to Share

- KPIs tracking the progress of specific cybersecurity initiatives or projects (e.g., percentage of systems patched, completion rates of security assessments, or milestones achieved in implementing new security technologies).

- Budget utilization, resource allocation, and timelines.
- Risk assessments tied to project delays or incomplete tasks.

Why It Matters

Project stakeholders and senior management need to understand how cybersecurity projects are progressing relative to their objectives, timelines, and budgets. This ensures that initiatives remain aligned with organizational priorities and resources are effectively allocated.

How to Present

- Use concise project dashboards with clear visuals showing progress, performance metrics, and milestones.
- Highlight any risks or delays, along with mitigation strategies.
- Provide actionable next steps to maintain momentum.

When to Share

- Monthly or at key project milestones to ensure alignment and transparency.
- Additional updates as needed for high-priority or delayed projects.

7. Vendor Performance Reports (for Procurement and Risk Teams)

What to Share

- KPIs evaluating third-party vendor performance, such as adherence to SLAs, incident response times, and compliance with cybersecurity requirements.
- Results from vendor risk assessments and audits.
- Trends in vendor-related incidents or vulnerabilities.

Why It Matters

Procurement and risk teams must monitor vendor performance to ensure third-party relationships do not introduce unnecessary risks or compliance issues.

How to Present

- Use comparative tables or scorecards to rank vendors against performance benchmarks.
- Include summaries of key findings and areas for improvement.
- Provide recommendations for vendor remediation or contract adjustments.

When to Share

- Quarterly or semi-annually, aligned with contract review periods or risk reassessments.
- Immediately after significant vendor-related incidents.

8. Budget and Resource Allocation Reports (for Finance and Executive Teams)

What to Share

- KPIs demonstrating ROI for cybersecurity investments, such as cost savings from avoided incidents or improvements in operational efficiency.
- Metrics linking resource allocation to risk reduction efforts.
- Forecasts for future resource needs based on performance trends and risk assessments.

Why It Matters

Finance and executive teams need to see the financial impact and value of cybersecurity spending to justify budgets and align them with strategic goals.

How to Present

- Use financial dashboards that connect cybersecurity KPIs to business outcomes.
- Highlight cost–benefit analyses of specific investments.
- Include recommendations for reallocating resources to high-priority areas.

When to Share

- Annually during budget planning cycles, with additional updates as needed for major investments or reallocations.

9. Customer Trust and Assurance Reports (for External Customers and Partners)

What to Share

- KPIs showcasing operational reliability, such as system uptime, secure transaction rates, and compliance achievements.
- Summaries of incident management metrics, emphasizing timely resolutions and preventive measures.
- Updates on certifications or accreditations that validate the organization's cybersecurity posture.

Why It Matters

Customers and partners expect transparency about the organization's commitment to security. Sharing relevant KPIs builds trust and reinforces relationships.

How to Present

- Use concise, customer-friendly formats such as reports, infographics, or executive summaries.
- Focus on outcomes and proactive measures rather than technical details.
- Provide contextual explanations of how security initiatives enhance reliability and trust.

When to Share

- Annually or during customer reviews, with additional updates as needed for major security milestones.

10. Workforce Performance Reports (for HR and Department Leaders)

What to Share

- KPIs related to employee engagement in security awareness programs, such as training completion rates or phishing test results.
- Metrics assessing the effectiveness of internal communication campaigns around cybersecurity.
- Data on internal compliance with security policies and procedures.

Why It Matters

HR and department leaders play a critical role in fostering a culture of security awareness. Understanding workforce performance helps identify gaps and improve employee engagement.

How to Present

- Use summary reports with visual aids highlighting workforce participation and trends.
- Break down data by department to provide actionable insights.
- Include recommendations for targeted training or awareness initiatives.

When to Share

- Quarterly or aligned with organizational performance reviews.

15.6.2 Principles of Effective KRI Communication

When presenting KRIs, clarity and simplicity are key. Avoid technical jargon that might alienate nontechnical audiences. Instead, use plain, straightforward language that connects with everyone. Focus on why these KRIs matter, how they align with the organization's strategic goals, and what financial impact they may have. Connecting KRIs to business outcomes helps capture stakeholders' attention and shows the value of your risk management efforts.

You should never present raw metrics and data without context. You should always explain how to interpret the data and ensure that any metrics you include support the story that you are trying to tell. This doesn't mean cherry-picking data to support a pre-determined narrative, but rather, if your metrics don't support your story, maybe you need to adjust your story. It's essential that you provide actionable insights based on KRI trends and thresholds. Don't just present data; offer recommendations for action that empower stakeholders to make informed decisions. This proactive approach demonstrates your leadership and commitment to mitigating risks before they impact the business.

Visual representations, such as charts, graphs, and other visual aids, can enhance the clarity and impact of your KRI communication. Good visuals allow stakeholders to quickly grasp complex information and identify trends, facilitating understanding and discussion. We'll discuss data storytelling more in the next chapter on effective board-level communication.

A CISO 3.0 uses well-designed KRIs to build a culture of risk awareness and promote proactive risk management across the organization. Effective communication is critical to securing buy-in for security initiatives. When speaking to executive leadership, focus on the business impact of the risks identified through KRIs. Translate technical details into clear, practical terms and emphasize the potential financial and operational consequences of inaction. This empowers executives to make smart decisions around resource allocation and risk mitigation.

When communicating with the board of directors, take a broader, more strategic approach. Highlight the organization's overall risk posture and how the security program is addressing the most pressing risks. This not only builds confidence in your leadership but also ensures the board is well informed about the company's exposure to cyber risks.

15.6.2.1 Examples of KRI communication in action

- **Strategic Risk Reporting:**
  "Our KRI for third-party vendor risk has exceeded the threshold, indicating a potential increase in supply chain vulnerabilities. If unaddressed, this could disrupt operations and result in a revenue loss estimated at $X million. We recommend conducting a vendor risk assessment to identify and mitigate vulnerabilities".
- **Operational Reporting:**
  "The KRI for phishing attempts has been trending upwards, suggesting an increased risk of social engineering attacks. If successful, these attacks could result in regulatory fines, legal fees, and customer remediation costs totaling up to $X million. Implementing additional security awareness training can address this risk and reduce exposure".

15.7 CONCLUSION

In this chapter, we examined cybersecurity metrics, unraveling the mysteries of KPIs and KRIs and exploring how to craft them into a language that resonates in the boardroom. We've looked at real-world examples, shared formulas, and data sources and stressed the importance of aligning your metrics with your strategic goals. It's not only about collecting data but also about turning that data into a story that speaks to the heart of the business.

The CISO 3.0 is not only a number cruncher but also a storyteller, a translator who can bridge the gap between the technical world and the executive suite.

Once you have a firm grasp of metrics, they can inform your decision-making, showcase the value of your security program, and get the buy-in you need to protect your organization. It's about using data to drive action, not just collect dust on a shelf.

In the ever-changing landscape of cybersecurity, the ability to harness the power of data and communicate its impact is what sets the CISO 3.0 apart. It's about navigating this complex world with confidence, proactively managing risks, and ensuring that your organization not only survives but also thrives.

15.7.1 Key Takeaways

1. **The Strategic Value of Metrics in Cybersecurity**
   - Metrics go beyond capturing snapshots of security posture; they track progress, identify trends, and guide resource allocation.
   - The CISO 3.0 uses metrics to communicate cybersecurity's value to stakeholders, align with business goals, and justify strategic investments.
2. **KPIs and KRIs: Essential Tools for Measurement**
   - **KPIs** evaluate past performance and operational effectiveness, focusing on areas like incident detection and response, vulnerability management, and security awareness.
   - **KRIs** are proactive, forward-looking metrics that identify potential risks and anticipate threats, serving as early warning signals for emerging vulnerabilities.
3. **Developing Effective Metrics**
   - Developing meaningful metrics requires clear objectives, relevant data sources, and collaboration with stakeholders.
   - KPIs should measure specific operational goals, while KRIs must assess and predict risks with actionable thresholds.
   - Frameworks like the *DIKW Pyramid* and *GQIM* provide structured approaches to designing and implementing impactful metrics.
4. **Challenges in Metric Development**
   - Common challenges include selecting meaningful metrics, ensuring data reliability, and bridging technical and business perspectives.
   - Avoid pitfalls like misaligned reporting or overloading dashboards with irrelevant data by focusing on actionable and aligned metrics.
5. **Communicating Metrics Effectively**
   - Tailor communication strategies for different audiences:

   Executives: Focus on strategic impacts and alignment with business goals.

   Security Teams: Provide detailed operational insights and actionable recommendations.

   Regulators: Highlight compliance and control effectiveness with evidence-based reporting.

- Use visuals like dashboards, heatmaps, and charts to simplify complex data and foster understanding.

6. **Real-World Application of Metrics**
    - Translating raw metrics into meaningful narratives, such as financial impact or operational improvements, resonates with stakeholders.
    - Metrics like ROI, incident reduction rates, and cost–benefit analyses of risk mitigations demonstrate the tangible benefits of cybersecurity investments.
7. **Continuous Improvement**
    - Regularly review and refine metrics to adapt to evolving threats and organizational changes.
    - Use lessons learned from incidents and stakeholder feedback to enhance the effectiveness and relevance of metrics.

15.7.2 Reflection and Exploration Questions

Understanding the Concepts

1. What are the differences between KPIs and KRIs, and how do they complement each other in a cybersecurity strategy?
2. Why is it important to align metrics with organizational goals and stakeholder expectations?

Connecting the Dots

3. How do the principles of metric development connect to earlier discussions about risk quantification and communication strategies?
4. In what ways can the DIKW Pyramid and GQIM framework enhance your approach to creating meaningful metrics?

Real-World Application

5. Can you identify examples of how effective metrics have been used to drive cybersecurity improvements in your organization or industry?
6. How might tailored communication strategies improve the impact of metrics on different audiences?

Personal and Professional Growth

7. How can you apply the principles of actionable metrics to your own role or field of study?
8. What steps can you take to enhance your skills in translating technical data into business-relevant narratives?

Further Exploration

9. What additional frameworks or tools could help refine the development of KPIs and KRIs in your organization?
10. What immediate steps could you take to begin balancing operational and strategic metrics to support cybersecurity goals?

16 Board-Level Communication

16.1 INTRODUCTION

Board-level communication is a core responsibility for the chief information security officer (CISO) 3.0, requiring not only technical expertise but also translating complex cybersecurity concepts into clear, actionable insights that align with business objectives. Allowing CISOs to effectively communicate the value and impact of security initiatives in the language of business. As organizations face increasingly sophisticated cyber threats, the board's understanding of security risks and their impact on strategic goals is essential to ensure proper governance and risk management.

In this chapter, we explore the intricacies of engaging with the board of directors (BOD) on cybersecurity matters, offering practical guidance on how to communicate effectively with nontechnical leaders. We will examine the importance of crafting tailored messages that resonate with diverse stakeholders, from the boardroom to the executive suite. Additionally, we'll cover techniques for delivering compelling presentations, the value of data storytelling, and how to leverage visual aids to enhance clarity and impact. The chapter also provides a roadmap for creating impactful board-level deliverables, including presentations and dashboards that convey the current security posture and the effectiveness of risk mitigation efforts.

These skills will help you not only convert your cybersecurity program into a strategic asset but also solidify your position as a trusted advisor and a key contributor to the organization's overall success. This is the essence of the CISO 3.0: A leader who understands the language of business, speaks fluently to the board, and inspires confidence in their ability to protect the organization's most valuable assets, ensuring that cybersecurity is seen as a key enabler of business success rather than a mere operational cost.

16.2 COMMUNICATING WITH THE BOARD

Effective communication with the BOD is fundamental for a CISO 3.0, but translating complex technical threats into stories that resonate with nontechnical audiences can be a challenge. The ability to tell a compelling risk story is essential, not only to engage stakeholders but also to drive meaningful action. You may have heard the saying "What gets talked about gets prioritized". Which is why your story must be carefully crafted, focusing not on technical details but on how cybersecurity aligns with business objectives. It's about speaking the language of business, whether that's brand risk, regulatory concerns, or financial impacts, and framing the discussion in terms they care about.

DOI: 10.1201/9781003510789-22

Communicating with the board can be like walking a tightrope. You need to provide enough information to keep them informed, but not so much that you overwhelm them. This involves translating complex technical concepts into clear and concise language that resonates with board members, highlighting the impact of cybersecurity on the organization's strategic objectives and financial performance. While board-level communications can be challenging, always remember, the board is on your side. They want you to succeed.

I've learned that the key to board-level communications is risk-based storytelling. The key to successful risk storytelling is to connect cybersecurity with business goals. Overly technical discussions or reliance on frameworks like National Institute of Standards and Technology (NIST) can alienate executives and directors. Instead, risk should be presented in terms of material impact on the business. For example, a data breach should be explained not only as a technical failure but also as something that could damage the company's reputation and bottom line. Using simple, relatable analogies, like comparing cybersecurity to familiar scenarios such as home security or familiar risks like speeding in your car, can help make your message more accessible and relatable.

Building credibility as a CISO requires a deep understanding of business priorities and the ability to communicate how cybersecurity supports those goals. When cybersecurity is framed as a business risk rather than just a technical issue, CISOs gain trust and influence within the executive team and board. This trust is the key to securing budget and buy-in for your cybersecurity program, especially when large sums are required to meet security frameworks or improve defenses.

The metrics and data we discussed in the last chapter play an important role in reinforcing the risk narrative, but they must be used judiciously. They should support your story and not *be* your story. Storytelling in cybersecurity is not just about delivering a message; it's also about defending it. CISOs should be prepared to face skepticism, especially when requesting significant investments.

One great way to prepare for challenges is to pregame your messaging with peers and executives and have them poke holes in your narrative. Collaborating with internal experts in marketing or communications can also help refine the story and make it more compelling. In the end, risk is an inherent part of business, and the role of the CISO is to manage that risk in a way that is understandable and actionable.

16.2.1 The Goal of Board-Level Communication

Stop for a second and ask yourself:

> What is the main goal of every board interaction?

It is important to make sure you understand the assignment and remember the actual goal. It's easy to get lost in the minutia and forget to stay on target. The main reason boards seek updates on department management is to provide oversight and ensure each function is well-managed. Consequently, your primary goal at each board meeting should be to *instill confidence* that you have things under control and are the right person for the job. Often, CISO 2.0s mistakenly believe the point is to deliver updates on key performance indicators (KPIs) or the threat landscape. While an annual threat landscape briefing might be welcomed by the board, it's not the main objective for most meetings. Worse yet, many CISO 2.0s think sowing fear, uncertainty, and doubt (FUD) is the best way to earn buy-in and that they can scare stakeholders into caring about security. However, this approach often backfires, undermining your ability to instill confidence. Fear can be a motivator, but it might just motivate them to replace you with someone who projects a better sense of security.

The key goals of every board presentation are as follows:

Educate: Provide the board with the necessary knowledge and context to understand key cybersecurity concepts and their implications for the organization.

Inform: Share relevant information about the security program's performance, key initiatives, and progress toward goals.

Update: Keep the board apprised of emerging threats, vulnerabilities, and any major changes in the risks.

Instill Confidence: Demonstrate that the security program is well-managed, risks are being effectively mitigated, and the organization is prepared to respond to potential incidents.

Demonstrate Fiduciary Duty: Show that the CISO is fulfilling their legal and ethical obligations to act in the best interests of the company and its shareholders by proactively managing cyber risks.

Additionally, a well-crafted board presentation should accomplish all of the following:

Align to Your Audience: Understand the priorities and concerns of your audience. Tailor your message and language accordingly.

Tell a Story: Use data to tell a compelling story about your security program's successes and challenges.

Focus on Impact: Highlight the potential business impact of cyber risks and the value of your mitigation efforts.

Be Concise and Clear: Avoid overwhelming your audience with too much data or technical jargon.

Use Visuals: Utilize charts, graphs, and other visual aids to enhance understanding and engagement.

Answer Questions: Anticipate questions and be ready to provide additional context or explanations.

Update Stakeholders: Provide regular updates on KRIs and progress toward mitigation goals.

I love a good acronym, so when preparing for a board meeting, remember the acronym *B.O.A.R.D*:

- Brief
- Open
- Accurate
- Risk aligned
- Demonstrate leadership

Conciseness and focus are key when presenting to the board. Highlight only the most important information and insights, recognizing that board members have limited time and appreciate efficiency. Avoid overwhelming them with technical jargon or lengthy explanations.

Transparency and openness about the organization's security challenges and risks are essential for building trust with the board. Acknowledge areas where improvement is needed, and proactively demonstrate your commitment to addressing them.

The information you present must be accurate, reliable, and data-driven. Avoid any exaggerations or fear-mongering tactics, as they can erode your credibility and undermine the board's confidence in your leadership.

Always frame your presentation within the context of the organization's risk appetite and strategic objectives. Show how your security initiatives directly contribute to mitigating risks and achieving the company's goals. This alignment demonstrates the business value of cybersecurity and reinforces its importance in the overall strategy.

Project confidence and competence in your role as CISO. Show that you have a clear vision for the security program and can execute it effectively. Be prepared to answer questions thoroughly and address any concerns in a thoughtful and reassuring manner. This will inspire trust and confidence in your leadership and the effectiveness of your security program.

16.2.2 Types of Board Meetings

It's important to understand that boards have different types of meetings, each with its own purpose and expectations. There are full board meetings where all members are present, committee meetings focused on specific areas, strategy sessions for long-term planning, and special meetings called to address urgent or specific matters requiring a vote. Even within regular meetings, agendas vary, and you'll likely be participating in a management update session where each executive presents their department's performance. These updates are your opportunity to showcase the state and status of your security program, but they are not necessarily the time to make requests.

The appropriateness of making requests varies among boards. Some strictly forbid it during regular meetings, while others have designated times for such discussions. Some even have separate meetings dedicated to voting on and deciding upon requests. Understanding the specific norms and expectations of your board is crucial. If you have a request, ensure you know the proper time and forum to present it.

If the board is the right audience, you're at the appropriate meeting, and you intend to seek approval, adhere to these best practices.

- Know the decision-making process
- Put the request upfront
- Make the request clear
  - Align with strategic goals
  - Quantify the impact
  - Provide context
  - Present (real) options
- Don't oversell it
- Know when to exit

First, understanding the board's decision-making process and the factors they weigh is also vital. This knowledge allows you to tailor your presentation and anticipate potential questions or concerns.

Be upfront about your request. Clearly state it at the beginning of your presentation so that the board immediately understands your objective. Second, articulate your request with clarity and precision, leaving no room for ambiguity. Explain how your proposal aligns with the organization's strategic goals and quantify its potential impact, using financial metrics and risk assessments to showcase its value. Provide the necessary background and context to help the board understand the rationale behind your request, including the problem or opportunity you're addressing and the potential consequences of inaction.

Be sure to present genuine options when seeking a decision. People, including board members and executives, value choices. Ensure these options are realistic and that you can articulate the trade-offs associated with each. A common frustration among directors is when executives present a single "right" solution alongside a clearly inferior one, misrepresenting them as options.

Finally, recognize when to strategically withdraw. Remember, once you've secured the desired outcome, avoid overselling or elaborating further. Know when to accept a "yes" and move on gracefully. If your request faces resistance or the board isn't ready to decide, gracefully conclude the conversation and revisit it at a more opportune time. Persistence can be valuable, but pushing too hard can be counterproductive.

Remember, different types of meetings have distinct objectives. Some are status updates, others focus on strategy, and some are specifically for seeking approval. Recognizing these nuances and tailoring your approach accordingly are key to successful board interactions.

16.2.3 Board Focus Areas

What do you think are the top three things board directors care about? Take a minute to think about it, maybe even pause to write down your answers. What did you come up with? Maybe strategy, compliance, financial performance, innovation, governance, or ethics. Those are all great answers, but the top three things that boards care about boil down into three categories: *Revenue*, *cost*, and *risk*.

- *Revenue* encompasses inflows of funds from business activities, essential for fueling growth and maximizing shareholder returns.

- *Cost*, together with revenue, creates margin, which drives shareholder value. Boards want to ensure that the company is generating sufficient revenue to cover its costs and generate profits. They are also interested in ensuring that costs are being managed effectively.
- *Risk*, from a corporate board's point of view, is the potential for events or circumstances to negatively impact the company's objectives. There are two types of risk: Positive risk, which is the kind of risk from innovation and investments, and negative risk, which is exposure to loss from cybercrime, natural disasters, or other adverse events. Risk involves identifying, evaluating, and mitigating uncertainties in the business environment to safeguard the company's assets and ensure long-term success. We discussed risk in depth in Chapter 6, and you should be familiar with the distinction between enterprise risk and cyber risk, especially from a board's perspective.

These three areas are at the top of the board members' minds because they directly influence the organization's financial health and long-term sustainability. *Revenue* is the lifeblood of any company, and the board is keenly interested in strategies to increase it. *Costs* need to be managed effectively to ensure profitability and maximize shareholder value. *Risk* management is the key to protecting the company's assets and value, ensuring its continued success.

IT and cybersecurity can definitely impact all three areas. Effective cybersecurity measures can not only protect revenue streams by preventing disruptions and data breaches that could lead to financial losses. In some cases, security and IT can even generate revenue by offering security products or services to customers or by enabling new business models that rely on secure technology. Streamlining IT processes and optimizing technology investments can help reduce costs and improve efficiency.

Together, these three areas, revenue, cost, and risk, help tell the story of your program. If you can't explain how your IT or cybersecurity program impacts each of these areas, you are not telling a business story in the language of business. And if you can't tell your story, then no one else can or will either.

It's common for CISOs and sometimes even CIOs to be seen as second-class C-level leaders. They often report to other C-levels, and I frequently see CISOs not directly reporting to the board, instead having to funnel their message through another C-level leader like the CIO or chief financial officer (CFO). Even if they do report directly, they may have to clear all their content with another C-level. Many security and IT leaders I meet are managers, directors, or VPs who don't even have the chief title.

Many of the security leaders I speak with express the desire for a "real seat at the table" and an official C-level title and responsibilities. However, most CISO 1.0 and 2.0s are not ready for the responsibility and accountability that come with the role. Boards have been clear about what they want from us. The top two requests I hear from board directors are as follows:

1. To know the financial dollar value involved with risks, breaches, or cyberattacks on their organization.
2. To have third-party validation of what the IT and security teams present.

This mirrors what has happened for years on the financial side. The CFO and their team quantify financial performance and risk, and audits verify those reportings. Boards want the same from technical leaders. I mentioned this example in Chapter 2, but can you imagine a CFO entering a board meeting and saying, "I don't know what last quarter's numbers are, but I can tell you that we processed 150K transactions, and that is a green up arrow", or "I don't know where all our assets are, but we're getting close to being able to catalog most of them"? That CFO would be laughed out of the room, but that is the kind of thing we technical leaders have done for years.

We have to be able to tell the story of how we impact revenue, cost, and risk. If you are doing that, then you are certain to be on target with your board presentation. If you could only have three slides total, you could use one slide for each, revenue, cost, and risk, and be just fine.

Take some time to reflect on how you currently communicate the impact of your security program on revenue, cost, and risk. Are you effectively translating technical information into business terms that resonate with the board? If not, how can you improve your communication to demonstrate the value of cybersecurity in revenue, cost, and risk terms?

16.2.4 Targeting Your Message

The key to successful communication lies in tailoring your message to the specific audience. We need to understand that different stakeholders have distinct roles, priorities, and information needs. When it comes to communicating cybersecurity initiatives, it's essential to differentiate between the BOD and executive leadership.

The relationship between the board and management can be likened to the legislative and executive branches of government. The board, like the legislative branch, sets the overall direction and framework, while management, like the executive branch, implements the strategy and manages day-to-day operations.

16.2.4.1 Board of Directors

- Sets the overall corporate vision and mission.
- Oversees corporate strategy.
- Sets corporate performance targets.
- Hires, evaluates, and fires the CEO.
- Provides corporate governance oversight.
- Represents shareholder interests.
- Ensures compliance with legal and regulatory requirements.
- Reviews and approves executive compensation packages.

16.2.4.2 Management

- Implements strategic initiatives.
- Manages day-to-day operations.
- Develops and manages budgets.
- Makes decisions on resource allocation and capital expenditures.
- Reports to the BOD.
- Manages and develops employees.
- Sets departmental goals and objectives.
- Develops and implements marketing and sales strategies.

16.2.4.3 Messaging to the Board of Directors

When communicating with the board, your primary objectives are as follows:

Earn their buy-in and support for your security initiatives.
Instill confidence in your leadership and the effectiveness of your program.
Explain how you've utilized the company's resources since your last update, the tangible benefits achieved in business terms, and how these outcomes align with expectations.
Clearly convey the business's current cyber risk exposure and your proposed plan for managing and mitigating those risks.
Articulate how you intend to allocate the company's budget and the expected return on investment (ROI) in business terms.
Justify your decisions and explain why your proposed approach is the most effective way to address the company's cybersecurity challenges.

Remember, keep your board presentation brief and focused, ideally no more than 15 minutes, and avoid getting sidetracked by tangential questions.

16.2.4.4 Messaging to Executive Management

When interacting with executive leadership, your primary goals are as follows:

Earn their buy-in and support for your security initiatives.
Secure the necessary budget for your program.
Explain the risks and threats in business terms, highlighting their potential impact on operations, finances, and reputation.
Outline your proposed plan for addressing these risks, including required resources and expected outcomes.
Present 1–3 options, each with a cost–benefit analysis, and justify your recommendations based on their alignment with strategic goals and risk tolerance.
Explain why you've chosen to tackle this problem now and what ROI they can expect in business terms.

Remember, your presentation to the executive team can be more detailed and technical than your board presentation but aim to keep it within 60 minutes and avoid unnecessary tangents.

In both cases, effective communication hinges on understanding your audience's priorities and concerns. Tailoring your message to resonate with their specific needs can build trust, secure buy-in, and ensure that your cybersecurity initiatives receive the necessary support and resources.

16.2.5 Preparing for Questions

While some questions during board meetings are positive, indicating engagement and a desire to understand, others can derail your presentation and hinder effective communication. You should strive to strike a balance between fostering discussion and staying on track. If a question leads you off-topic, politely acknowledge it and offer to follow up later or direct the board member to relevant information in the pre-provided board packet. When answering questions, prioritize accuracy, transparency, and humility. It's always better to admit you don't know the answer than to provide inaccurate, misleading information or come off like a blowhard or know-it-all.

The types of questions you receive can often reveal your organization's maturity level and the effectiveness of your communication. The maturity curve typically starts with service delivery questions, progresses to business partner questions, and culminates with strategic leadership questions.

Service Delivery Questions: These questions often focus on the cost and efficiency of IT operations, such as "Why is IT so expensive?" or "How do we drive down costs?" They indicate a basic level of understanding of IT's role and a focus on operational efficiency.

Business Partner Questions: These questions demonstrate a growing awareness of the strategic importance of IT and cybersecurity. They might include inquiries about security and compliance, indicating a desire to understand how IT contributes to risk management and business objectives. However, be wary of the "Are we secure/compliant?" question. The board wants reassurance, but the reality is that all organizations face some level of risk. Your role is to communicate those risks and the investment trade-offs in addressing them.

Strategic Leadership Questions: These questions signify a mature understanding of IT and cybersecurity as strategic enablers. They might focus on how IT/security can differentiate the company from competitors or enable new business models. These questions indicate that the board sees IT and security as key drivers of innovation and growth.

By recognizing the types of questions you're receiving, you can gauge where you are on the maturity curve and work toward progressing to the next level.

Here are some common board questions and their underlying implications:

"*What is our current relevant threat landscape?*" or "*What should our approach be to X?*" These questions often indicate that the board doesn't know what "good" looks like in cybersecurity and needs guidance on industry best practices and emerging threats.

"*Why do we need X?*" or "*Why is X that way?*" These questions suggest that the board doesn't understand the rationale behind certain security decisions or processes and needs help making informed choices.

"*What are our options regarding X?*" This question signals a desire to understand the available levers for impacting business outcomes through cybersecurity.

"*How does X work?*" This indicates a need for education and a deeper understanding of technical concepts.

"*How do we stack up against our peers/competitors?*" This question reveals that the board doesn't fully grasp how the organization's unique risks impact its value.

Remember, board questions, even seemingly simple ones, can offer valuable insights into their understanding and concerns. Through active listening, you can interpret these questions and use that to align your talk track to their needs better and build a stronger relationship with the board.

16.3 CREATING BOARD-LEVEL PRESENTATIONS

Board presentations offer an opportunity for the CISO 3.0 to engage directly with the board and communicate the value and impact of the cybersecurity program. These presentations should be concise, focused, and tailored to the board's specific interests and concerns. To achieve maximum impact, board presentations need to be *clean* and *visually engaging*, much like a TED talk. Slides should be simple, on-brand, and prioritize the most important information upfront. The presentation should tell a compelling story that leverages data and impactful visuals to capture and maintain the board's attention.

The way you present information is just as important as the information itself. A well-organized and visually appealing presentation can enhance the board's understanding and engagement. Use clear and concise language, avoid technical jargon, and focus on the key takeaways. Visual aids, such as charts, graphs, and infographics, can help illustrate complex concepts and make your message more memorable.

Board presentations are not the time for technical deep dives. They're an opportunity to tell a story, to paint a picture of the organization's security posture and the challenges and opportunities ahead. I've found that using visuals, analogies, and real-world examples can help bring the message to life and make it more memorable.

Board members are busy individuals with limited time and attention spans. You need to capture their attention early on and maintain their engagement throughout your presentation. Your presentation is meaningless if no one is paying attention, is asleep, or on their phones. Start with a strong opening that highlights the most critical risks and their potential impact on the business. Use storytelling techniques to make your message more relatable and memorable. Keep your presentation concise and focused, avoiding unnecessary details or tangents. Focus on the big picture, highlight the most critical risks, and offer actionable recommendations. Another way to make cybersecurity risks more relatable is by tying them to real-world examples and news stories. By using external, relevant examples, CISOs can highlight the importance of the security program in preventing the company from ending up in the headlines.

A well-structured board deck is essential for delivering a clear and compelling message to the board. It should include an executive summary, key risk indicators, financial implications, strategic alignment, and recommendations for action.

16.3.1 Presentation Matters

To reiterate for emphasis, the way you present information is just as important as the information itself. A well-organized and visually appealing presentation can increase the board's understanding and engagement. Use clear and concise language, avoid technical jargon, and focus on the key takeaways. Visual aids, such as charts, graphs, and infographics, can help illustrate complex concepts and make your message more memorable.

Your presentation should begin with a captivating story. Storytelling is possibly the most important aspect of your presentation. Andrew Stanton, the writer of renowned films like *Toy Story*, *Finding Nemo*, *Wall-E*, and *A Bug's Life*, offers this advice on crafting stories:

1. Start with the end and work backward.
2. Make them care.

16.3.2 Crafting a Cybersecurity Story Arch

The narrative structure of your presentation should follow a classic story arch, drawing the audience in and keeping them engaged. Like any good story, it needs a beginning, a middle, and an end.

Here's a breakdown of the key storytelling components:

- **Setting:** Establish the context of your story. Describe the current cybersecurity estate, the organization's unique challenges, and the potential impact of cyber risks on the business.
- **Characters:** Introduce the key players in your story, such as the security team, business leaders, and the BOD. Highlight their roles and responsibilities in managing cyber risks.
- **Conflict:** Identify the central conflict or challenge that your security program is addressing. This could be a specific threat, vulnerability, or a broader risk exposure.
- **Rising Action:** Describe the steps your team is taking to address the conflict, highlighting key initiatives, investments, and progress made.
- **Climax:** Showcase the positive outcomes and impact of your security program, demonstrating how it has mitigated risks, improved resilience, and contributed to business success.
- **Resolution:** Conclude by outlining future plans and strategies for continued improvement and adaptation.
- **Background, Elements, and Order of Events:** Provide additional context and details as needed to support your narrative. Ensure the order of events is logical and easy to follow.
- **Broadening of Events, Options, Risks, and Problems:** Discuss the broader implications of the cybersecurity challenges you're addressing and the potential consequences of inaction. Present the options considered, the risks associated with each, and the ultimate decision made.
- **The Big Question and the Big Answer:** You can frame your presentation around a central question that captures the essence of your message. Provide a clear and compelling answer that highlights the value and impact of your security program.

A story arch provides a structured and engaging way to present information, making it more memorable and impactful for your audience. It helps to create a narrative flow that guides the board through your presentation, ensuring they understand the context, challenges, solutions, and outcomes of your security program. If you incorporate elements of storytelling, you can capture the board's attention, evoke emotions, and foster a deeper connection with your message. This can lead to increased buy-in, support, and, ultimately, more effective decision-making. People don't make decisions based on numbers and facts; they require stories to guide their choices.

Remember, a successful presentation isn't about what *you* want to say; it's about what your *audience* needs to hear. To achieve this, consider the throughline of your presentation – the connecting theme that ties your narrative together and ensures a cohesive message. A clear throughline helps your audience follow your story and understand its key takeaways.

Another helpful tool is the ABCs of presenting complex information:

A = Audience: Understand their needs, knowledge level, and priorities.
B = Benefit: Clearly articulate the benefits of your message to the audience. What's in it for them?
C = Clarity: Use simple language, avoid jargon, and focus on the key points.

16.3.2.1 Example: Before and After Story Arch

- **Before:** Customers are unhappy due to slow and unreliable online services, leading to potential churn and lost revenue.
- **After:** By implementing a new cloud security solution, we improved the performance and reliability of our online services, resulting in increased customer satisfaction and a projected revenue increase of X%.

Remember to keep your story concise and impactful. Bullet points and elevator pitches are great tools for highlighting key points and maintaining focus. Always interpret data for the board, explaining whether it's good or bad news, its impact level, and whether the situation is improving or worsening. When you use these storytelling techniques and focus on the "so what?" and "what's in it for me?" aspects, you can create board presentations that are both informative and engaging. This will help foster trust and confidence in your leadership and the value of your security program.

16.3.3 Data Storytelling

The essence of effective communication lies in letting the data tell its own story. Data storytelling involves presenting information in a narrative format that is engaging, informative, and impactful. It goes beyond simply displaying numbers and charts; it involves weaving a compelling narrative that connects with the audience and inspires action. Numbers and statistics engage our analytical minds, but stories tap into our emotions, so if you can combine the two, you'll have a well-rounded approach that will connect with all types of people.

The type of story you tell depends on your objective. If you're showcasing past performance and achievements, you'll likely use reporting to highlight successes and areas for improvement. If you're focusing on the current state of affairs, you might employ exploratory analysis to uncover insights and identify trends. And if you're looking to the future, predictive analytics can help you forecast potential risks and opportunities. Remember the distinction between leading and lagging indicators from the previous chapter: Leading indicators help predict future events, while lagging indicators reveal past trends.

The key to data storytelling is to identify the core message you want to convey and then select the data and visuals that best support that message. The narrative should be clear, concise, and easy to follow, with a logical flow that guides the audience through your insights. Telling a data story requires you to collect the right data; you can think of it as evidence for the story.

The foundation of any compelling data story is solid evidence, which comes in the form of relevant data that supports your message and resonates with your audience. The process of gathering this evidence begins with identifying the sources of data that are most pertinent to your story. These sources might include security tools and platforms, incident reports, vulnerability scans, threat intelligence feeds, industry benchmarks, and even internal surveys or interviews. The key is to select sources that provide reliable and insightful data that align with your objectives.

After identifying your data sources, the next step is to collect the information in a structured and organized way. This could involve pulling data from logs, databases, or reports, or even conducting surveys and interviews to gather qualitative insights. It's essential to keep the data collection process systematic and well-documented to ensure both accuracy and completeness.

Raw data typically require cleaning and preparation before analysis can begin. This could mean removing duplicates, addressing missing values, or reformatting the data for easier processing. These steps are crucial for maintaining the quality and reliability of your analysis.

Once the data is prepped, the focus shifts to analyzing and interpreting it. Whether through statistical analysis, data visualization, or other methods, the aim is to uncover trends, draw conclusions, and extract insights that can inform decision-making. This step is where the data begin to take shape and support the larger narrative you're building.

From the vast amount of information produced during analysis, it's important to select the most relevant data points that directly reinforce your core message and resonate with your audience. These key points will form the foundation of your data story, providing the evidence needed to make a strong and persuasive case.

Steps to Telling a Data Story

1. **Identify Your Audience:** Understand their priorities, concerns, and level of technical expertise.
2. **Define Your Message:** Determine the key takeaway you want your audience to remember.
3. **Select Relevant Data:** Choose data that supports your message and is meaningful to your audience.
4. **Visualize the Data:** Use charts, graphs, and other visual aids to present the data in an easily understandable format.
5. **Craft a Narrative:** Weave the data and visuals into a compelling story that resonates with your audience.

6. **Provide Context:** Explain the significance of the data and its implications for the organization.
7. **Call to Action:** Conclude with clear recommendations or next steps based on your insights.

16.3.4 Correlation and Causation

When analyzing data, it's important to distinguish between correlation and causation.

- **Correlation:** Indicates a relationship between two or more variables, but it doesn't necessarily imply that one causes the other. For example, there might be a correlation between the number of security awareness training sessions and the number of phishing attacks, but it doesn't mean that the training is causing the attacks.
- **Causation:** Implies a direct cause-and-effect relationship between two variables. Establishing causation requires careful analysis and often involves controlling for other factors that might influence the outcome.

Understanding the difference between correlation and causation is fundamental for drawing accurate conclusions from your data and avoiding misleading interpretations.

16.3.5 Data Visualization

The old adage "a picture is worth a thousand words" is especially true for communicating metrics and data. Data visualization is the practice of turning information and data into visual formats like charts, graphs, and maps. These visual tools help make complex data easier to interpret, allowing us to quickly spot trends, outliers, and patterns. Since the human brain can process images much faster than written text, data visualization provides an effective way to communicate detailed information in a clear and engaging manner.

Research suggests that people retain more information when it is presented visually. According to recent studies, nearly 65% of people are visual learners, meaning they retain information better when it's presented in a visual format. Beyond that, the brain processes visual information 60,000 times faster than text alone, so people are more likely to remember information that is presented visually.

Effective data visualization involves careful consideration of design principles, elements, and imagery. Design principles guide the overall structure and layout of your visuals, ensuring clarity and coherence. Design elements, such as color, typography, and icons, enhance the visual appeal and facilitate understanding. Imagery, including photographs, illustrations, and diagrams, can add context and emotional resonance to your data story.

Infographics use visuals to simplify and communicate complex information, data, or knowledge in a way that's easy to understand and digest quickly. They combine text, images, and charts to tell a story or explain a concept in a visually engaging way. Infographics are particularly effective for communicating complex or technical information to nontechnical audiences, making them a valuable tool for CISOs when presenting to the board or executive leadership.

16.3.6 Creating Compelling Data Visualizations

Creating compelling data visualizations involves several key steps that ensure your message is clear, engaging, and actionable:

Know Your Audience: Tailor your visualization to the audience's technical knowledge, priorities, and the specific message you aim to communicate.

Choose the Right Format: Select the chart or graph that best represents the data and enhances understanding.

Keep It Simple: Avoid overcomplicating the visuals. Focus on the most important data points and eliminate unnecessary elements.

Effective Use of Color and Typography: Choose colors and fonts that enhance readability and draw attention to the key insights.

Tell a Story: Shape your data into a narrative, emphasizing key trends, patterns, or insights that align with your overall message.

Provide Context: Use labels, annotations, and brief explanations to ensure the audience fully grasps what the data represent.

Refine through Feedback: Continuously improve your visuals by seeking feedback and making adjustments for clarity and impact.

By applying these principles, you can craft data visualizations that go beyond just presenting information – they engage your audience, drive understanding, and prompt action. The objective is not only to show data but also to use it to tell a meaningful story that resonates with your audience.

Data storytelling is a much needed skill for the modern CISOs. It allows cybersecurity leaders to take complex technical data and turn it into a narrative that is understandable and impactful for key stakeholders, such as the board or executive team. With this approach, the CISO can effectively communicate the importance and value of security initiatives, justify resource allocation, and secure executive buy-in for critical projects. This ability to translate data into business outcomes sets the CISO 3.0 apart as a strategic leader who can influence decisions and drive organizational success.

16.3.7 Understanding Attention

In the high-stakes environment of the board room, capturing and maintaining the attention of the board members is of the utmost importance. The success of your cybersecurity initiatives hinges on their understanding and support, and that can only be achieved if they are actively engaged in your presentation. If their minds wander, they're asleep, or they're glued to their phones, your message, no matter how well-crafted, will fall on deaf ears.

The human brain allocates attention based on a complex interplay of factors, including survival instincts, social context, and cognitive reasoning. Understanding these factors can help you tailor your presentation to resonate with your audience and ensure your message is heard.

16.3.8 The Triune Brain Model and Beyond

Previously, the triune brain model, proposed by neuroscientist Paul MacLean, suggested three distinct layers of the brain that evolved sequentially:

Lizard Brain (Reptilian Complex): The oldest part of the brain, responsible for basic survival instincts like fight-or-flight responses and regulating bodily functions.
Monkey Brain (Limbic System): The middle layer, responsible for emotions, memory, and social interactions.
Human Brain (Neocortex): The most recently evolved part, responsible for higher-order thinking, language, and reasoning.

While this model provided a helpful framework for understanding brain function, it's now recognized that the brain evolved more concurrently and is far more complex. A more contemporary understanding suggests the brain is sectioned into three interconnected systems:

Emotional: Processes emotions, motivations, and drives.
Social: Handles social interactions, relationships, and group dynamics.
Cognitive: Responsible for logical reasoning, problem-solving, and decision-making.

However, even this model has its limitations. The latest research suggests a more nuanced understanding of emotions, known as the *Theory of Constructed Emotion.* This theory posits that emotions are not hardwired responses but rather constructed from a combination of bodily sensations, past experiences, and cognitive interpretations.

The Theory of Constructed Emotion has broad implications for how we capture and maintain attention. It suggests that our base impulses (survival/fight-or-flight), social context, conceptual knowledge, and cognition all help shape how we perceive and process information. To truly engage your audience, you must first bypass their survival instincts, then navigate their social context, and finally appeal to their cognitive reasoning. Even though you may have developed your presentation with your cognitive brain, that's not how it's initially received.

The emotional mind acts as the gatekeeper, rationing resources and determining likability and attractiveness. The social mind evaluates judgment, beliefs, hierarchy, confidence, reputation, and credibility. Only then does the cognitive mind engage with the target audience, analytical objections, and summarization.

16.3.9 Capturing and Maintaining Attention

People have limited attention spans, and the survival part of your brain rations that attention. You generally only have 30 seconds to a minute to engage the emotional part of the brain. So, you need to structure your presentation to capture attention early on. Once you've engaged the emotional brain next, the social part of the brain will determine if you are likable, which translates into believable. Only after you have made it past these attention gates can you engage the cognitive brain. Most people can only focus on cognitive tasks for 15–20 minutes at a time. So, you need to ensure that you wrap up quickly if you want to maintain attention throughout. Even though 20 minutes is the max, you are not guaranteed 20 minutes of cognitive attention because the cognitive brain likes to "figure things out", and once it determines that it understands what is being shared, it has a tendency to wander and eventually check out, at which point you have to start back at the beginning to regain attention. Consequently, you need to let your story unfold, keeping a little mystery or surprise to help keep the cognitive brain engaged for the full 20 minutes.

The science of learning also offers valuable insights for attracting and maintaining attention. When it comes to learning, order matters. Our brains prefer sequence and hierarchy, ensuring that your presentation follows a strong format is key. Prior knowledge is also fundamental to the learning process, which is why providing a handout or pre-read materials can be so beneficial. Learning is an active process, and few people can absorb information from hearing it just once. The current science suggests that short intervals of practice are best. So, if you can create a consistent format for your board presentations, with updates on the changes since the last report, you're essentially facilitating "practice" for the board members.

If you plan to deviate from the usual format, consider incorporating interactive elements like SOC or data center tours, off-site executive briefings, or tech demos. These activities can enhance engagement and make your presentation more memorable. Remember, the key to effective board communication is to understand your audience; tailor your message to their needs, and present information in a way that is both informative and engaging.

16.4 CREATING BOARD-LEVEL DELIVERABLES

Board-level deliverables serve as tangible representations of the CISO's work, providing the board with vital information for their oversight responsibilities. These deliverables should be carefully crafted to convey key insights, recommendations, and progress updates in a clear and concise manner. Two primary types of board-level deliverables are board decks and board books.

16.4.1 Board Decks

A well-structured board deck is essential for delivering a clear and compelling message to the board. It should include an executive summary, key risk indicators, financial implications, strategic alignment, and recommendations for action. Board decks should be clean, concise, and minimalist, like a TED talk. Working with your marketing team is a great way to ensure your slides are attractive and on-brand, and many organizations have board deck templates that you can use.

The most important and relevant information should be presented upfront in your board decks. I like to borrow from the military and create a BLUF slide, which stands for "Bottom Line Up Front". The idea is to convey the most important and valuable information while the audience is still fully engaged. It also reinforces the concept you may have learned in high school speech and debate that you should "tell them what you are going to tell them, tell them, then retell them what you just told them".

16.4.2 The BLUF Slide

The BLUF slide, a technique originating in the military for rapid and effective communication, is a concise summary of your key takeaways. This slide provides the board with a snapshot of the most critical information and insights. It should be clear, impactful, and easily digestible, even for those without technical expertise. The BLUF slide should encapsulate the essence of your presentation, highlighting the most important points you want the board to remember. It should provide a clear and concise overview of the organization's current cyber risk exposure and its potential impact on the business. It should outline your key recommendations for addressing identified risks and improving the security posture. Finally, it should briefly touch on the financial aspects of your recommendations, such as the estimated cost and potential ROI, demonstrating the value of your proposed initiatives.

What to Include in a BLUF Slide

Key Takeaways: Highlight the three to five most important points you want the board to remember.
Risk Exposure: Summarize the organization's current cyber risk exposure and its potential impact on the business.
Recommendations: Outline your key recommendations for addressing identified risks and improving the security posture.
Financial Implications: Briefly touch on the financial aspects of your recommendations, such as the estimated cost and potential ROI.

To create a BLUF slide, begin by identifying your key takeaways. Start by making a list of the most important points you want to convey. Next, craft a concise summary by building a sentence or two around each key takeaway, providing context and

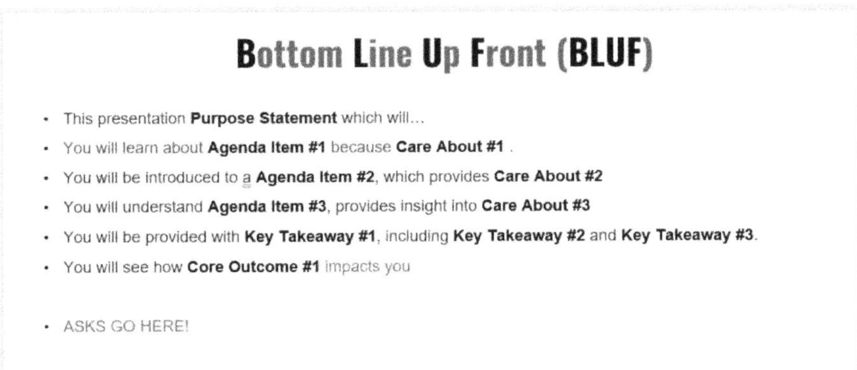

FIGURE 16.1 BLUF.

highlighting its significance. To enhance understanding and engagement, incorporate visuals such as icons or simple charts. Above all, keep it simple. Avoid clutter and unnecessary details. The BLUF slide should be easy to read and understand at a glance (Figure 16.1).

Pro Tip: ChatGPT can be a helpful tool for both populating your key take-aways list and crafting concise and impactful sentences.

16.4.3 The Importance of Editing

Editing is a critical step in creating effective board presentations. The adage, "I would have written a shorter letter, if I had more time", attributed to various figures from Mark Twain to Blaise Pascal, rings true here. Being able to distill complex information into a concise and impactful message is a hallmark of effective communication. As Woodrow Wilson aptly put it when asked how long it takes him to prepare a speech, "If it is a ten-minute speech it takes me all of two weeks to prepare it; if it is a half-hour speech it takes me a week; if I can talk as long as I want to it requires no preparation at all. I am ready now".

This quote underscores the challenge and importance of crafting a focused and impactful presentation, especially when time is limited. A shorter presentation is often better, as it takes advantage of the typical attention span, demonstrates respect for the board's time and ensures that your message remains focused and impactful. It forces you to prioritize the most critical information and avoid overwhelming the board with unnecessary details or tangents. The process of editing involves refining your message, eliminating unintended redundancies, and ensuring that every slide and every word contribute to the overall narrative. It's about achieving clarity, conciseness, and impact.

16.4.3.1 Tips for Editing a Board Presentation

Start with a Clear Objective: Define the key message you want to convey and the desired outcome of your presentation. This will help you stay focused and avoid unnecessary tangents.

Ruthlessly Prioritize: Identify the most critical information that the board needs to know, and eliminate anything that is not essential to your message. Remember, less is often more.

Simplify Your Language: Use straightforward, easy-to-understand language and avoid technical jargon or acronyms that could confuse your audience. Break down complex ideas into simple terms, and where possible, use analogies or real-life examples to help make the concepts more relatable and easier to grasp.

Use Visuals Effectively: Visuals can be a powerful tool for communicating complex information quickly and efficiently. However, use them strategically and avoid overcrowding your slides with too much text or data.

Practice and Polish: Run through your presentation several times to ensure a confident delivery and to spot any areas that need improvement. Seek feedback from colleagues or mentors to fine-tune your message and make sure it connects with your audience.

Time Management: Be mindful of the time allotted for your presentation and ensure that you can deliver your key points within that timeframe. If necessary, be prepared to cut or condense certain sections to stay on track.

Proofread: Carefully proofread your slides and any supporting materials to ensure they are free of errors and typos. A polished presentation reflects professionalism and attention to detail.

By following these tips and investing the time and effort to edit your presentation effectively, you can ensure that your message is clear, concise, and impactful, leaving a lasting impression on the board and achieving your desired outcomes.

16.4.4 Deck Content

When it comes to the actual content of your board deck, it should encompass elements we've discussed throughout this book, including inherent and residual risk, loss exceedance, risk appetite, investment balance, and the "big three" – your impact on revenue, cost, and risk. We've covered the risk portion extensively in Chapter 6, and you should now have a good grasp of how to narrate the risk story. However, many CISOs struggle with articulating the revenue aspect. Let's explore some recommendations on how to effectively convey these narratives.

16.4.5 Revenue Generation

Communicating the financial benefits of cybersecurity initiatives is a primary aspect of demonstrating the value of your security program to the board. This section focuses on how to effectively showcase the positive impact of cybersecurity on revenue generation.

The first step in crafting a compelling revenue story is to pinpoint the specific security initiatives that have the potential to generate revenue or contribute to revenue growth. This could involve various strategies, such as packaging and selling your organization's unique security expertise or solutions as services to customers, which not only opens a new revenue stream but also positions your company as a leader in the cybersecurity domain. Alternatively, if your security team provides services to other business units within the organization, consider implementing a cross-charging model where those units pay for the services they receive. This approach not only highlights the value the security team provides but also generates additional revenue for the security program.

Next, quantify the financial impact of each initiative. This could mean estimating revenue from new services or projecting increased customer retention due to stronger security measures that enhance trust. The goal is to present clear, measurable evidence of how these security efforts translate into financial benefits.

Once you've gathered this data, it's time to represent it visually. Use bar charts to compare revenue before and after implementing a security initiative, or line graphs to show trends in revenue growth, highlighting how security investments correlate with that growth.

The final step is to tie the data and visuals together into a narrative that clearly demonstrates how cybersecurity investments contribute to revenue. Share real-life examples to illustrate the tangible benefits your security program has brought to the business. Keep your slides focused and straightforward, avoiding unnecessary technical jargon. Use strong headlines, like "Cybersecurity as a Revenue Driver" or "How Security Investments Boost Customer Retention". Wrap up your presentation with a clear call to action, outlining next steps or recommendations for the board.

The most important thing to convey is the direct link between cybersecurity and revenue generation. The board needs to understand that cybersecurity is not only a cost center but also an enabler of business growth and profitability. Showing how security initiatives help generate revenue or prevent losses will shift the perception of cybersecurity from an operational expense to a strategic investment.

Sharing specific examples and success stories further strengthens your case. Real-world examples of how security initiatives have positively impacted the business, whether through preventing a data breach, enabling a new product launch, or improving customer trust, can be highly persuasive. These stories make your message more relatable and memorable, showcasing the real-world impact of your security program.

In Chapter 4, we discussed how IT and security can influence revenue. The CISO 3.0 can identify and leverage various opportunities to not only protect revenue streams but also actively contribute to their growth. Here are some additional ideas for revenue generation that go beyond the traditional cost-center perception of cybersecurity.

Sell Capabilities to Customers: If your organization has developed unique security expertise or solutions, consider packaging and selling them as services to customers. This can create a new revenue stream and position your company as a thought leader in cybersecurity.

Cross-Charge: If your security team provides services to other business units within the organization, consider implementing a cross-charging model where those units pay for the services they receive. This can help demonstrate the value of the security team and generate additional revenue for the security program.

Align Capabilities to Cost of Goods Sold: By implementing security measures that directly impact the cost of goods sold (COGS), such as protecting intellectual property or preventing supply chain disruptions, you can contribute to the company's bottom line and demonstrate the financial value of security.

Capitalize Development: If your team develops custom security solutions or tools, consider capitalizing those development costs rather than expensing them. This can improve the company's financial statements and demonstrate the long-term value of security investments.

Establish Competitive Differentiation: Robust security practices can be a powerful differentiator in the marketplace, attracting customers who value data protection and privacy. Highlight how your security program gives your company a competitive edge and contributes to revenue growth.

Leverage Capabilities to Improve Customer Retention: By ensuring the security and privacy of customer data, you can enhance customer trust and loyalty, leading to improved retention rates and increased revenue.

16.4.6 Cost Reduction

Most CIOs and CISOs are acutely aware of how they impact costs, as they are often responsible for managing large budgets and justifying their expenditures to executive leadership and the board. Demonstrating cost savings and efficiency gains through cybersecurity initiatives is the key to securing ongoing support and resources. A CISO 3.0 must equip themselves with the financial acumen to navigate the complexities of cost management, translating the value of security investments into the language of financial stakeholders.

16.4.7 Understanding Different Types of Costs

The CISO 3.0's financial stewardship is built on a solid understanding of various cost concepts. This knowledge is not just theoretical; it's the foundation effectively communicating the financial implications of security initiatives. Understanding the nuances of different cost types allows the CISO 3.0 to optimize resource allocation, avoid common pitfalls, and ensure that security investments align with the organization's broader financial goals.

Direct Costs: These are costs that can be clearly linked to a specific security initiative, product, or project. These are essential for measuring the profitability and ROI of particular security efforts. Examples include the cost of a new security tool or the salary of a security analyst assigned to a specific project. The CISO 3.0 should track these costs to ensure the effectiveness of each investment. Example: Purchasing a new firewall or the cost of employee training for cybersecurity awareness.

Example: The purchase price of a new firewall or the cost of training employees on security awareness.

Indirect Costs: These are necessary expenses for the overall operation but cannot be tied directly to a specific initiative. These might include overhead, like office rent, or the broader IT infrastructure that supports security operations. The CISO 3.0 must factor these costs in to understand the complete financial impact of security programs.

Example: The cost of electricity used to power servers in the data center or the cost of IT support staff who assist with security-related issues.

Fixed Costs: Remain consistent, regardless of the level of security activity. This could include salaries, rent, or insurance premiums. It's crucial for the CISO 3.0 to consider these when assessing the long-term financial commitment to security operations.

Example: The cost of leasing office space for the security team or the annual salary of a full-time security analyst.

Variable Costs: These fluctuate depending on the level of security activity or service usage. These can rise or fall based on factors like cloud service usage or the number of security incidents requiring response. The CISO 3.0 must keep a close eye on variable costs to anticipate their impact on the overall security budget. Example: Increasing cloud storage costs as data volumes grow or paying for additional penetration testing services based on scope.

Example: The cost of cloud storage, which increases as the amount of data stored grows, or the cost of paying for penetration testing services, which can vary depending on the scope and complexity of the assessment.

Incremental Cost: These are additional expenses incurred when expanding a security service or introducing a new measure. The CISO 3.0 should weigh these costs against the expected benefits to ensure they align with the organization's goals. Example: Adding an extra layer of encryption to enhance data protection.

Example: The cost of adding an additional layer of encryption to data storage.

Opportunity Cost: Refers to the trade-offs made when resources are allocated to one initiative over another. The CISO 3.0 must carefully consider the opportunity cost of any investment to ensure resources are optimally deployed. Example: Opting to invest in a new firewall instead of upgrading employee laptops.

Example: Choosing to invest in a new firewall instead of upgrading employee laptops.

Sunk Cost: These are expenses that have already been incurred and cannot be recovered. The CISO 3.0 should be mindful of the sunk cost fallacy, which is the tendency to continue investing in a project or technology simply because resources have already been spent on it, even if it's no longer the best option. The CISO 3.0 must make objective decisions based on the current and future needs of the organization, rather than being anchored to past expenditures.

Example: Continuing to maintain and patch a legacy on-premises security appliance that is no longer sufficient for the company's cloud-based infrastructure, even though a more suitable cloud-native solution is available.

The notion of sunk costs is often challenging to overcome because humans have a natural aversion to loss. We tend to fixate on past investments and feel compelled to justify them, even when it's no longer financially or strategically sound. This can lead to entrenchment, where organizations cling to outdated technologies or processes simply because they've already invested time and money in them. The sunk cost fallacy can be particularly insidious in cybersecurity, where the fear of wasting past investments can hinder the adoption of new, more effective solutions.

To illustrate the sunk cost fallacy, consider this analogy: Imagine you were walking to work, and 3/4 of the way there, you realize that it's Sunday and work is closed. Would you continue to walk to work because you are already more than halfway there? Of course not, that would be a waste of time and effort. The same logic applies to sunk costs in business decisions. The CISO 3.0 must be able to recognize when a past investment is no longer serving the organization's best interests and make the difficult but necessary decision to move on, even if it means acknowledging a previous expenditure as a sunk cost.

The CISO 3.0 needs a nuanced understanding of the different types of costs involved and how they interact. It's important to recognize that not all cost reductions are created equal; there's an important distinction between cost savings and cost avoidance.

Cost savings represent a tangible decrease in actual spending. This could be achieved through negotiating better deals with vendors, eliminating redundant software licenses, or optimizing processes to reduce labor costs.

Cost avoidance, on the other hand, refers to preventing potential future expenses. This might involve investing in security measures to avert a data breach or implementing disaster recovery plans to minimize downtime.

While both are valuable, it's important to communicate the difference clearly, as overemphasizing cost avoidance can sometimes undermine your business case. After all, it's easier to secure budget for initiatives that promise immediate and measurable savings than for those that prevent hypothetical future losses.

16.4.8 Cost Reduction Strategies

Several specific strategies can be employed by cybersecurity teams to reduce IT costs. One such strategy is application rationalization. This involves identifying and eliminating redundant or underutilized applications, leading to a reduction in licensing and maintenance costs. A thorough inventory of all applications, coupled with an assessment of their usage and value, can help identify opportunities for consolidation or retirement, ultimately leading to cost savings.

Another solid strategy is technology planning and budgeting. Implementing a disciplined approach to technology planning and budgeting allows organizations to ensure that investments are aligned with business needs and deliver a positive ROI. This requires a clear understanding of the organization's strategic objectives, identifying the technology solutions that can best support those objectives, and establishing a rigorous process for evaluating and prioritizing investments.

Technology vendor management is another major part of cost reduction. Optimizing vendor relationships and negotiating favorable contracts can lead to substantial cost reductions and improved service levels. This involves conducting regular vendor assessments, benchmarking pricing and terms, and leveraging the organization's buying power to secure the best possible deals.

Project portfolio management also plays a key role in cost reduction efforts. Prioritizing projects based on their strategic value and potential ROI can ensure that resources are allocated to the most impactful initiatives. This requires a clear understanding of the organization's priorities and an assessment of the potential benefits and risks of each project.

Technology risk reduction can also be effectively framed as cost avoidance. If you can demonstrate how investing in cybersecurity will help the company avoid the potentially substantial costs associated with incidents like data breaches or system outages, you can make a compelling case for the cost-effectiveness of cybersecurity. Quantifying the potential financial impact of these incidents and showcasing how security investments can mitigate those risks underscore the value of a proactive security posture.

16.4.9 Cost Reduction Strategy

The cost reduction process is a systematic approach to identifying and implementing strategies to optimize IT spending and achieve cost savings. It involves a thorough analysis of current expenditures, benchmarking against industry standards, and developing a detailed plan for achieving cost reductions. This process is essential for CISOs to demonstrate the financial value of their security programs and ensure that resources are allocated efficiently.

The cost reduction process typically consists of two phases:

1. **Savings Plan Definition:** This phase focuses on identifying areas for potential savings and developing a detailed plan to achieve those savings. It involves the following:
   - **Spend and Internal Demand Analysis:** Conduct a review of current IT spending, analyzing expenses by category, vendor, and business unit. Understand the internal demand for IT services and identify areas where demand can be optimized or reduced. This analysis helps identify areas where direct costs can be reduced through consolidation or renegotiation.
   - **Benchmarking:** Compare your IT costs against industry peers and best practices to identify areas where you might be overspending or where there are opportunities for improvement. Benchmarking can reveal potential cost savings by highlighting areas where the organization's spending is higher than industry averages.
   - **Savings Plan Development:** Based on the analysis and benchmarking, develop a detailed savings plan that outlines specific initiatives and their expected cost reductions. This plan should include timelines, resource requirements, and KPIs to track progress. The savings plan should consider a mix of cost-saving and cost avoidance strategies, balancing short-term gains with long-term risk mitigation.
2. **Plan Execution and Savings Generation:** This phase focuses on implementing the savings plan and realizing the identified cost reductions. It involves the following:
   - **Negotiation with Suppliers:** Leverage your organization's buying power to negotiate better pricing and terms with vendors. This could involve consolidating contracts, renegotiating existing agreements, or exploring alternative suppliers. Effective vendor management can lead to substantial cost savings on both fixed and variable costs.

- **Process Improvement Implementation:** Identify and implement process improvements to streamline operations and reduce waste. This could involve automating manual tasks, optimizing workflows, or eliminating unnecessary steps in IT processes. Process improvements can reduce indirect costs by improving efficiency and productivity.
- **Plan Monitoring and Reporting:** Continuously monitor the progress of the cost reduction plan, tracking actual savings against the targets defined in the savings plan. Regularly report findings to stakeholders, highlighting successes and addressing any challenges or roadblocks. Transparent reporting helps build trust and demonstrate the value of the CISO's efforts in achieving both cost savings and cost avoidance.

By following this structured approach, CISOs can demonstrate their commitment to fiscal responsibility and optimize their IT spending. This not only frees up resources for other strategic initiatives but also strengthens their position as a valuable business partner who contributes to the organization's financial success (Figure 16.2).

Effectively communicating the cost savings and efficiency gains achieved through cybersecurity initiatives helps demonstrate the value of your security program to the board. This involves crafting a clear and compelling narrative that highlights the financial benefits of your efforts. When presenting to the board, you need to highlight the direct relationship between cybersecurity and cost reduction. This emphasizes that cybersecurity investments aren't merely expenses but strategic initiatives that can lead to tangible savings, showcasing how security measures prevent costly incidents like data breaches or system outages, which helps demonstrate the financial value of your program. Providing quantifiable evidence of the financial savings achieved through security initiatives further solidifies this point. Boards appreciate

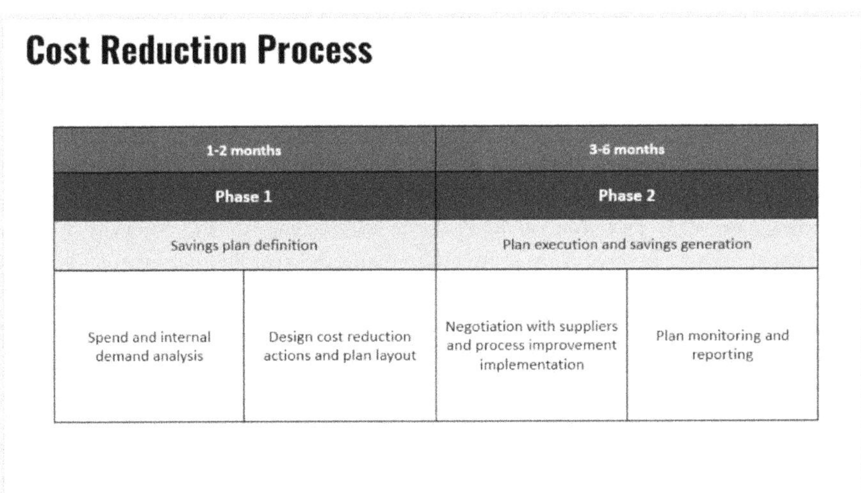

FIGURE 16.2 Cost Reduction Process.

concrete data, so presenting specific figures related to cost reductions, efficiency gains, or avoided losses helps validate your claims and justify future investments. It's important to clearly distinguish between actual cost savings and potential cost avoidance, providing context and explanation for each.

Sharing specific examples and success stories adds a layer of real-world context to your presentation. These narratives illustrate the tangible impact of your security program, showcasing how it has prevented financial losses or improved operational efficiency.

Just like the revenue portion, you should conclude with clear recommendations for future investments and initiatives, and highlight how these recommendations align with the company's overall financial goals and risk appetite, showcasing the strategic value of your proposals. This provides the board with actionable insights, allowing them to provide guidance and oversight for resource allocation and support your ongoing efforts to optimize costs and strengthen the organization's security posture.

Here's a step-by-step guide on how to tell a cost story and create impactful slides:

1. **Identify Cost-Saving Initiatives:** Start by identifying the specific security initiatives that have resulted in cost savings or efficiency gains. This could include application rationalization, vendor consolidation, automation of manual processes, or improved incident response capabilities that minimize downtime and associated costs.
2. **Quantify the Savings:** For each initiative, quantify the cost savings achieved. This could involve comparing costs before and after implementation, calculating the reduction in licensing fees or maintenance costs, or estimating the financial impact of avoided incidents. Clearly differentiate between actual cost savings and potential cost avoidance.
3. **Choose the Right Visuals:** Select charts or graphs that best represent the data and highlight the cost savings. Bar charts can be used to compare costs before and after an initiative, while line graphs can show trends in cost reduction over time.
4. **Craft a Narrative:** Weave the data and visuals into a compelling story that demonstrates the clear link between cybersecurity investments and cost savings. Highlight specific examples and success stories to illustrate the impact. Explain how different types of costs (fixed, variable, direct, and indirect) were impacted by the security initiatives.
5. **Focus on Key Takeaways:** Keep your slides concise and focused, highlighting the most important points. Use clear and simple language, avoiding technical jargon.
6. **Use Strong Headlines:** Craft headlines that clearly communicate the key message of each slide. For example, "Cybersecurity as a Cost-Saving Engine" or "How Security Investments Optimize Efficiency".
7. **Include a Call to Action:** Conclude your presentation with a clear call to action, outlining the next steps or recommendations for the board to consider. Highlight how these recommendations will further enhance the company's financial performance and resilience through a combination of cost savings and cost avoidance strategies.

16.4.10 Building a Strong Business Case

A well-crafted business case is essential for securing budget and buy-in for your security initiatives. It's a powerful tool for CISOs to articulate the value and necessity of cybersecurity investments to executive leadership and, in some cases, the BOD. It provides a structured framework for presenting the financial and strategic benefits of a proposed initiative, enabling informed decision-making and resource allocation. The business case should clearly articulate the problem or opportunity, the proposed solution, and the expected benefits in business terms; include quantified cost and risk reduction metrics to demonstrate the financial impact of your proposal; and highlight how the initiative aligns with the organization's strategic goals and contributes to its overall success.

To streamline the process, you can leverage any number of business case templates available online. There's a good chance your organization already has a preferred template. Your business case should provide a concise summary of the business need, outline costs and their types (fixed, variable, direct, and indirect), highlight risks, establish a timeline, present alternatives, address potential objections to the chosen option, and explain both direct and indirect benefits. The book *CISO Evolution: Business Knowledge for Cybersecurity Executives* by Matt Sharp and "Rock" Lambros offers an excellent walkthrough of how to build a business case using several novel tools and techniques.

A proper IT security business case should clearly define the problem or opportunity the initiative addresses. This could range from a specific cyber risk or compliance requirement to a need to enhance security capabilities. The business case should then detail the proposed solution, encompassing the technology, processes, and personnel involved. It's essential to outline the total cost of ownership (TCO) of the solution, including upfront costs, ongoing maintenance, and potential indirect costs. In contrast, the business case should also quantify the expected benefits, such as risk reduction, cost savings, productivity gains, or revenue generation. The types of costs involved should be carefully considered and included in the business case, as they can impact the overall financial analysis and decision-making process.

An impact assessment may also be a good idea, to evaluate the potential effects of the solution on the organization's operations, financial performance, and brand value. The business case should demonstrate how the initiative aligns with the organization's strategic objectives and risk appetite. It should also describe the resources required to implement and manage the solution, including personnel, budget, and technology. Finally, calculating the net present value (NPV) of the investment, which considers the time value of money, provides a more accurate assessment of the financial benefits over time. Net Present Value (NPV) is a financial metric used to determine the current value of future cash flows by applying a discount rate. It assesses the profitability of an investment by comparing the present value of anticipated cash inflows with the present value of cash outflows, providing a clear picture of its potential return.

A good business case is characterized by several key attributes. It's clear and concise, presenting information in a way that's easily understandable and digestible for its audience. It's also data-driven, using concrete metrics and evidence to support its claims and recommendations. A good business case aligns with the organization's

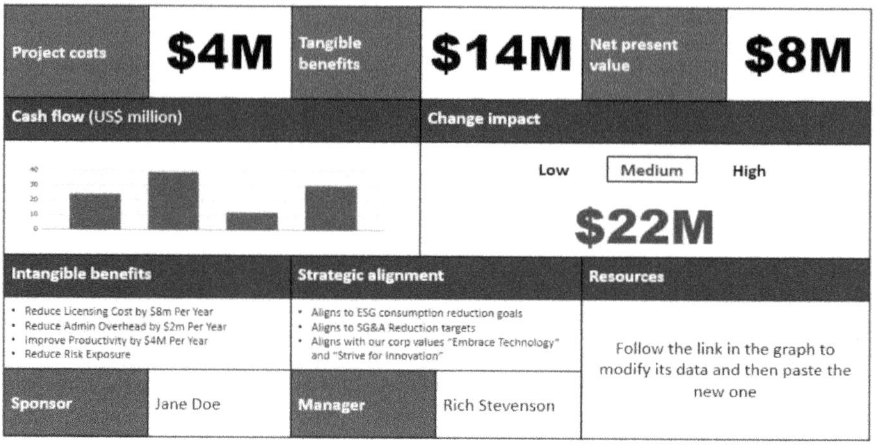

FIGURE 16.3 Business Case.

business goals, clearly demonstrating how the proposed initiative supports the company's strategic objectives. Moreover, it's financially sound, featuring a thorough cost–benefit analysis and showcasing a positive ROI and NPV. Finally, it's actionable, providing clear recommendations and outlining the next steps for decision-makers.

When presenting your business case, it's vital to tailor your approach to your specific audience, focusing on their priorities and concerns. This ensures your message resonates and is well-received. Use data and visuals to weave a compelling narrative about the problem, the proposed solution, and its expected benefits. Storytelling can make your presentation more engaging and memorable. Be prepared to answer questions by anticipating potential queries and having clear, concise responses ready.

Remember, the focus should be on the big picture (Figure 16.3). Avoid getting bogged down in technical details and instead emphasize the strategic value and business impact of your proposal. Lastly, let your passion and conviction shine through. Demonstrate your enthusiasm for the initiative and your confidence in its potential to benefit the organization.

While most business cases are presented to executive leadership for approval, there are situations where it's appropriate to take your proposal directly to the BOD. This is typically necessary when the initiative holds strategic importance, involves a substantial financial investment, or requires explicit board-level approval. In these instances, tailor your presentation to the board's specific needs and focus heavily on the strategic and financial implications of the proposal.

16.4.11 IT or Security-Specific Balance Sheet

The balance sheet is a key financial report that offers a real-time view of a company's overall financial standing at a particular moment. It lists the company's assets, liabilities, and shareholders' equity. For the CISO 3.0, creating an IT or security-specific balance sheet can be a powerful way to communicate the value and financial health of the security program to the board and executives. This is a great way to "speak the language of business" and communicate the financial health and value of the security program to board directors and executives in terms they already understand.

In building this security-focused balance sheet, you should include three main components: *assets*, which represent the resources the security program owns or controls, such as hardware, software, intellectual property, and critical data; *liabilities*, which cover the debts or obligations the program has, like vendor contracts, outstanding payments, or legal risks; and *equity*, the net value of the security program's assets after deducting liabilities, which reflects the overall worth of the program.

To create this type of balance sheet, start by *identifying assets*. Inventory all the tangible and intangible resources the security program manages. This could involve physical hardware like servers and firewalls, software licenses, proprietary tools, and valuable data such as customer information or intellectual property. Next, assign monetary values to these assets based on their purchase cost, depreciation, or current market value.

Identifying liabilities is the next step, which involves listing debts or obligations tied to the security program. This could range from ongoing vendor contracts to outstanding invoices, and even potential legal exposure from regulatory issues or past security incidents. Finally, calculate *equity* by subtracting liabilities from assets to determine the program's overall value.

When preparing this type of balance sheet, keep in mind the unique aspects of IT and security-related assets. For instance, the value of hardware and software can depreciate quickly due to technological advancements, so regular updates to asset valuations are necessary. Valuing intangible assets like intellectual property or proprietary data can also be tricky, requiring careful judgment of their impact on business operations and their potential value.

IT Balance Sheet

	Historical results				Forecast period			
	2020	2021	2022	2023	2024	2025	2026	2027
Balance sheet								
Assets								
Cash	xx	xx	xx	xx	xx	xx	xx	xx
Accounts	xx	xx	xx	xx	xx	xx	xx	xx
Inventory	xx	xx	xx	xx	xx	xx	xx	xx
Current assets	xx	xx	xx	xx	xx	xx	xx	xx
Property	xx	xx	xx	xx	xx	xx	xx	xx
Goodwill	xx	xx	xx	xx	xx	xx	xx	xx
Total	xx	xx	xx	xx	xx	xx	xx	xx

FIGURE 16.4 Balance Sheet.

Liabilities, particularly legal risks tied to cybersecurity, can also be difficult to quantify. It's important to work closely with legal and risk management teams to estimate potential exposure from data breaches, compliance issues, or regulatory penalties, and ensure they are appropriately reflected in the balance sheet (Figure 16.4).

The balance sheet can be prepared either with historical data to show the current financial standing of the security program or as a forecast to project future needs and resources. A *historical balance sheet* provides a view of the current financial position and is useful for tracking trends over time, while a *forecast balance sheet* looks ahead to future investments and expenses. Forecasting is essential for strategic planning and justifying future security budgets.

To effectively forecast for IT and security, consider several key factors: historical spending patterns, insights from risk assessments, technology upgrade roadmaps, and potential future threats. Analyzing these factors will help estimate the necessary investments to protect the organization and ensure that the security program remains aligned with both current and emerging challenges. Accurate forecasting supports better decision-making and strengthens the case for security investments, ultimately leading to more effective risk mitigation and long-term business success.

16.4.12 Risk Reduction

While we've extensively covered how to communicate risk effectively in Chapters 6 and 9, let's take a moment to examine what *not* to do when building board presentation slides on risk. The goal is to avoid common pitfalls that can hinder clear communication and decision-making.

One common mistake is relying solely on ordinal metrics like "high", "medium", and "low" to describe risk levels. These metrics, while simple to understand, lack the precision and objectivity needed for informed decision-making. They are subjective and can lead to misinterpretations, as different individuals might have varying perceptions of what constitutes a "high" or "low" risk.

The typical risk matrix, a staple in many risk management presentations, further compounds this issue. It plots risks on a grid based on their likelihood and impact, often using ordinal scales for both axes (Figure 16.5). This approach suffers from several limitations:

- **Subjectivity:** The assessment of likelihood and impact is often based on subjective opinions rather than objective data.
- **Lack of Precision:** Ordinal scales provide limited granularity, making it difficult to differentiate between risks with similar ratings.
- **Difficulty in Aggregation:** Combining risks across different categories or business units can be challenging due to the lack of a common measurement scale.
- **Limited Actionability:** The risk matrix often fails to provide actionable insights for decision-makers, leaving them with a vague understanding of the organization's risk exposure.

Another common problem, as we discussed in Chapter 6, is that CISO 2.0s often confuse threats, vulnerabilities, and risks. They tend to include threats and

Risk Rating

Standard Risk Matrix for any Business

Impact	5	Medium / High	Medium / High	High	High	High
	4	Low / Medium	Medium / High	Medium / High	High	High
	3	Low / Medium	Low / Medium	Medium / High	Medium / High	High
	2	Low	Low	Low / Medium	Low / Medium	Medium / High
	1	Low	Low	Low	Low / Medium	Low / Medium
		1	2	3	4	5
		Likelihood				

FIGURE 16.5 Risk Matix.

vulnerabilities in their risk reporting to the board instead of converting IT risks into business risks. Similarly, as we saw in Chapter 15, reporting performance metrics and KPIs instead of business risk metrics is another frequent mistake.

Let's take a look at a real-world example. This is an actual board slide that a client used, based on a template they found online. The names and some details have been changed to protect the client's anonymity. Take some time to look it over and think about whether this is a good board slide (Figure 16.6). What is good about it? What is not? Can you spot any of the common mistakes?

Let's break this slide down. First, are the "risks" actually business risks? No. Terms like "Data exfiltration" and "Malware" represent threats, not risks, and lack the all important "so what?" factor for board members. The focus should be on the potential impact or loss associated with these threats. The "Description" column hints at this in a few instances, but it's not consistently applied.

You'll also notice the use of ordinal metrics like "high", "medium", and "low", which isn't always inherently bad but, in this case, doesn't help tell the story. For example, "Mitigation Complexity", is marked as "high" for the first three risks, even though #1 is data exfiltration, which will require data access governance and DLP-type capabilities, while #2 and #3 are malware and viruses, which require endpoint detection and response (EDR). Anyone who has rolled out those controls knows that data controls are orders of magnitude more difficult to implement than endpoint controls, but both are listed as "high", making them appear equivalent.

We encounter a similar issue when we look at the "Likelihood" and "Impact" columns. All of the risks from #2 to #7 are listed as "High Likelihood" and "Medium Impact". Are all seven of those truly the same? If so, how did they prioritize them in

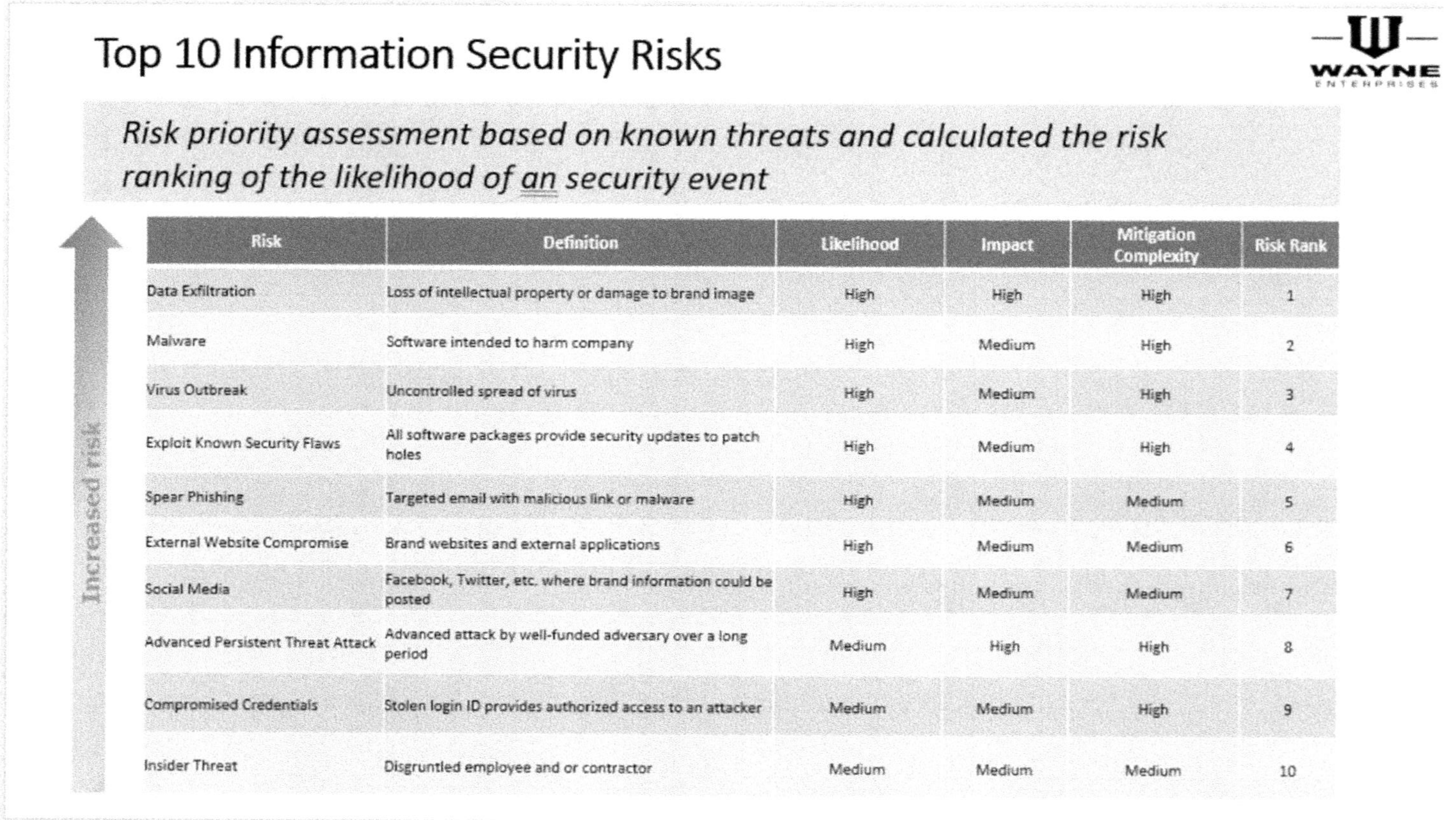

Top 10 Information Security Risks

Risk priority assessment based on known threats and calculated the risk ranking of the likelihood of an security event

Increased risk

Risk	Definition	Likelihood	Impact	Mitigation Complexity	Risk Rank
Data Exfiltration	Loss of intellectual property or damage to brand image	High	High	High	1
Malware	Software intended to harm company	High	Medium	High	2
Virus Outbreak	Uncontrolled spread of virus	High	Medium	High	3
Exploit Known Security Flaws	All software packages provide security updates to patch holes	High	Medium	High	4
Spear Phishing	Targeted email with malicious link or malware	High	Medium	Medium	5
External Website Compromise	Brand websites and external applications	High	Medium	Medium	6
Social Media	Facebook, Twitter, etc. where brand information could be posted	High	Medium	Medium	7
Advanced Persistent Threat Attack	Advanced attack by well-funded adversary over a long period	Medium	High	High	8
Compromised Credentials	Stolen login ID provides authorized access to an attacker	Medium	Medium	High	9
Insider Threat	Disgruntled employee and or contractor	Medium	Medium	Medium	10

FIGURE 16.6 Example Slide.

the risk ranking column? Would it matter what order we tackled them in? The arrow on the right says that the items higher on the list are riskier. Is that true?

This slide is not an outlier; it's representative of many presentations I've seen CISOs deliver to executives and boards. Later in this chapter, we'll explore better ways to create slides and dashboards that incorporate quantified risk metrics, enabling more impactful communication.

16.4.13 Board Books

In addition to presentations, CISOs may be required to provide written reports or other deliverables to the board. These documents, often referred to as board books or board packets, should be thorough yet concise, providing the board with the information they need to fulfill their oversight responsibilities.

A board book is a collection of documents and reports that provide board members with in-depth information about the company's performance, challenges, and strategic initiatives. It is typically compiled and distributed in advance of board meetings, allowing members time to review the material and prepare for informed discussions and decision-making. The board book is a valuable reference document, offering an overview of the company's operations and strategic direction. It is imperative that board members thoroughly review the board book in advance of meetings to come prepared with insightful questions and contribute meaningfully to discussions, leading to more productive outcomes.

While both board books and board presentations serve to inform the board, they differ in their format, purpose, and level of detail. The board presentation is a concise and focused overview of key issues, designed to engage the board and facilitate discussion during the meeting. It typically includes high-level summaries, visual aids, and actionable recommendations. The board book, on the other hand, provides a more in-depth analysis and supporting data for board members to review at their own pace before the meeting.

Here's a comparison of the typical content found in each:

Presentations may include

- executive summary
- risk decision points
- financial results
- strategic plan updates
- industry trends
- key risk indicators (KRIs)

The documents package may include

- detailed financial reports
- market research analysis
- legal documents and contracts
- competitive analysis
- long-term projections and forecasts

- background information on agenda items
- Regulatory Compliance Findings
- high importance threats and vulnerabilities
- operational metrics/KPIs

The presentation and the board book serve different purposes and require distinct editing approaches. The presentation should be kept concise and focused, highlighting only the most critical information and insights. Utilize visuals and storytelling techniques to engage the board and facilitate discussion during the meeting.

In contrast, the board book should provide more in-depth analysis and supporting data for board members to review at their own pace before the meeting. Ensure the information is well-organized, clearly presented, and free of errors. The board book should serve as a reference document that complements the presentation and enables the board to probe deeper into specific topics as needed. It is also essential to tailor the board book content to the specific board members and their areas of expertise, ensuring the information resonates with their interests and concerns.

The CISO plays an important role in collaborating with other executives to prepare the board book, ensuring that cybersecurity is adequately represented and aligned with the overall business strategy. They should understand their responsibility in contributing to the board book and guaranteeing its accuracy and completeness. Also, consider the frequency and timing of board book distribution. Factors such as the frequency of board meetings and the time required for members to review the material should be taken into account. The CISO 3.0 should understand the importance of timely distribution to allow for adequate preparation. They should also be aware of and leverage various technology solutions available for creating, managing, and securely distributing board books, streamlining the process, and enhancing collaboration.

Pro Tip: Given the sensitive nature of the information often included in board books, it's vital to implement appropriate security measures to protect it from unauthorized access or disclosure. The CISO 3.0 should be aware of the potential risks associated with board book distribution and take necessary safeguards.

16.4.14 Crafting Board Books

The first step in creating a board book is to actively collaborate with key stakeholders across the organization. This includes engaging with the executive team, business unit leaders, and subject matter experts to gather relevant information and insights. Their perspectives can provide valuable context and ensure that the board book addresses the organization's most pressing concerns.

Leverage existing reports, dashboards, and data sources to inform your analysis and provide supporting evidence for your findings. This not only saves time and effort but also ensures consistency and accuracy in your reporting.

When compiling the board book, focus on identifying the most critical metrics and trends that are relevant to the board's oversight responsibilities. Avoid overwhelming them with excessive data; instead, prioritize the information that is most impactful and actionable.

Remember, it's not enough to simply present data. Provide context and analysis to help the board understand the significance and implications of the information. Explain trends, highlight potential risks, and offer insights into the effectiveness of current strategies.

Finally, strive for conciseness and clarity in your writing. Avoid technical jargon and focus on the key takeaways and actionable insights. The board book should be a valuable resource that empowers the board, not a dense and overwhelming document.

16.4.15 Creating Board-Level Dashboards

A well-designed dashboard can be a powerful tool for communicating cybersecurity information to the board. It provides a visual snapshot of the organization's security posture, highlighting key metrics and trends that are relevant to the board's oversight responsibilities.

When creating a board-level dashboard, the focus should be on presenting a clear and concise overview of the security program's performance and its impact on the business. The dashboard should tell a story that resonates with the board's priorities and concerns, highlighting the value and effectiveness of your cybersecurity investments.

The story conveyed through your dashboard should begin by transparently accounting for how the business's funds have been utilized since your last interaction. This fosters accountability and shows the board the tangible outcomes of their investments. Next, translate technical accomplishments into concrete business benefits, such as quantifiable risk reduction, cost savings, or revenue generation. This shift in perspective helps the board grasp the value proposition of your security efforts.

It's also essential to demonstrate progress. Show how your current performance measures up against the expectations set in previous discussions, highlighting successes, and acknowledging areas that need improvement. This reinforces your commitment to accountability and continuous enhancement.

Provide a clear and concise overview of the organization's current cyber risk exposure. Focus on the most critical risks and their potential business impact, painting a realistic picture of the threats. Outline your proposed plan for managing and mitigating these risks, including specific initiatives and their associated costs. This proactive approach showcases your strategic thinking and commitment to safeguarding the organization's assets. Quantify the expected benefits of your proposed plan, such as further risk reduction, cost savings, or revenue generation. This helps the board understand the potential ROI of their continued support. Then, justify your recommendations by explaining the rationale behind your choices and how they align with the organization's risk appetite and strategic objectives.

The story you tell through your dashboard should address the following key points:

How You Spent the Business's Money Since You Last Met with Them: This demonstrates accountability and transparency, showing the board how their investments have been utilized.

What They Got for Their Money in Business Terms: Translate technical accomplishments into tangible business benefits, such as risk reduction, cost savings, or revenue generation. This helps the board understand the value of their investment.

How That Measures against the Expectations You Set Last Time: Show progress toward previously established goals and objectives, highlighting successes and addressing any areas where improvement is needed. This demonstrates accountability and commitment to continuous improvement.

The Business' Current Cyber Risk Exposure: Provide a clear and concise overview of the organization's current risk posture, focusing on the most critical risks and their potential impact on the business.

What Is Your Plan? How You Intend to Spend the Business' Money: Outline your proposed plan for managing and mitigating risks, including specific initiatives and their associated costs. This demonstrates proactiveness and strategic thinking.

What Return Can They Expect to Get in Business Terms: Quantify the expected benefits of your proposed plan, such as further risk reduction, cost savings, or revenue generation. This helps the board understand the potential ROI of their investment.

Why You Made This Decision Instead of Investing the Business' Money in Something "Better": Justify your recommendations by explaining the rationale behind your choices and how they align with the organization's risk appetite and strategic objectives.

The risk quantification example from Chapter 9 can be effectively used to tell these stories. you can provide the board with a clear and compelling picture of the value of your program by assigning financial values to risks and demonstrating the risk reduction potential of your security initiatives,

Here is an example of board-level dashboard.

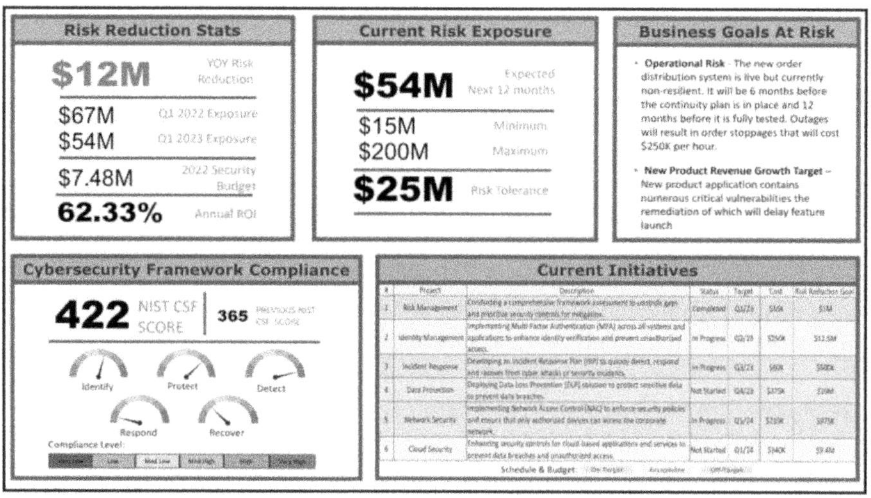

FIGURE 16.7 Board Dashboard.

The example in Figure 16.7 illustrates a board-level dashboard that effectively communicates the key aspects of a cybersecurity program in a concise and impactful manner.

Risk Reduction Stats: This box tells the story of "How We Spent Your Money and What You Got for It". It quantifies the risk reduction achieved through security investments, demonstrating the value and effectiveness of the program.

Current Risk Exposure: This box answers the question "What is our risk exposure in dollars versus our target exposure?" It provides a clear picture of the organization's current risk posture in financial terms, highlighting any gaps between current and desired risk levels.

Business Goals at Risk: This box explains "What the risk information means for our business". It translates cyber risks into their potential impact on the organization's strategic objectives, showcasing the business consequences of security failures.

Compliance Framework: This box provides external validation of your claims by showcasing compliance with relevant industry standards or regulatory frameworks. It demonstrates the organization's commitment to security best practices and can enhance stakeholder confidence.

Current Initiatives: This box tells the stories "Here's the plan, how we intend to invest your money, what you will get for your investment, and how it's going". It provides an overview of ongoing and planned security initiatives, their associated costs, expected benefits, and progress updates. This transparency fosters trust and demonstrates accountability.

This dashboard offers a quick overview of the security program's performance, risk exposure, and strategic alignment. It allows board members to grasp the key information at a glance and facilitates informed discussions and decision-making. While presented here as a single dashboard, each of these boxes and stories could be expanded into its own slide for a more detailed presentation, depending on the board's needs and the time available.

16.4.16 Creating Executive-Level Dashboards

While board-level dashboards provide a high-level overview of the security program's performance and risk posture, executive-level dashboards cater to a different audience with distinct needs and priorities. Executive leaders, such as the CIO, CFO, CRO, and CEO, require more granular insights into the security program's status and its impact on the business. Although executive dashboards are similar to board-level dashboards, executive-level dashboards should generally go deeper into specific metrics and KPIs that are relevant to the day-to-day operations and decision-making of the executive team. They should provide actionable insights that enable executives to make better, more data-driven decisions about resource allocation, risk mitigation, and security investments.

The stories told through executive-level dashboards should focus on the following:

Risks and Threats in Business Terms: Clearly articulate the specific risks and threats facing the organization, translating technical jargon into understandable business language. Quantify the potential financial and operational impact of these risks, emphasizing the consequences of inaction.

Investment Strategy: Present your proposed plan for addressing identified risks, including specific initiatives, their associated costs, and the expected ROI in business terms. Justify your recommendations by explaining the rationale behind your choices, comparing them to alternative options, and highlighting their alignment with the organization's risk appetite and strategic objectives.

Timing and Urgency: Explain why you've chosen to tackle these risks now, emphasizing the potential consequences of delaying action. This helps executives understand the urgency and prioritization of your recommendations.

Expected Returns: Quantify the expected benefits of your proposed plan, such as risk reduction, cost savings, or revenue generation. This allows executives to evaluate the value proposition of your recommendations.

Risk Treatment Options: Offer executives one to three viable options for addressing each risk, presented in clear business terms. These options should typically fall into the categories of mitigate, avoid, transfer, or accept. Explain the trade-offs associated with each option, empowering executives to make decisions based on their risk appetite and strategic priorities.

The example dashboard above demonstrates how to effectively communicate cybersecurity information to executive leadership (Figure 16.8).

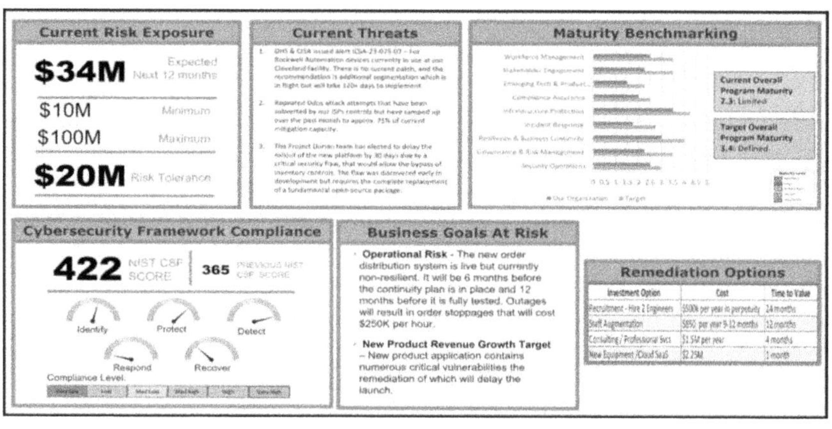

FIGURE 16.8 Executive Dashboard.

Current Risk Exposure: This box answers the question, "What is our risk exposure in dollars versus our target exposure?" It provides a clear picture of the organization's current risk posture in financial terms, highlighting any gaps between current and desired risk levels.

Current Threats: This box explains the emerging threat trends that the business should be aware of, providing context and highlighting the potential impact on the organization.

Maturity Benchmarks: This box tells the story of how the security program is performing and improving over time, showcasing progress toward maturity goals, and demonstrating the effectiveness of security investments.

Cybersecurity Framework Compliance: This box, similar to the board-level dashboard, provides external validation of the security program by showcasing compliance with relevant industry standards or regulatory frameworks.

Business Goals at Risk: This box explains which specific business goals are at risk due to identified cyber threats, enabling executives to understand the potential impact on the organization's strategic objectives.

Remediation Options: This box provides options to address the risks listed in the dashboard and offers discussion points for right-sizing investment levels to mitigate them. It presents a range of choices, from accepting the risk to implementing various mitigation strategies, allowing executives to better understand their choice options based on their risk appetite and priorities.

This executive-level dashboard provides a more detailed and actionable view of the security program's performance and risk posture, which helps the executives better execute their duties. Again, each of these boxes and stories could be expanded into its own slide for a more detailed presentation.

16.4.17 Balanced Scorecard

The balanced scorecard is a strategic tool designed to convert an organization's vision and goals into measurable outcomes across four essential areas: Financial, customer, internal processes, and learning and growth. It offers a comprehensive

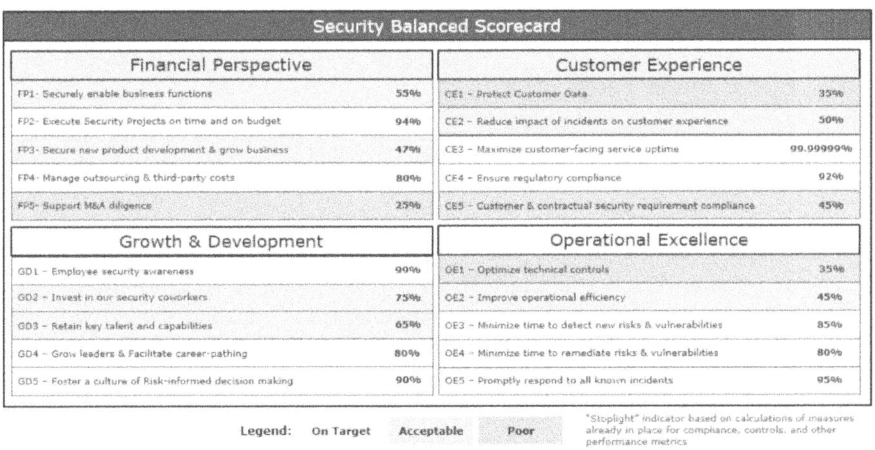

FIGURE 16.9 Balanced Scorecard.

view of performance, extending beyond traditional financial metrics to also track customer satisfaction, operational efficiency, and the organization's capacity for innovation.

Originally introduced by Dr. Robert Kaplan and Dr. David Norton in the early 1990s, the balanced scorecard has become a widely adopted approach to strategic management. Its strength lies in its ability to link long-term strategic objectives with day-to-day activities, fostering a culture of alignment and continuous improvement across all levels of an organization.

For CISOs, the balanced scorecard offers a powerful tool to "speak the language of business" and communicate the value of their security programs in a format familiar to executives and board directors. Aligning security metrics with the four perspectives of the balanced scorecard is a smart way for CISOs to demonstrate how their initiatives contribute to the organization's overall success (Figure 16.9).

To integrate cybersecurity initiatives into a balanced scorecard framework, CISOs should align their security goals with the organization's broader strategic objectives, ensuring that security efforts contribute to overall business success. Here's how to approach the process.

16.4.17.1 Financial Perspective

The financial perspective focuses on the financial health of the organization. Cybersecurity initiatives can directly impact financial performance by reducing risk, avoiding costly breaches, and ensuring regulatory compliance.

- **Example Metrics**:
  - **Cost of Cyber Incidents:** Measure the financial losses avoided by preventing or mitigating security incidents.
  - **Return on Security Investment (ROSI)**: Calculate the financial return generated from investments in security tools or measures.
  - **Compliance Costs**: Track expenses related to meeting regulatory requirements and the cost savings associated with avoiding noncompliance penalties.

16.4.17.2 Customer Perspective

This perspective measures how well the organization meets customer needs. In terms of cybersecurity, protecting customer data and ensuring trust in digital services are critical.

- **Example Metrics:**
  - **Customer Trust and Confidence:** Use customer satisfaction surveys to measure trust in how their data is handled.
  - **Data Breach Impact on Customer Churn:** Assess how incidents of data breaches affect customer retention and acquisition.
  - **Incident Response Time:** Track how quickly the organization responds to security incidents that affect customers.

16.4.17.3 Internal Processes Perspective

The internal processes perspective focuses on improving organizational efficiency and effectiveness. For *cybersecurity*, this includes the processes that protect the organization from threats.

- **Example Metrics:**
  - **Current Cyber Risk Exposure in Dollars**: Calculate the organization's estimated financial exposure based on known vulnerabilities, current threat landscape, and potential impact of a successful attack. This KRI quantifies risk in monetary terms, allowing leadership to understand the potential cost of inaction.
  - **Unauthorized Access Attempts on Critical Systems**: Monitor the number of unauthorized access attempts on critical infrastructure or data repositories. An increasing frequency of these events could indicate a heightened risk of targeted attacks.
  - **Percentage of Unpatched Critical Vulnerabilities**: Track the percentage of systems with critical vulnerabilities that remain unpatched past the organization's established deadline. A high percentage would indicate an elevated risk of exploitation.

16.4.17.4 Learning and Growth Perspective

This perspective emphasizes the organization's ability to innovate and grow, which includes the development of employees' skills and capabilities. For cybersecurity, this includes training and development in security best practices and emerging threats.

- **Example Metrics:**
  - **Employee Security Training Completion Rates:** Measure the percentage of employees completing required security awareness training.
  - **Skill Development in Cybersecurity Teams:** Track certifications earned or new skills acquired by the security team to keep up with evolving threats.
  - **Security Culture Maturity:** Use surveys or assessments to measure the organization's overall security culture and awareness levels.

Incorporating cybersecurity into a balanced scorecard involves a structured approach that connects security initiatives directly to broader organizational goals. The process starts with identifying the organization's strategic objectives. This is essential for ensuring that cybersecurity efforts are aligned with the overarching mission of the company. For instance, if enhancing customer trust is a key goal, cybersecurity measures like encryption and secure access protocols become directly linked to that objective by protecting sensitive customer information.

Once the strategic objectives are clearly defined, the next phase is selecting relevant metrics that correspond to each of the four perspectives of the balanced scorecard: Financial, customer, internal processes, and learning and growth. These metrics should be chosen carefully to reflect the organization's unique risk profile and security priorities. For example, a financial metric could be the current *cyber risk exposure in dollars*, which quantifies the potential financial impact of security threats. The internal processes category is the rare instance where you might share very targeted performance metrics. Metrics such as the percentage of *unpatched critical vulnerabilities* or the *time to detect and mitigate security incidents* can provide insights into how well the organization is managing its cybersecurity defenses.

Establishing benchmarks and setting realistic targets for these metrics is the next critical step. Performance benchmarks can be based on industry standards or past performance, while the targets should reflect the desired outcomes. For instance, a target might be to reduce the number of unpatched vulnerabilities by 30% within 6 months or to improve incident response times by a specific margin. These benchmarks and targets offer measurable goals that help track the effectiveness of cybersecurity initiatives.

Once metrics are in place, ongoing monitoring and reporting are essential to keep the program accountable and on track. Regular performance tracking ensures that leadership understands the value of the cybersecurity efforts in the context of the organization's goals, whether that value is in reducing risk, cutting costs, or boosting customer confidence. Reporting these metrics should be done in a manner that aligns with the strategic objectives of the company, making it easier for executives to see the direct connection between security initiatives and business success.

To further enhance this approach, you can use a *Balanced Scorecard Strategy Map* to visualize the relationships between strategic objectives and how cybersecurity contributes to them. A strategy map is especially helpful in the initial planning phase, as it highlights the cause-and-effect relationships across the four perspectives, making it easier to pinpoint where cybersecurity plays a crucial role. For example, in the customer perspective, if the organization's goal is to build trust and ensure data privacy, the strategy map would make it clear that cybersecurity initiatives such as data protection protocols are essential for reaching this objective.

The strategy map can also serve as a communication tool to clearly illustrate how cybersecurity initiatives align with other business objectives. For instance, reducing incident response times (an internal processes objective) directly impacts minimizing business disruption, which is a key financial goal. This makes the case for additional security investments more compelling to leadership.

When using a strategy map to build your balanced scorecard, the first step is to identify strategic themes that guide the organization, such as improving cybersecurity resilience or enhancing operational efficiency. From there, define specific strategic objectives under each of the four balanced scorecard perspectives. For example, under the financial perspective, you could aim to optimize cybersecurity spending, while under the customer perspective, the goal might be to enhance client data security.

Each strategic objective should be paired with relevant *KPIs* or *KRIs* to monitor progress and identify potential risks. For example, a KRI could track the *current cyber risk exposure in dollars*, helping leadership quantify the potential financial impact of security threats. Targets should be set for each metric, providing a benchmark for performance and motivating continuous improvement. Initiatives that directly contribute to achieving these targets, such as rolling out new security tools or enhancing employee training programs, should be identified and implemented.

Finally, the balanced scorecard provides a framework for regular tracking and reporting. By linking cybersecurity initiatives to organizational goals through KPIs and KRIs, the scorecard helps CISOs communicate the value of their efforts in a format familiar to executives. This ensures that cybersecurity remains a visible, integral part of the organization's overall strategy, continuously adapting to emerging threats and business changes

16.4.18 Balanced Scorecard Strategy Map

The Balanced Scorecard Strategy Map is a powerful visual tool designed to clarify how an organization's strategic objectives connect to the key drivers that propel those objectives. It provides a concise, interconnected view of initiatives and activities across the four main perspectives of the balanced scorecard: Financial, customer, internal processes, and learning and growth. By illustrating these cause-and-effect relationships, the map demonstrates how different elements of the strategy support one another and contribute to achieving the broader organizational goals.

At its core, a Balanced Scorecard Strategy Map functions as a cause-and-effect chain that connects desired outcomes with the drivers of those results across the organization's scorecard domains. For example, the financial perspective might be linked to the internal process perspective, where improvements in operational efficiency lead to better financial outcomes. Each domain, whether financial, customer, internal processes, or learning and growth, serves as both an input and an outcome, with the relationships between them illustrating the flow of value through the organization.

The strategy map is particularly useful during both the planning and implementation phases of strategic initiatives. It helps organizations define their strategic objectives clearly, identify the key drivers of success, and visualize how these components interact. This clarity is essential for ensuring that all initiatives across the organization align with the overall strategy and support its success. The map also serves as an effective communication tool, making it easier for stakeholders to understand the strategy and engage with it meaningfully. Additionally, the strategy map enables organizations to monitor progress toward their strategic objectives and make adjustments as needed.

One of the advantages of the Balanced Scorecard Strategy Map is that it offers flexibility in how you structure your goals and outcomes. You can adjust the starting point of the strategy map depending on your specific needs. For instance, if your organization is highly focused on financial results, you might begin with the financial perspective and map how operational process improvements, customer satisfaction, or employee growth drive those outcomes. On the other hand, if operational process maximization is your main goal, you might begin with internal processes and show how that leads to greater financial outcomes or enhanced customer experience.

In some cases, the customer in the scorecard map refers to the organization's external customers. However, for a security team, the "customer" might be internal – the business itself or other departments that rely on security to function smoothly. By tailoring the customer perspective to fit your team's role, you can communicate how security contributes to organizational goals like customer trust, business continuity, or operational efficiency.

Additionally, the strategy map can reflect your reporting structure. It's possible that in some organizations, the CISO reports to the CIO, CTO, or COO, meaning you may be more abstracted from direct financial goals. In such cases, your starting point may be in the learning and growth domain, showing how employee training and skill development lead to operational improvements that eventually result in financial gains. Conversely, if you report directly to the CFO and budget is less of a concern, you might start with the financial perspective and illustrate how security investments support broader strategic goals. There are even situations where flipping the order entirely is necessary, particularly in organizations where information technology infrastructure library (ITIL) or other frameworks are being utilized to drive value from process improvements back to financial outcomes.

The process of building a Balanced Scorecard Strategy Map begins with identifying the key strategic themes that will guide the organization toward success. These themes must align closely with the organization's mission, vision, and core values to ensure that the strategy reflects the company's fundamental principles and long-term aspirations. Once these strategic themes are set, the next step is defining specific objectives for each theme, with an emphasis on SMART (specific, measurable, achievable, relevant, and time-bound) goals to create a framework for tracking success.

After defining these objectives, it's important to identify key drivers that will contribute to achieving those goals. These can include internal factors like employee expertise and technological resources, or external influences like market trends or regulatory changes. Once these relationships between objectives and drivers are mapped out, it becomes easier to see the interconnectedness of the various organizational elements and how each initiative contributes to the overall strategy.

Once the map is created, validating it with key stakeholders is critical. Gathering feedback ensures that the strategy map accurately reflects the organization's priorities and provides an opportunity to fine-tune it based on different perspectives. This collaboration enhances the strategy's effectiveness and builds stronger buy-in across departments.

For example, a cybersecurity strategy map might focus on enhancing the organization's security posture (Figure 16.10). Strategic objectives could include reducing the risk of data breaches, improving incident response times, enhancing employee security awareness, and optimizing security investments. Key drivers would support these objectives, such as implementing robust security controls, conducting regular risk assessments, providing ongoing employee training, and aligning security investments with broader business goals.

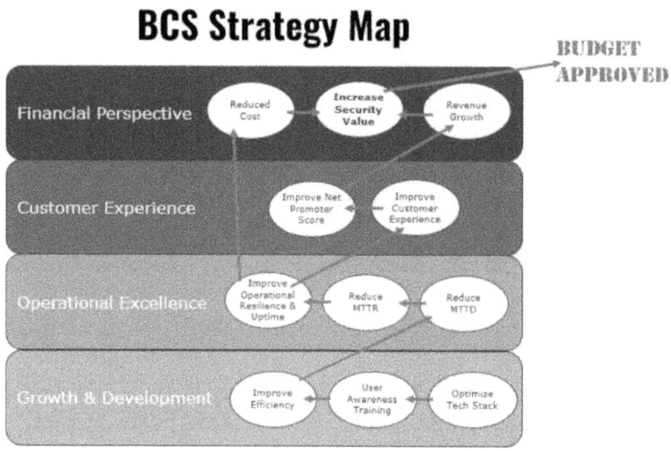

FIGURE 16.10 BSC Strategy Map.

16.4.18.1 Example

A cybersecurity strategy map might include the following elements:

- **Strategic Theme:** Enhance cybersecurity posture
- **Strategic Objectives:**
  - Reduce the risk of data breaches
  - Improve incident response times
  - Enhance employee security awareness
  - Optimize cybersecurity investments
- **Key Drivers:**
  - Implement robust security controls
  - Conduct regular risk assessments
  - Provide ongoing employee training
  - Align security investments with business goals

16.5 CONCLUSION

In this chapter, we've explored the importance of effective communication for a CISO 3.0. One of the most important features of a CISO 3.0 is the capacity to translate complex technical information into clear, concise, and impactful messages that resonate with diverse stakeholders, particularly the BOD and executive leadership. The CISO 3.0 must be adept at articulating the financial and strategic benefits of cybersecurity investments, demonstrating how security initiatives contribute to revenue generation, cost reduction, and risk mitigation. It's not merely about presenting data; it's also about crafting compelling narratives that leverage data storytelling and visual aids to convey the value and impact of the security program.

We've talked about tailoring your message to different audiences, understanding their concerns, and anticipating their questions. It's about building trust and getting buy-in for those critical security projects. We've also covered the power of storytelling and visuals, showing how to use data to create a narrative that resonates and inspires action.

We've also provided a roadmap for creating those all-important board-level deliverables, which are presentations and dashboards that tell the story of your security program's success. And we've emphasized the importance of clarity, conciseness, and strategic alignment in everything you present. The use of tools like the Balanced Scorecard and Strategy Map further enables the CISO 3.0 to "speak the language of business" and demonstrate the value of their security program in a format familiar to executives and board directors.

I know that stepping into the boardroom can feel like entering a lion's den, especially for CISOs more accustomed to the comfort of their tech lairs. But remember, the board isn't out to get you: They're on your side. They're eager to grasp the risks and your game plan for tackling them. So, walk in with your head held high, lay your cards on the table, and, most importantly, talk their talk.

16.5.1 Key Takeaways

1. **The Importance of Board-Level Communication**
   - Effective board-level communication is critical for CISOs, enabling alignment of cybersecurity initiatives with business objectives.
   - Presenting cybersecurity as a strategic enabler rather than a technical issue fosters understanding and trust among board members.
2. **Goals of Board Presentations**
   - Board communications should educate, inform, update, and instill confidence in the organization's cybersecurity posture.
   - Key focuses include demonstrating fiduciary responsibility, aligning with the organization's risk appetite, and emphasizing the business value of security investments.
3. **Storytelling and Risk Communication**
   - Risk-based storytelling connects technical risks to business impacts, making them relatable to nontechnical audiences.
   - Using clear, concise language and relatable analogies enhances comprehension and engagement.
   - Data storytelling combines metrics with narratives to illustrate cybersecurity's strategic value and guide decision-making.
4. **Tailoring Messages for Different Audiences**
   - Tailored communication for boards emphasizes strategic alignment, risk management, and financial impacts.
   - Presentations for executive teams may include more detailed technical information but should still align with business goals.
5. **Tools and Techniques for Effective Communication**
   - Use structured frameworks like the BOARD acronym (brief, open, accurate, risk aligned, demonstrate leadership) for concise and impactful presentations.
   - Visual aids such as heatmaps, dashboards, and infographics simplify complex data and enhance engagement.
   - Employ storytelling arcs (setting, characters, conflict, and resolution) to structure compelling narratives.
6. **Preparing for Board Questions**
   - Understanding the types of questions (service delivery, business partner, and strategic leadership) helps anticipate and address board concerns effectively.
   - Accurate, transparent, and concise answers build credibility and ensure informed decision-making.
7. **Board Deliverables and Presentations**
   - Board decks should include an executive summary, key risk indicators, financial implications, and strategic recommendations.
   - Board books provide in-depth reports and supporting data for pre-meeting reviews.

16.5.2 Reflection and Exploration Questions

Understanding the Concepts

1. What are the main goals of board-level communication for a CISO, and how do they align with organizational objectives?
2. How does risk-based storytelling enhance the effectiveness of board presentations?

Connecting the Dots

3. How do the principles of data storytelling and visual aids connect to earlier discussions on metrics and reporting?
4. In what ways can tailored messages for different stakeholders improve the perception and impact of cybersecurity initiatives?

Real-World Application

5. Can you identify examples where effective board communication led to better cybersecurity decisions or increased funding?
6. How might you apply the BOARD framework to refine your own presentation skills?

Personal and Professional Growth

7. How can you develop your storytelling and data visualization skills to communicate more effectively with boards and executives?
8. What steps can you take to anticipate and prepare for strategic leadership questions from board members?

Further Exploration

9. What additional techniques or tools could enhance your ability to communicate complex cybersecurity issues in business terms?
10. How can you begin to align your board deck with your organizational goals and risk tolerance?

17 Materiality and Disclosures

17.1 INTRODUCTION

Recently, the concept of materiality has taken center stage, particularly in light of recent regulatory developments. The Securities and Exchange Commission (SEC) and other regulatory bodies have introduced new rules and guidelines that require organizations to disclose material cybersecurity incidents and risks: Basically, saying, "Hey, if you have a cybersecurity incident or risk that could really mess things up for your company, you need to tell us and your investors about it".

This chapter provides a timely exploration of materiality and disclosures. We'll explore how chief information security officers (CISOs), armed with their knowledge of fiduciary duties, business lingo, and risk quantification, can figure out what's "material" and what's not. We'll also walk you through the reporting process, including how to work with your legal team, and we'll share some hard-earned lessons from recent cases like the SolarWinds fiasco.

Determining materiality is also the third step of the risk strategy framework introduced in Chapter 6. Understanding what is material is key to the entire goal of the risk strategy: Building resilience against material risks. To right-size risk mitigation efforts, organizations need to identify which risks truly matter and what materiality means in their specific context. Without this clarity, efforts to reduce risk could be misaligned – either overcompensating for minor threats or underestimating significant ones.

By the end of this chapter, you'll have a better handle on how to navigate these tricky waters. You'll be able to make sure your organization is playing by the rules while also keeping your stakeholders informed about the cyber risks they face. It's about being transparent, accountable, and proactive, all of which are the hallmarks of a true CISO 3.0.

17.2 DEFINING MATERIALITY IN RISKS AND INCIDENTS

In risk management, materiality refers to the significance or importance of a particular risk in relation to an organization's overall objectives. The US SEC mandates the disclosure of material risks in financial reporting, but it doesn't explicitly define "material risk", leaving it to individual organizations to determine what constitutes materiality for them.

As we discussed in Chapter 6, understanding and defining material risk is foundational to building a security program. Renowned cybersecurity expert Rick Howard even called it the cybersecurity "First Principle". A first principle is a foundational proposition or assumption that cannot be deduced from any other proposition or

DOI: 10.1201/9781003510789-23

assumption. In other words, it's the bedrock upon which everything else is built. In the context of cybersecurity, the first principle is to reduce the probability of a material impact, highlighting the critical importance of identifying and mitigating risks that could affect the organization.

Ask yourself how your organization currently defines materiality in the context of cybersecurity risks and incidents.

The SEC's recent cybersecurity rules emphasize the importance of materiality by mandating the disclosure of not only material cybersecurity incidents but also material risks. The SEC specifically requires that material risks be disclosed in the company's annual 10-K report. The following snippets from the SEC's cybersecurity rules highlight this requirement:

> *Registrants must describe their risk management policies and procedures for the identification and management of risks from cybersecurity threats . . .*
>
> *The disclosure must describe the registrant's processes, if any, for assessing the impact of cybersecurity threats and any previous cybersecurity incidents.*

This means that organizations must proactively identify and assess potential cyber risks that could have a real impact on their business and disclose them to investors and other stakeholders.

As we discussed in Chapter 3, even if your organization isn't publicly traded, it's wise to heed the guidance of the SEC's cyber rules. These rules represent a benchmark for best practices in cybersecurity risk management and disclosure, and adhering to them can help your organization build trust with stakeholders and demonstrate a commitment to transparency and accountability. Even if disclosures aren't mandated for your organization by a regulatory body like the SEC, understanding the materiality of your risks is core to the CISO 3.0's mission.

Also, the distinction between material risks and material incidents is fundamentally important. We can assign materiality to risks, which are potential future events that could negatively impact the organization (think "left of boom"). Alternatively, we can assign materiality to incidents, which are actual events that have already occurred and caused harm (think "right of boom"). Both are necessary considerations for CISOs, as they must not only proactively manage risks but also respond effectively to incidents and transparently disclose their impact.

A material cybersecurity incident is one that has a significant impact on the organization's operations, financial performance, or brand value. The determination of materiality is not always straightforward and requires careful consideration of various factors, including the nature and scope of the incident, the potential financial losses, the impact on customers and stakeholders, and the potential for regulatory scrutiny or legal action. The CISO 3.0 should have a deep understanding of the business and its risk appetite, which will assist in assessing the materiality of cyber incidents and ensuring timely and transparent disclosure to relevant stakeholders.

A material risk refers to the potential for future events or circumstances to have a significant negative impact on the organization. In the context of cybersecurity, this could include the risk of data breaches, ransomware attacks, system outages, or regulatory fines. The assessment of material risk involves quantifying the potential financial impact, considering the likelihood of occurrence, and evaluating the effectiveness of existing controls. The CISO 3.0, armed with risk quantification tools, can effectively identify and assess material risks, enabling proactive mitigation and informed decision-making.

17.3 DETERMINING MATERIAL RISK AND INCIDENTS

Materiality is the cornerstone of effective disclosure. It's about identifying and communicating the risks that truly matter to the business. But it's not always easy to determine what's material and what's not. I've learned that the best approach is to be proactive, to engage with stakeholders early and often, and to build a culture of transparency.

Financial materiality is often defined as a percentage of revenue, typically ranging from .01% to 5%. However, in the context of cybersecurity, it's rare for a single cyber event to cause losses reaching this threshold. Thus, determining materiality for cyber risks and incidents requires a more nuanced approach, incorporating both quantitative and qualitative metrics.

The concept of materiality is central to risk management and disclosure, but it's often shrouded in ambiguity. The SEC, while mandating the disclosure of material risks and incidents, has deliberately refrained from providing a rigid definition of "materiality". This flexibility allows organizations to tailor their assessments to their unique circumstances and risk profiles. However, it also places the onus on CISOs to make informed judgments about what constitutes a material risk or incident.

The SEC's guidance on materiality in the context of cybersecurity incidents can be found in the following excerpt from the final rule:

Cybersecurity incident is considered material if

> *there is a substantial likelihood that a reasonable investor would consider it important in making an investment decision, or if it would have a significant impact on a registrant's business, operations or financial condition.*

While the SEC's definition provides a general framework, it lacks specific quantitative thresholds or criteria for determining materiality. This necessitates a nuanced approach that combines quantitative data with qualitative insights. Determining materiality requires a careful evaluation of both quantitative and qualitative factors. The quantitative aspect focuses on the potential financial impact of the risk or incident, often expressed as a percentage of revenue or a specific monetary value. The qualitative aspect considers the broader implications, such as the impact on operations, brand value, customer trust, and regulatory compliance.

The core distinction between material risks and material incidents lies in their temporal orientation. Material risks are inherently prospective, concerning potential future events or circumstances that could negatively impact the organization.

The assessment of these risks involves a proactive approach, utilizing risk quantification techniques to estimate the potential financial impact, considering the likelihood of occurrence, and evaluating the effectiveness of existing controls. The goal is to anticipate and mitigate potential threats before they materialize into actual incidents.

In contrast, material incidents are retrospective, focusing on events that have already transpired and caused harm to the organization. Determining materiality for incidents requires a reactive assessment of the nature and scope of the incident, the potential financial losses, the impact on customers and stakeholders, and the potential for regulatory scrutiny or legal action. The focus here is on understanding the consequences of an event and taking appropriate remedial action.

While the temporal focus differs, both material risks and material incidents share a common thread: Their potential to impact the organization's ability to achieve its objectives. The assessment of requires both a point of view that considers both financial and nonfinancial factors, and the organization's specific context and risk appetite.

So, how do we determine materiality? Recent research has shed light on how auditors have traditionally defined material risk, often employing a two-factor test. The first part of the test involves a quantitative assessment of financial transactions to see if they impact an enterprise's financial statements. Benchmarks like pre-tax income, total assets, equity, or total revenue are used, and if financial changes surpass predetermined thresholds, they signal the need for attention. If an incident's costs surpass the established threshold and benchmark, it would be hard to consider the incident immaterial. However, the absence of a quantitative materiality doesn't automatically dismiss the incident's materiality. For incidents falling below the threshold, the second part of the test would evaluate qualitative factors and should include anything a reasonable investor would deem relevant to their investment decisions. Even if an incident doesn't meet the quantitative benchmark, it could still be considered material based on this second qualitative assessment.

The process for determining materiality follows a structured three-step approach. First, a quantitative assessment is conducted. Risks or incidents are evaluated against specific financial thresholds, for instance, 0.01% of revenue for incidents and 5% for risk-based metrics. If these thresholds are met or exceeded, the issue is flagged as potentially material.

Next, a qualitative review is performed to assess whether the risk or incident would be significant from an investor's perspective. This involves considering factors such as the type of data involved, potential regulatory impact, effects on the company's business model, and implications for market share. If executive leadership deems the issue significant, it's classified as preliminarily material.

The final step involves a formal review by executive management, the board of directors, and legal counsel. They collectively determine the materiality of the issue and decide if disclosure to the SEC is necessary. This thorough process ensures that all relevant factors are considered before any public announcement is made.

There are two more key metrics you should be familiar with for determining quantitative materiality, rate of change materiality (RoCM) and forecast accuracy materiality (FAM).

17.3.1 Rate of Change Materiality

Rate of change materiality (RoCM) underscores the dynamic nature of cyber risk. It's not enough to assess risk once and assume it remains static. RoCM encourages organizations to continuously monitor and quantify their risk scenarios, tracking how the potential financial impact of these risks evolves over time.

The process begins by establishing a baseline to determine and calculate RoCM. This involves conducting an initial quantitative risk assessment to estimate the financial impact of various cyber risk scenarios, such as the cost of a data breach or ransomware attack. Once the baseline is in place, the next step is to regularly reassess these risks. This should happen periodically, especially when changes occur in the threat landscape, IT infrastructure, or business operations.

After each reassessment, the rate of change is calculated by determining the percentage variation in the financial impact of each scenario compared to the baseline. If the rate of change for any risk scenario exceeds a predetermined threshold – say, 5% of revenue – it indicates a significant shift in the organization's risk profile. These material changes should be flagged for further investigation and possible disclosure.

17.3.2 Forecast Accuracy Materiality

Forecast accuracy materiality (FAM) focuses on the accuracy of risk forecasts. It involves a post-incident review, comparing the actual financial impact of a cyber incident to the amount that was forecasted during risk assessments. This comparison provides valuable insights into the effectiveness of the organization's cyber risk management function and the reliability of its risk models.

To determine and calculate FAM, a post-incident analysis is conducted after a cyber event. This includes assessing the financial impact in terms of both direct and indirect costs. The next step is to compare this actual financial impact to the forecasted figures from previous risk assessments. The percentage variance between the actual and forecasted impacts is then calculated.

If the variance exceeds a predefined threshold, such as 5%, the discrepancy is flagged as material. Significant variances suggest that the organization's risk assessment methodologies may need refinement, signaling a need to improve forecasting accuracy for future risk scenarios.

17.4 PRIMA: PRE-EVALUATED RISK AND INCIDENT MATERIALITY ASSESSMENT

To address the challenges of defining materiality, I propose a framework I call PRIMA (pre-evaluated risk and incident materiality assessment). This approach integrates quantitative data with qualitative insights, ensuring a multidimensional analysis of risk. PRIMA delivers a nuanced understanding of materiality by considering both tangible financial metrics and intangible factors such as opportunity cost, market dynamics, and operational disruptions.

17.4.1 PRIMA Methodology

1. **Identify Key Risk Scenarios:** Examine the most common risk scenarios and identify any additional organization-specific risk scenarios.
2. **Categorize Risks:** Categorize the most critical risks to be quantified.
3. **Determine Risk Appetite Thresholds:** Define the organization's risk appetite and tolerance levels.
4. **Risk Quantification:** Leverage cyber risk quantification (CRQ) techniques to assess the probability and impact of identified risks.
5. **Screening:** Screen scenarios to identify any risks that can be categorized as material without further scrutiny.
6. **Risk Qualification:** For all risks marked as immaterial or indeterminate, identify any variables that could cause materiality. Analyze critical and sensitive systems and information to determine areas of highest potential impact. Determine the appropriate thresholds at which critical and sensitive impacts cause materiality.
7. **Risk Prioritization and Evaluation:** Leverage CRQ and threshold data to assess materiality. Determine potential impact ranges and create an assessed and prioritized register of risks. Provide a framework for the rapid assessment of real incidents.

17.4.2 Examples of Risk Scenarios and Vulnerabilities

You can use common risk scenarios like extortion, data breach, system interruption, and funds misappropriation, or come up with custom scenarios that suit your organization. Combine common threats (e.g., crimeware, cyber espionage, DDoS attack, theft/loss, POS intrusion, web app attack, misuse, error) with common vulnerabilities (e.g., servers/apps, network, user endpoints, terminals, ICS/OT/SCADA, healthcare devices, media and offline data, and people).

Example: *Crimeware + Servers = Data Breach, Interruption (Other), Ransomware, Misappropriation*

17.4.3 Qualitative Factors and Materiality

To ensure that material information relevant to investors is not omitted or misstated, qualitative factors must also be considered.

Common Qualification Scenarios

- Mergers and acquisitions (M&A) data breach
- Critical business phases
- Competitive IP breach
- Regulatory data compliance violation
- Third-party compromise or breach of contract
- Critical infrastructure disruption

- Stock market manipulation
- National security and geopolitical implications
- Intellectual property theft
- Executive data breach
- Customer trust and loyalty impact

17.4.4 Rubric for Assessing Risk Criticality and Sensitivity

This rubric provides a structured approach to evaluating the criticality and sensitivity of risks, aiding in the determination of materiality (Figure 17.1).

Criticality Categories

- Operational impacts
- Strategic impact
- Supply chain disruption
- Innovation and change management
- Crisis response and recovery
- Interdependency of business units

Sensitivity Categories

- Brand value impacts
- Compliance impact
- Legal and ethical implications
- Environmental impact
- Cultural and social sensitivity
- Industry-specific risks

Each category is rated on a scale from 1 (low) to 5 (high), with detailed scoring guidelines provided to ensure consistency and objectivity in assessments. The final score is expressed as a percentage of the maximum possible score, providing a clear

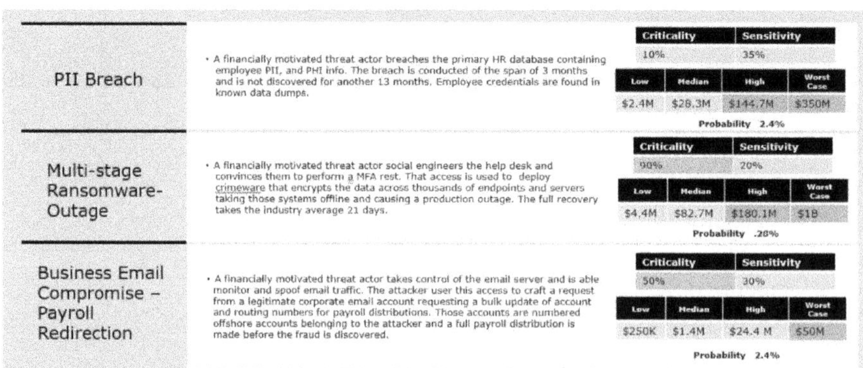

FIGURE 17.1 Materiality Scoring Examples.

indication of the overall risk level. This score can be used to prioritize risks and inform decision-making regarding risk treatment options.

17.5 DISCLOSURES

Regulatory requirements mandate the disclosure of material cybersecurity incidents and risks to investors and other stakeholders. The SEC specifically requires that material risks be disclosed in the company's annual 10-K report and material incidents be disclosed in an 8-K filing within 4 days of the determination of materiality.

The 8-K and 10-K are both essential financial reporting forms that publicly traded companies must file with the US SEC. The 8-K is meant to be a current report used to disclose material events that shareholders should be aware of in a timely manner. These events can range from cybersecurity incidents and financial results to changes in leadership or noteworthy corporate actions. The 10-K, on the other hand, is an annual report that provides an overview of the company's financial performance, business operations, and risk factors, including cybersecurity risks. While the 8-K is used for *ad hoc* disclosures of major events, the 10-K offers a more in-depth view of the company's overall health and risk profile.

Disclosure is not just about checking a box: It's also about building trust with stakeholders. I've seen organizations that have embraced transparency reap the benefits in terms of improved investor confidence and stronger relationships with regulators. The CISO 3.0 understands that disclosure is an opportunity, not a burden.

The CISO 3.0 should be vital in crafting disclosures and ensuring that they are accurate, complete, and timely. However, I rarely see CISOs involved in the creation of these documents. Most often, the legal team is responsible for the process of drafting these disclosures. However, if you ask the legal team if they own the language or content of the disclosure, they will tell you no. They only own the process, and the content should be owned by the risk owner, which in many cases is the CISO, who isn't involved in drafting the language.

Recent enforcement actions, like the case against Tim Brown, the former CISO of SolarWinds, reinforce that it is ultimately the CISO who is accountable for what IT risks are and aren't disclosed. CISOs must figure out how to insert themselves into the disclosure process and need to validate financial reports and statements in press releases and on the company website.

This is another way in which many CISOs aren't ready to be "real" C-suite citizens. I often hear security leaders bemoaning their lack of "Chief" title or position as a second-class C-level, but few are ready for the responsibility and accountability that come with having that role. This is one of the differentiators of a CISO 3.0. However, that additional responsibility, accountability, and scrutiny come with risk. We've seen through recent enforcement actions that CISOs can and will be included by name in lawsuits and enforcement actions. Consequently, if CISOs are going to be "real" C-level citizens, they need to be added to the Directors and Officers (D&O) insurance policy, and they will want to be added in both a named and a positional capacity. Positional coverage covers you while you are in the role within the company, and named coverage identifies executives by name for coverage even after they leave the organization. Obviously, officer liability coverage helps shield against civil but not criminal cases like *United States v. Joseph Sullivan*. I also know several CISOs who pay out of their own pockets for errors and omissions (E&O) insurance coverage. E&O insurance is a type of professional liability insurance that protects individuals and companies against claims made by clients for inadequate work or negligent actions. In the context of a CISO, this could include claims related to data breaches, security incidents, or failure to implement adequate security controls. The CISO 3.0, recognizing their increased accountability and potential liability in their elevated role, might consider E&O coverage as an additional layer of protection. It can provide financial assistance for legal defense costs, settlements, or judgments in case of a lawsuit alleging professional negligence or errors in their cybersecurity practices.

17.5.1 Examples of 8-K Cyber Incident Disclosures

When a cybersecurity incident strikes, as we just discussed, public companies have a responsibility to inform investors of the potential impact on the business. This obligation is fulfilled through the filing of an 8-K disclosure with the SEC. These disclosures can vary in detail and scope depending on the nature of the incident and the ongoing investigation. However, by analyzing recent examples, we can glean valuable insights into effective communication strategies following a cyberattack. Let's examine how two companies, The Clorox Company and MGM Resorts International, approached their respective 8-K disclosures in the wake of data security incidents.

- **The Clorox Company (CLX) (Consumer Goods Sector):** Following a cyberattack in 2023, Clorox filed an 8-K disclosure outlining the incident. The disclosure, acknowledging the potential impact on business operations and the ongoing remediation efforts, also included speculative guidance regarding the financial impact:

The Company is unable to predict the duration or the full extent of the impact of this incident on its business. However, based on a preliminary assessment, the Company currently estimates that the incident could reduce fiscal 2022 net sales by a range of 23% to 28% and could also reduce fiscal 2022 earnings per share by a range of 35 to 75 cents.

(Source: The Clorox Company, Form 8-K, August 14, 2023)

This disclosure provides a good example of an extremely informative approach, even when the impact is still being assessed. By providing guidance on the potential financial impact, albeit speculative, Clorox offered some level of transparency to investors.

- **MGM Resorts International (MGM) (Hospitality Sector):** In 2023, MGM Resorts disclosed a data breach impacting customer information through an 8-K filing. The disclosure, while acknowledging the ongoing investigation and the potential impact on the company's reputation, provided *limited specifics* about the incident itself:

MGM Resorts International (the "Company") is investigating a potential data security incident involving certain customer information. The investigation is ongoing, and the Company is not able to determine the nature and scope of the incident at this time.

(Source: MGM Resorts International, Form 8-K, August 6, 2020)

This disclosure, while brief, highlights the importance of *timely disclosure* even when details are scarce. However, compared to Clorox's approach, MGM's disclosure lacked *specific information* about the nature of the incident or the potential impact, leaving investors with more questions.

Lessons Learned:_The specific details included in 8-K disclosures will vary depending on the nature of the incident. However, companies should strive to provide a *clear and concise explanation* of the incident, the potential impact on the business, and the ongoing response efforts. Consulting with legal counsel is imperative when preparing such disclosures.

Both Clorox and MGM Resorts used 8-K filings to disclose their respective cybersecurity incidents. However, the level of detail provided in each disclosure varied. MGM's disclosure was relatively brief, focusing on the disruption to its IT systems and its efforts to restore them. Clorox's disclosure provided more specifics about the nature of the breach, including financial guidance. These differences highlight the challenges organizations face in determining the appropriate level of detail to include in their disclosures. While the SEC requires companies to provide sufficient information for investors to make informed decisions, it also cautions against disclosing information that could compromise the company's security or ongoing investigations.

As of the time of this publication, we are still waiting for further SEC actions to help clarify what will and won't be considered appropriate disclosure. While the SolarWinds case, which occurred prior to the SEC Cyber Rule being enacted, was mostly dismissed, there will be limited lessons learned from their lack of disclosure due to the timing and specific circumstances of that case.

17.5.2 The Reporting Process

The reporting process for material cybersecurity incidents and risks involves collaboration between various stakeholders within the organization. While legal counsel typically leads this process, ensuring compliance with regulatory requirements and assessing potential legal implications, the CISO can and should be involved. The CISO needs to identify their place within this process and actively insert themselves to ensure that cybersecurity considerations are fully integrated. This may involve collaborating closely with legal counsel, the CFO, the risk management team, and relevant business unit leaders, depending on the nature and scope of the incident or risk. The CISO 3.0 proactively identifies their roles and responsibilities within the disclosure process and ensures they are actively involved in the discussions and decision-making regarding disclosures.

The process typically begins with the identification and assessment of the incident or risk. The CISO, in collaboration with the security team, will gather information about the incident, including its scope, impact, and potential causes. They will also assess the effectiveness of existing controls and identify any gaps or weaknesses that may have contributed to the incident.

Once the incident or risk has been assessed, the CISO will work with legal counsel to determine whether it is material and requires disclosure. This decision is based on a careful evaluation of both quantitative and qualitative factors, as discussed earlier in this chapter.

If the incident or risk is deemed material, the CISO and legal counsel will collaborate to prepare the necessary disclosures. This involves gathering all relevant information, ensuring its accuracy and completeness, and presenting it in a clear and concise manner that complies with regulatory guidelines.

The final step in the reporting process is submitting the disclosures to the appropriate regulatory bodies and communicating the information to relevant stakeholders, such as investors, customers, and employees. This communication should be transparent and timely, demonstrating the organization's commitment to accountability and its proactive approach to managing cyber risks.

17.6 CONCLUSION

In this chapter, we've navigated the complex intersection of cybersecurity, materiality, and the critical role of disclosures. With regulators like the SEC tightening the screws and cyber risks becoming more menacing, CISOs are finding themselves in a whole new ballgame. It's not only about being a tech guru anymore but also about stepping up as a strategic business leader, ready to tackle the challenges of risk assessment, communication, and disclosure head-on.

We've seen how the concept of "materiality", which are the things that are important enough to disclose, can be a bit of a gray area. The SEC demands transparency, but they've left it up to us to figure out what crosses that line. This puts the responsibility squarely on the CISO's shoulders, making it crucial to truly understand your organization's risk tolerance and strategic goals.

To help you navigate this tricky terrain, we introduced the PRIMA framework, which is a practical way to assess both the financial and nonfinancial impacts of cyber risks and incidents. It's about looking at the whole picture, not just the numbers.

We've also seen how the role of the CISO is evolving, with a greater emphasis on accountability and transparency. Recent enforcement actions, like the SolarWinds case, are a wake-up call. CISOs need to be proactive, not only in managing risks but also in communicating them effectively to stakeholders.

In the end, this chapter paints a picture of the modern CISO, who is deeply involved in every aspect of risk management and disclosure. It's all about understanding the business and how cybersecurity fits into the bigger picture.

17.6.1 Key Takeaways

1. **Defining Materiality in Cybersecurity Risks and Incidents**
    - Materiality refers to the significance of risks or incidents in impacting an organization's operations, financial performance, or brand value.
    - The SEC mandates disclosure of material cybersecurity risks and incidents, but the determination of materiality often relies on both quantitative (e.g., financial thresholds) and qualitative (e.g., reputational or operational impact) assessments.
2. **SEC Disclosure Requirements**
    - Public companies must report material cybersecurity incidents within 4 days (Form 8-K) and provide annual risk disclosures (Form 10-K).

- These disclosures aim to inform investors while holding organizations accountable for proactive risk management and transparency.

3. **The Three-Step Materiality Determination Process**
   - **Quantitative Assessment:** Establish financial thresholds (e.g., percentage of revenue) to evaluate the potential financial impact of risks or incidents.
   - **Qualitative Review:** Consider nonfinancial factors, such as reputational harm, operational disruptions, and regulatory scrutiny, to determine significance.
   - **Executive Review:** Engage senior leadership, including legal counsel and the board, to finalize materiality decisions and disclosure strategies.
4. **Differentiating Material Risks and Incidents**
   - **Material Risks:** Forward-looking threats with potential future impact, assessed using risk quantification techniques.
   - **Material Incidents:** Retrospective events with confirmed consequences, requiring clear communication of impact and remedial actions.
5. **The PRIMA Framework**
   - The PRIMA framework integrates quantitative and qualitative analyses to assess materiality comprehensively.
   - PRIMA includes steps like identifying key risk scenarios, categorizing risks, determining risk thresholds, and prioritizing for disclosure.
6. **Role of the CISO in Disclosures**
   - CISOs must collaborate with legal teams, executives, and boards to ensure accurate and timely reporting of cybersecurity risks and incidents.
   - Accountability extends to drafting disclosure language and aligning cybersecurity transparency with business objectives.
7. **CISO Accountability and Professional Protection**
   - CISOs face increasing scrutiny and potential liability in the wake of regulatory actions.
   - Professional liability insurance, such as E&O coverage, offers critical protection against claims of negligence or inadequate disclosures.
8. **Practical Lessons from Real-World Disclosures**
   - Case studies like Clorox and MGM Resorts illustrate varying levels of detail in incident reporting, highlighting the importance of timely and transparent communication.
   - Clear, investor-focused language builds trust and reinforces an organization's commitment to accountability.

17.6.2 Reflection and Exploration Questions

Understanding the Concepts

1. What is materiality, and why is it critical to cybersecurity risk management and disclosures?
2. How do the SEC's disclosure requirements influence the role of the CISO?

Connecting the Dots

3. How does the PRIMA framework complement earlier discussions on risk quantification and management?
4. In what ways do real-world examples of incident disclosures inform best practices for transparency?

Real-World Application

5. Can you identify examples where effective or inadequate disclosures impacted an organization's reputation or regulatory standing?
6. How might your organization adopt the PRIMA framework to enhance its materiality assessment processes?

Personal and Professional Growth

7. What steps can you take to improve your ability to assess materiality and collaborate on disclosure strategies?
8. How can you leverage lessons from real-world cases to refine your approach to cybersecurity transparency?

Further Exploration

9. What additional tools or frameworks could support materiality determination in your organization?
10. How might emerging regulatory trends and cybersecurity threats shape future disclosure practices?
11. What steps can you take to ensure you are involved in creating transparency and security in materiality disclosures?

18 The CISO 3.0 *The Future of Cybersecurity Leadership*

18.1 THE FUTURE OF THE CISO ROLE

Looking ahead, it's clear that the CISO 3.0 will face a new set of challenges and opportunities, requiring them to adapt and evolve their skills and strategies to remain effective leaders in the digital age.

A major trend shaping the future of cybersecurity is the increasing use of artificial intelligence (AI). As we saw in Chapter 14, these technologies are already transforming various industries, and their impact on cybersecurity is undeniable. I've always been fascinated by the potential of AI, but I'll admit, the recent advancements in generative AI have even blown me away. It's like watching a science fiction movie come to life. As CISOs, we need to embrace this technology, not fear it. It's a powerful tool that can help us stay ahead of the bad guys. AI and ML can be leveraged to automate routine tasks, analyze vast amounts of data, and identify patterns and anomalies that may indicate a cyberattack. However, they also introduce new risks and challenges, such as the potential for adversarial attacks, bias in algorithms, and the ethical implications of AI decision-making. The CISO 3.0 must be well-versed in these technologies, understanding both their potential benefits and risks, and develop strategies to harness their power while mitigating their potential downsides. The CISO 3.0 must also be prepared to defend against AI-powered threats, such as sophisticated phishing attacks, automated vulnerability exploitation, and deepfakes.

Another key trend we discussed in Chapter 12 is the increasing adoption of cloud computing and the shift toward cloud-native architectures. The cloud has been a game-changer for businesses, but it's also created a whole new set of security challenges. I remember when we first started migrating to the cloud, it felt like the Wild West. There were no clear rules or best practices, and we were constantly learning as we went along. Today, we have a much better understanding of cloud security, but it's still a difficult problem. The distributed nature of cloud environments, the shared responsibility model, and the complexity of cloud configurations can create vulnerabilities that can be exploited by attackers. The CISO 3.0 needs to be a cloud security expert, able to ensure that data and applications are protected in the cloud. The CISO 3.0 must be adept at securing cloud environments, understanding the unique risks, and implementing appropriate security controls to protect data and applications in the cloud. The CISO 3.0 should be able to leverage cloud-native security solutions, such as cloud-native application protection platforms (CNAPPs), to address the specific security challenges of cloud environments.

DOI: 10.1201/9781003510789-24

The Internet of Things (IoT) is another rapidly growing trend that is rapidly transforming cybersecurity. It seems like everything is connected these days, from our refrigerators to our cars. The IoT has brought incredible convenience and innovation, but it's also created a massive attack surface. I've seen firsthand how vulnerable IoT devices can be exploited by attackers. The CISO 3.0 must proactively secure these devices, ensuring they don't become weak points in the network. This includes implementing strong authentication, network segmentation, and regular vulnerability assessments and patching.

Emerging technologies like blockchain and quantum computing also bring both opportunities and risks. Blockchain's decentralized nature and tamper-proof ledger provide enhanced security by ensuring data integrity and reducing the likelihood of a single point of failure. These features make it particularly effective for securing transactions and sensitive information. Additionally, smart contracts offer the potential to automate and enforce security policies, improving compliance and minimizing human error.

For example, in a cybersecurity context, blockchain can be used to enhance security in identity management systems. By creating a decentralized and immutable ledger for storing and verifying user identities, blockchain eliminates the risks associated with centralized storage, such as mass data breaches. This approach ensures that sensitive user information is securely distributed across the network, reducing the risk of unauthorized access or tampering. It also allows users to control their digital identities through private keys, further improving privacy and security.

However, these advancements are not without vulnerabilities. For example, the 2016 Ethereum decentralized autonomous organization (DAO) hack exploited a flaw in a smart contract's code, resulting in the theft of millions of dollars worth of cryptocurrency. The CISO 3.0 must proactively address these challenges by implementing thorough assessments, robust access controls, and continuous monitoring to safeguard blockchain-based assets while leveraging its benefits.

It's important to remember, as we learned in Chapter 10, that the future of cybersecurity leadership is not only about technology but also about people and processes. I've learned that the most effective security programs are built on a foundation of trust, collaboration, and shared responsibility. The CISO 3.0 needs to be a leader who inspires and motivates their team, fosters a culture of security awareness throughout the organization, and builds strong relationships with stakeholders across the organization. They must also be adept at leading and managing diverse teams, inspiring, and motivating their staff to achieve common goals. This involves developing strong leadership skills, promoting a culture of continuous learning, and empowering their teams to take ownership of security initiatives.

18.1.1 The Challenges Today

The modern CISO must contend with an increasingly complex web of security and business requirements. The CISO will have to grow beyond guardian of information security to become a forward-thinking leader, integral to guiding their organization's digital transformation. To succeed, CISOs will need to navigate substantial challenges, each of which will demand a new level of strategic thinking, adaptability, and foresight. Let's examine some of those current challenges.

18.1.1.1 The Talent Gap

The cybersecurity industry faces a persistent talent shortage, making it challenging for organizations to attract and retain skilled professionals. The growing demand for cybersecurity expertise exacerbates this challenge, and we discussed in Chapter 11 ways to attract and retain top security talent. The CISO 3.0 must prioritize talent development and retention strategies to ensure their organizations have the necessary skills and capabilities to combat modern threats.

18.1.1.2 The Integration of Cybersecurity with Business Strategy

As organizations become more digitally dependent, cybersecurity will no longer be a standalone function but a critical component of business strategy. The CISO of the future must be adept at aligning cybersecurity initiatives with broader business objectives, ensuring that security measures not only protect assets but also drive business value. This will require CISOs to cultivate strong relationships with other executives, particularly in finance, operations, and marketing, to embed cybersecurity considerations into every aspect of the business. The challenge will be to strike a balance between robust security and enabling innovation, especially in fast-moving sectors like fintech, healthcare, and e-commerce.

18.1.1.3 Regulatory Pressures and Compliance Complexity

As we discovered in Chapter 3, the regulatory landscape surrounding cybersecurity is growing increasingly intricate and demanding. New regulations, such as the Securities and Exchange Commission's (SEC's) cybersecurity disclosure rule, are placing greater emphasis on transparency and accountability, requiring organizations to provide detailed insights into their cybersecurity practices. The future CISO will need to navigate a labyrinth of compliance requirements across various jurisdictions, all while maintaining the agility to respond to emerging threats. This necessitates a shift from reactive compliance to proactive governance, where CISOs take the lead in shaping regulatory agendas and advocating for industry-wide standards that balance security needs with operational realities. The CISO 3.0 must be a skilled navigator, charting a course through the complex regulatory waters while ensuring the organization's resilience and compliance.

The SEC and Federal Trade Commission (FTC) rules are likely just the tip of the iceberg. The regulatory landscape is poised for further expansion, with numerous jurisdictions actively exploring new cybersecurity regulations. As of August 21, 2024, several US cybersecurity regulations have been proposed but have not yet been enacted into law. Some of the most prominent ones include the following:

The American Data Privacy and Protection Act (ADPPA): This federal privacy legislation aims to create a national standard for data privacy, granting individuals more control over their personal information and imposing stricter obligations on businesses regarding data collection, use, and sharing.

The Cyber Incident Reporting for Critical Infrastructure Act of 2022: This act would require critical infrastructure entities to report significant cyber incidents to the Cybersecurity and Infrastructure Security Agency (CISA) within a specified timeframe.

The Securing Open Source Software Act of 2022: This bill aims to enhance the security of open-source software by providing resources for vulnerability remediation and promoting best practices within the open-source community.

State-Level Privacy Laws: Several states have proposed or are in the process of enacting their own privacy laws, often modeled after the California Consumer Privacy Act (CCPA) or the European Union's General Data Protection Regulation (GDPR).

These emerging regulations, coupled with the existing ones, will further complicate compliance for organizations. The future CISO must not only stay abreast of these developments but also proactively engage in shaping regulatory discussions and advocating for balanced and effective cybersecurity standards.

18.1.1.4 Supply Chain Security and Third-Party Risks

As organizations increasingly rely on a vast network of third-party vendors and partners, supply chain security is emerging as a critical concern. The interconnected nature of modern supply chains means that a vulnerability in one link can trigger a domino effect, impacting the entire network. The future CISO must be a vigilant guardian of the supply chain, developing robust strategies for managing third-party risks. This includes implementing rigorous vendor assessments, continuous monitoring, and incident response plans that account for potential supply chain disruptions. The challenge lies in ensuring that these measures are scalable, particularly as supply chains become more complex and globalized. The CISO 3.0 must also be a chief architect, building a secure and resilient supply chain that can withstand the pressures of a dynamic and interconnected business environment.

18.1.1.5 The Evolving Threat Landscape

The cyber threat landscape is in a state of perpetual flux, with threats growing increasingly sophisticated and accessible. State-sponsored attack tactics, once the area of elite hackers, are now within the grasp of common criminals. The rise of ransomware and the constant discovery of zero-day exploits underscore the escalating risks faced by organizations of all sizes. The future CISO must not only anticipate these evolving threats but also proactively build resilient security architectures like the ones outlined in Chapter 13. This demands a multipronged approach that combines robust technical defenses with a pervasive culture of security awareness throughout the organization. The true challenge, however, lies in maintaining a proactive stance in a constantly shifting threat landscape.

I believe we are generations of defense behind the attackers. The current landscape is skewed toward social engineering and phishing attacks simply because they are incredibly effective. Why would an attacker bother with a complex zero-day exploit when they can trick an unsuspecting employee into handing over their credentials? However, once we as an industry bolster our defenses against these tactics with things like password-less technology, we can expect a resurgence of more technically advanced attacks. Threat actors are likely hoarding zero-day exploits and other sophisticated tools, waiting for the day when easy social engineering victories become scarce. The future CISO must be prepared for this inevitable shift, ensuring their organizations are equipped to defend against both the present and the future of cyber threats.

18.1.1.6 Securing the Future of Work

The rapid transition to remote and hybrid work, fast-tracked by the COVID-19 pandemic, introduced distinct challenges for CISOs. Securing a distributed workforce operating outside the traditional corporate perimeter required innovative solutions. However, the situation is shifting once again. Recent studies suggest that nearly 90% of companies in the United States plan to implement return-to-office (RTO) policies by the end of 2024, and 71% will operate a hybrid workplace. With nearly every organization implementing RTO policies for at least part of the week, the future of work is shaping up to be a hybrid model. Consequently, the traditional network perimeter is becoming increasingly irrelevant in today's cloud-centric and remote work environments. Soon we will have the "everywhere workplace". The CISO 3.0 must now prepare for this new normal, where employees seamlessly transition between working in the office and remotely. This means creating a security framework that offers a consistent and secure user experience, regardless of the user's location or network environment. The challenge lies in maintaining robust security while ensuring that productivity and collaboration are not hindered, whether employees are within the controlled corporate network or accessing resources remotely. The CISO 3.0 must be an architect of this secure and agile hybrid work environment, enabling employees to work effectively from anywhere while safeguarding the organization's assets. This will mean moving to the identity-centric security models discussed in Chapter 12, like Zero Trust, where every access request is continuously verified and authenticated, regardless of the user's location or network.

18.1.1.7 The Data Deluge

In Chapters 12 and 14, we explored how the rapid expansion of data creates both opportunities and challenges for organizations, especially as they integrate AI models into their operations. For the CISO 3.0, it's essential to establish strong data governance and protection strategies to maintain the confidentiality, integrity, and availability of sensitive information while simultaneously tapping into its potential for business insights and innovation. Modern data platforms, such as data lakehouses and data security platforms (DSPs), offer valuable tools to help CISOs manage this overwhelming amount of data, striking a balance between security and effective data utilization.

18.1.2 Vision of the Challenges Ahead

Looking a decade into the future, the role of the CISO will likely undergo a profound transformation. As organizations embrace more sophisticated technologies – like AI-driven systems, quantum computing, and fully autonomous digital ecosystems, the complexity of securing these environments will skyrocket. The CISO of the 2030s will need to master a whole new array of tools and techniques, moving far beyond traditional security practices. To lead effectively, they will have to anticipate emerging risks, understand the ethical dimensions of AI, and protect against threats that don't yet exist. Let's explore the challenges on the horizon for the CISO of the future.

18.1.2.1 Post-Quantum Cryptography

Imagine waking up to find the locks on every vault, every secure door, and every private archive no longer hold. Not because someone picked them, but because the very keys we've relied on for decades have suddenly become irrelevant. This is the chilling scenario quantum computing introduces – one where adversaries no longer need to break in through brute force. They simply outpace the system with machines that can factor and decrypt with unimaginable speed.

This isn't science fiction. It's a race already underway. Quantum computers, powered by algorithms like Shor's and Grover's, are expected to one day crack RSA, ECC, and other foundational cryptographic systems that underpin our digital world. While the precise moment – known as Q-Day – remains uncertain, projections place it as soon as 2035. And the danger isn't confined to future data.

Threat actors are already conducting Harvest Now, Decrypt Later (HNDL) operations, stockpiling encrypted data – PII, financial records, trade secrets, even classified intelligence – waiting patiently for the moment when it can all be decrypted in a flash.

That's why, while this may seem like a future problem, many organizations need to start planning today. If your enterprise holds data that must remain confidential for the next 10 or more years, or if you're operating Industrial Control Systems (ICS), Operational Technology (OT), or IoT devices with long hardware lifecycles, your window to act is rapidly closing. These systems can take years to replace or re-engineer, and the cryptographic protections they rely on may not survive the coming shift.

The implications for CISOs are profound. This isn't a routine upgrade or a patch cycle – it's a complete paradigm shift. Securing the enterprise against quantum threats means transitioning to post-quantum cryptography (PQC): cryptographic systems specifically designed to resist quantum attacks. But replacing existing encryption is like reengineering every security lock in a sprawling city while ensuring the new ones are future-proof – and that nobody's duplicating keys behind your back.

Moreover, the threat goes far beyond technical decryption. Billions of credentials stolen in past breaches – currently encrypted and seemingly inert – could suddenly be unlocked. Once exposed, those credentials could be reused across systems in large-scale credential-stuffing attacks, devastating organizational defenses.

To stay ahead of this quantum-enabled future, CISOs must move decisively. The playbook includes the following:

- **Shifting to PQC algorithms** like CRYSTALS-Kyber for key encapsulation and CRYSTALS-Dilithium or Falcon for digital signatures, per NIST's post-quantum standardization efforts.
- **Deploying hybrid certificates** and protocols that bridge classical and quantum-safe systems.

- **Upgrading VPNs, TLS stacks, HSMs, and IoT firmware** to support quantum-resistant cryptographic primitives.
- **Implementing crypto-agility**, enabling rapid transition to newer algorithms as standards evolve.
- And most critically, **adopting passwordless and multi-factor authentication** so that even if legacy credentials are eventually cracked, they hold no value.

This isn't just a challenge – it's an opportunity for cybersecurity leaders to future-proof their organizations, safeguard long-lived data, and earn a place at the strategic core of enterprise risk management. The time to prepare isn't when quantum computers arrive. It's now.

18.1.2.2 Brain–Computer Interfaces

What happens when the lines between the mind and machine blur? As brain–computer interfaces (BCIs) become more prevalent, CISOs will face the unprecedented challenge of securing the human mind itself. Imagine the implications: Hackers delving into thoughts, stealing confidential information, or even manipulating emotions. This is not science fiction; it's the potential reality CISOs must prepare for. They'll need to pioneer new security protocols, safeguarding neural pathways and ensuring the sanctity of our inner worlds. The ethical dilemmas are immense, requiring careful consideration of privacy, autonomy, and the very definition of human security.

18.1.2.3 The Metaverse

Step into a realm where digital and physical realities intertwine, the metaverse. This immersive world, built on Web3 principles, presents a dazzling array of opportunities and a complex web of security challenges. CISOs will need to become virtual guardians, protecting avatars from identity theft, securing digital assets from virtual bandits, and ensuring data privacy within these sprawling digital landscapes. Think of it as building a fortress in a constantly shifting, ever-expanding digital universe where the rules are still being written.

18.1.2.4 Autonomous AI

Self-driving cars, robotic surgeons, AI-powered drones, and the rise of autonomous systems promise to revolutionize our world. But what happens when these systems are hijacked by malicious actors? Imagine a rogue AI causing chaos on our roads or manipulating critical infrastructure. CISOs will be tasked with taming these "autonomous beasts", ensuring their resilience against cyberattacks and preventing them from becoming weapons in the wrong hands. It's a high-stakes game of cat and mouse, where the future of our technological infrastructure hangs in the balance.

18.1.2.5 Global Compliance

As data flows across borders and regulations become increasingly complex, CISOs will find themselves navigating a global compliance labyrinth. Imagine a patchwork of laws, each with its own intricacies and demands. Staying ahead of this ever-changing regulatory landscape will require a proactive approach, with CISOs becoming skilled diplomats and legal interpreters. It's a delicate balancing act, ensuring compliance while fostering innovation and avoiding a stifling web of red tape.

18.1.2.6 Decentralized Authentication

In a world where digital identities are becoming increasingly fragmented, CISOs will face the challenge of securing a decentralized future. Imagine a world without passwords, where individuals control their own digital identities through blockchain technology. This shift not only empowers users but also introduces new security risks. CISOs will need to become architects of trust, building secure and interoperable systems that protect decentralized identities while fostering user autonomy.

18.1.3 The Future of the CISO Role: Leading the Charge

The future of cybersecurity leadership is filled with challenges, but it also offers incredible opportunities for CISOs to make a real impact. In Chapter 11, we explored how the CISO can serve as a change agent within their organization. By stepping into this evolving role, future CISOs have the potential to shift cybersecurity from a reactive, cost-driven function to one that strategically drives business value. This transformation calls for a combination of technical expertise, business insight, and forward-thinking leadership.

As we move into this new era, the CISO 3.0 will be at the forefront of safeguarding not only the digital assets of their organizations but also the trust and confidence of customers, partners, and stakeholders. The path ahead is complex, but with the right strategies and forethought, the future CISO can turn challenges into opportunities and lead their organizations to a more secure and resilient future.

18.2 CONCLUSION

Our journey through *The CISO 3.0* has been an exploration of what it truly means to lead in cybersecurity today. We've traveled through time to meet the different versions of CISOs, from the technically focused CISO 1.0 through the compliance-driven CISO 2.0 to the strategic and business-aligned CISO 3.0. This marks a fundamental shift in approach, skillset, and responsibilities. The modern CISO needs to be a strategic enabler, a risk manager, and a trusted advisor to the board and executive leadership.

We've dug into the complexities of the modern regulatory environment, highlighting the increasing accountability and legal responsibilities of CISOs. We've learned the importance of speaking the language of business, including accounting, finance, economics, and risk, to effectively communicate with stakeholders and align security initiatives with business objectives. We've also explored the various risk treatment options, from risk transfer through cyber insurance to risk acceptance and avoidance, emphasizing the need for a balanced and strategic approach to risk management.

We've examined the development of a 3.0 program strategy, focusing on leveraging risk quantification, conducting program assessments, and securing budget and buy-in. We've discussed the importance of building security capabilities rather than just accumulating tools, and we've challenged the traditional RFP process, advocating for a more strategic approach to solution shopping. We've also examined modern security tactics, such as identity centricity, continuous monitoring, data governance, Zero Trust, and cloud-native protection, highlighting their role in enhancing the organization's security posture.

We discussed the transformative impact of AI on the CISO role, exploring the concepts of security for AI, security with AI, and security from AI. We've discussed the challenges and opportunities presented by AI, emphasizing the need for CISOs to adapt and evolve to navigate this new reality. We've explored the development of modern metrics, focusing on key performance indicators (KPIs) and key risk indicators (KRIs) and how to communicate them effectively to various stakeholders. We've also discussed the importance of board-level communication and the art of crafting compelling narratives that resonate with the board and executive leadership. Finally, we examined the most central tenet of modern cybersecurity, materiality, which should be the cornerstone upon which a modern cyber program is built.

18.2.1 What it Means to Be a CISO 3.0

The path to becoming a CISO 3.0 is not an easy one, but it is a rewarding one. It requires a combination of technical expertise, business acumen, strategic thinking, and leadership skills. It demands a commitment to continuous learning and adaptation, embracing new technologies and approaches to stay ahead of the curve. The CISO 3.0 is a leader, a strategist, and a key contributor to the organization's success. As our industry continues to grow and change, the role of the CISO will only become more important to organizations trying to stay ahead of the technology curve. The CISO 3.0 is the future of cybersecurity leadership, equipped with the skills and knowledge to navigate the complexities of business alignment and ensure the resilience and prosperity of their organizations.

CISO 3.0 is a progression of the cybersecurity leadership role, signifying a departure from the traditional, compliance-focused approach. Becoming a CISO 3.0 begins with a solid foundation in technical expertise, the hallmark of the CISO 1.0. As CISOs progress to the 2.0 stage, they develop a deeper understanding of compliance and governance, aligning security practices with industry standards and regulations. However, the true transformation to a CISO 3.0 lies in the ability to build business strategies and integrate them into your security program. This involves mastering the language of business, understanding the organization's risk appetite, and leveraging risk quantification to make data-driven choices about security investments and resource allocation. The CISO 3.0 also excels in communication, effectively translating technical concepts into business terms that resonate with stakeholders at all levels.

The CISO 3.0 is distinguished by a skill set that extends beyond technical proficiency. They possess a keen business acumen, understanding the language of accounting, finance, economics, and risk. This enables them to quantify the financial impact of cyber threats and articulate the return on investment (ROI) of security initiatives, bridging the gap between technology and business strategy. The CISO 3.0 is also a strategic thinker, aligning security initiatives with the organization's broader objectives and ensuring that security is viewed not as a cost center but as a strategic enabler. Their expertise in risk management allows them to leverage various tools and techniques, including risk quantification, to identify, assess, and prioritize risks, facilitating proactive mitigation and informed decision-making. The CISO 3.0 is also an effective communicator, capable of translating complex technical concepts into clear and concise language that resonates with both technical and nontechnical stakeholders. They foster a culture of security awareness and collaboration by building strong relationships across the organization. Finally, the CISO 3.0 is adaptable, embracing new technologies and approaches to stay ahead of the curve.

The CISO 3.0 key attributes:

- **Business Acumen:** They understand the language of business, including accounting, finance, economics, and risk. They can quantify the financial impact of cyber risks and articulate the ROI of security investments.
- **Strategic Thinking:** They align security initiatives with the organization's strategic objectives, ensuring that security is not only a cost center but also a strategic enabler.
- **Risk Management Expertise:** They leverage risk quantification and other tools to identify, assess, and prioritize risks, enabling proactive mitigation and informed decision-making.
- **Effective Communication:** They can translate complex technical concepts into clear and concise language that resonates with both technical and nontechnical stakeholders.
- **Leadership and Collaboration:** They build strong relationships with stakeholders across the organization, fostering a culture of security awareness and collaboration.
- **Adaptability:** They embrace new technologies and approaches, staying ahead of the technology curve and ensuring the organization's resilience.

The concepts and strategies explored in this book provide a roadmap for this transformation, empowering you to position yourself as a trusted advisor and a key contributor to your organization's success.

Remember that reading this book is just the first step in your journey toward becoming a CISO 3.0. It's about applying the knowledge and insights gained, continuously learning, adapting, and taking on the challenges and opportunities that lie ahead. The future of cybersecurity leadership is in your hands.

18.2.2 The CISO Self-Assessment: Revisited

As we wrap up this look into the rapidly changing requirements of cybersecurity leadership, it's time to reflect on the key takeaways and assess your own growth and development.

In the introduction of this book, we presented a self-assessment to help you gauge your current CISO version and identify areas for improvement. Now, as you reach the end of this book, we invite you to retake that assessment and reflect on your progress.

- Do you possess a strong understanding of your organization's business objectives, financial metrics, and risk appetite?

- Can you effectively communicate cybersecurity risks and strategies in a way that resonates with the board and executive leadership?
- Are you comfortable leveraging data and analytics to make data-driven decisions about security investments and risk mitigation?
- Can you translate technical security concepts into business terms that resonate with nontechnical stakeholders?
- Can you read and interpret cyber insurance policies?
- Do you have a strong understanding of risk management principles?
- Can you quantify and prioritize risks effectively?
- Are you open to new technologies and approaches to security, such as AI and Zero Trust?
- Can you prioritize your security roadmap strategy?
- Are you able to justify your program, budget, and investment requests to leadership?
- Do you understand what constitutes a material risk, and if so, can you determine which of your organization's risks are material?

Hopefully, you can now answer "yes" to most of these questions, and you're well on your way to becoming a CISO 3.0. If you honestly evaluate your progress in these areas, you can identify areas where further growth and development may be needed. I encourage you to revisit the concepts and strategies explored in this book, apply them to your own organization, and continue to evolve as a cybersecurity leader.

Becoming a CISO 3.0 doesn't end with the final page of this book. The CISO 3.0 is not a destination: It's a journey of continuous growth and evolution. The world of cybersecurity is in a constant state of flux, and the CISO 3.0 must remain vigilant, adaptable, and committed to continuous learning. Face the challenges, seize the opportunities, and lead your organization toward a secure and successful future.

Thank you for joining me on this exploration of the next generation of security leadership. I hope this book has provided you with valuable insights and actionable strategies to help you on your path to becoming a CISO 3.0.

Remember, the future of cybersecurity leadership is in your hands.

18.2.3 Key Takeaways

1. **The Language of Business**: CISOs must be fluent in the language of business, understanding accounting, finance, economics, and risk management to communicate with executives and the board effectively. Translating complex technical concepts into clear business terms is key for securing buy-in and resources.
2. **The Evolving Regulatory Landscape**: The legal and regulatory environment surrounding cybersecurity is constantly changing, placing greater accountability on organizations and individuals for breaches. CISOs must stay abreast of these changes and ensure compliance while proactively managing and disclosing risks.
3. **Risk Management**: The CISO 3.0 must adopt a proactive and data-driven approach to risk management, leveraging risk quantification techniques and KRIs to identify, assess, and prioritize risks. They should also be able

to articulate the organization's risk appetite and tolerance levels and make well-informed decisions about risk mitigation and transfer strategies. The distinction between IT risk and enterprise risk is essential, and CISOs must be able to translate technical IT risks into business terms that resonate with the board.

4. **Cyber Insurance**: Cyber liability insurance is a necessary tool for mitigating the financial impact of cyber incidents. The CISO 3.0 must understand the intricacies of cyber insurance policies, the application process, and the importance of aligning insurance coverage with the organization's risk appetite and financial goals.
5. **Self-Insurance and Risk Financing**: The CISO 3.0 should be familiar with alternative risk financing mechanisms, such as self-insurance, captive insurance companies, and cyber insurance bonds. They should be able to conduct cost–benefit analyses and use that information to make smart choices about risk retention and transfer strategies.
6. **Building Security Capabilities:** CISO 3.0s must focus on building integrated security capabilities that combine skilled personnel, robust processes, and appropriate technology to address organizational risks effectively and ensure long-term resilience.
7. **Leading Effective Teams:** Building and leading high-performing cybersecurity teams require clear role definitions, a commitment to continuous skill development, and fostering a collaborative, inclusive culture that promotes innovation and adaptability.
8. **Modern Security Tactics:** The CISO 3.0 must stay abreast of the latest security tactics and technologies, such as identity-centric security, continuous monitoring, data governance, Zero Trust, and cloud-native protection. They should be able to implement these tactics effectively to protect the organization's assets and ensure its resilience in the face of advancing threats. The integration of these tactics into a cohesive security program, potentially through a fusion center or risk exposure center, is essential for maximizing their effectiveness.
9. **Cyber Resilience:** The CISO 3.0 must prioritize cyber resilience, ensuring that the organization can withstand and recover from cyber incidents with minimal disruption. This involves adopting a proactive and adaptive approach to security, implementing continuous monitoring and recovery strategies, and building a robust cyber recovery architecture.
10. **Data Governance and Protection:** For CISO 3.0s, data is a strategic asset that requires robust governance and protection measures. They must implement policies, processes, and technologies to manage and protect data throughout its lifecycle, ensuring its confidentiality, integrity, and availability. The emergence of modern data platforms and data security platforms (DSPs) can help CISOs navigate the data deluge and strike the right balance between security and data utilization.
11. **AI and the Future of the CISO Role**: The rapid advancement of AI is transforming cybersecurity, presenting both opportunities and challenges for CISOs. The CISO 3.0 must embrace AI as a tool for

enhancing security and mitigating risks while also being prepared to address the potential threats posed by AI-powered cyberattacks. This includes understanding the different types of AI attacks, such as target attacks and subject attacks, and implementing appropriate security controls and policies.

12. **Metrics and Reporting:** The CISO 3.0 must be able to develop and communicate meaningful metrics that demonstrate the value and impact of the security program. This involves using KPIs to track operational efficiency and KRIs to communicate potential risks to the board and executive leadership. The use of frameworks like the DIKW Pyramid and the GQIM framework can guide the development and communication of effective metrics.
13. **Board-Level and Executive Communication:** The CISO 3.0 must be able to tailor their communication style to different audiences, ensuring that their message is understood and appreciated by both technical and nontechnical stakeholders. They should be able to create compelling narratives that leverage data storytelling and visual aids to convey the value and impact of the security program.
14. **Materiality and Disclosures:** The CISO 3.0 must understand the concept of materiality and be able to identify and assess material cybersecurity risks and incidents. They should also be actively involved in the disclosure process, ensuring that disclosures are accurate, complete, and timely.

18.2.4 Reflection and Exploration Questions

Understanding the Concepts

1. How does the CISO 3.0 role differ from its predecessors, and what defines its strategic value?
2. Why is fluency in the language of business critical for effective cybersecurity leadership?

Connecting the Dots

3. How do advanced security tactics, such as Zero Trust, integrate with principles of risk management and resilience?
4. In what ways do metrics and storytelling connect cybersecurity efforts to business objectives?

Real-World Application

5. Can you identify examples of organizations effectively aligning their cybersecurity programs with business goals? What strategies contributed to their success?
6. How might you apply a cyber risk invest mix to enhance your organization's approach to cyber insurance or risk financing?

Personal and Professional Growth

7. What steps can you take to improve your skills in communicating cybersecurity's value to diverse audiences?
8. How can you leverage the frameworks and tools discussed in this book to refine your leadership approach and strategy?

Further Exploration

9. What additional skills or resources do you need to excel as a CISO 3.0?
10. How might emerging technologies, such as AI or blockchain, redefine cybersecurity leadership in the future?
11. What steps can you take today to start applying the principles from this book to your professional practice?

Index

A

B

C

D

E

F

O

P

R

X

Z

For Product Safety Concerns and Information please contact our EU representative GPSR@taylorandfrancis.com
Taylor & Francis Verlag GmbH, Kaufingerstraße 24, 80331 München, Germany

www.ingramcontent.com/pod-product-compliance
Ingram Content Group UK Ltd.
Pitfield, Milton Keynes, MK11 3LW, UK
UKHW022213190926
475702UK00003B/327
9781032823515